FLORA ZAMBESIACA

Flora terrarum Zambesii aquis conjunctarum

VOLUME THREE: PART THREE

FLORA ZAMBESIACA

MOZAMBIQUE

MALAWI, ZAMBIA, ZIMBABWE

BOTSWANA

VOLUME THREE: PART THREE

Edited by
J.R. TIMBERLAKE, R.M. POLHILL, G.V. POPE & E.S. MARTINS

on behalf of the Editorial Board:

S.J. OWENS
Royal Botanic Gardens, Kew

M.A. DINIZ
Centro de Botânica, Instituto de Investigação
Científica Tropical, Lisboa

J.R. TIMBERLAKE
Royal Botanic Gardens, Kew

Published by the Royal Botanic Gardens, Kew,
for the Flora Zambesiaca Managing Committee
2007

PLANTS PEOPLE
POSSIBILITIES

First published in 2007 by
Royal Botanic Gardens, Kew
Richmond, Surrey, TW9 3AB, UK
www.org.uk

ISBN 978 1 84246 195 2

British Library Cataloguing in Publication Data
A catalogue record for this book is available from the British Library

Design and typesetting by Christine Beard,
Kew Publishing, Royal Botanic Gardens, Kew

Printed in the United Kingdom by Hobbs the Printers
This book is printed on paper that has been manufactured using wood from well managed forests certified in accordance with the rules of the Forest Stewardship Council.

For information on or to purchase all Kew titles please visit
www.kewbooks.com or e-mail publishing@kew.org

All proceeds go to support Kew's work in saving the world's plants for life

CONTENTS

ARRANGEMENT OF TRIBES IN VOLUME 3

LIST OF NEW NAMES PUBLISHED IN THIS PART

61. LEGUMINOSAE
Subfamily PAPILIONOIDEAE*

by R.K. Brummitt, D.K. Harder, G.P. Lewis, J.M. Lock, R.M. Polhill
& B. Verdcourt

Papilionoideae DC., Prodr. **2**: 94 (1825) as 'Papilionaceae' (alternative
name Faboideae).

Trees, shrubs, lianes or more often herbs, unarmed or (less often) spiny. Leaves most often pinnate or pinnately 3-foliolate, less often digitately 3-foliolate, occasionally 1-foliolate, simple, palmate or lacking; stipels sometimes present. Inflorescences racemose or paniculate, less often umbellate, capitate or spicate, or flowers solitary. Flowers zygomorphic. Calyx tubular, 2-lipped or regularly (4)5-toothed or -lobed, rarely spathaceous or opening irregularly; teeth or lobes imbricate or valvate. Petals normally 5, free or some of them united, imbricate, the adaxial petal (the vexillum or standard) almost always outermost, the 2 lateral (the alae or wings) parallel with each other, the 2 lower innermost and often cohering at the margin, forming the carina or keel; the most primitive genera (tribes Swartzieae and Sophoreae) sometimes with similar petals, some sometimes absent. Stamens 10, rarely fewer or very rarely (in Swartzieae) more, sometimes free, more usually joined in a sheath open on the upper side, the upper filament free, or all joined into a closed tube, rarely joined irregularly in bundles; anthers 2-thecous, usually opening by longitudinal slits, uniform or less often dimorphic, basifixed and dorsifixed. Seeds without areoles, normally with a distinct groove across the hilum; aril sometimes well developed; embryo usually with an incurved radicle.

A subfamily of about 450 genera and some 12,000 species, found in all parts of the world.

The subfamily is generally easily recognized by the characteristic papilionoid flowers with a distinct standard, wings and keel, but genera of the Swartzieae and some Sophoreae have flowers more like Caesalpinioideae and allowed for in the keys there. The seeds, if hard, are also characteristic, with a distinct hilar groove, which acts as a valve to dry out the seed as it ripens.

The genera are arranged here following Polhill, Classification of Leguminosae in Bisby, Buckingham & Harborne, Phytochemical Dict. of Legum. **1**: xxxv–lvii (1994), with some minor readjustments following Lewis, Schrire, Mackinder & Lock, Legumes of the World (2005). A considerable amount of evidence is accumulating from molecular studies to refine the overall classification and there is an ongoing need for more general revisionary work on individual genera and groups of genera. The present arrangement should be reasonably familiar and practical. In the sequence of genera, Nos. 64 and 65 are omitted as Loteae was transferred from Flora Zambesiaca volume 3 part 5 to part 7 after the numbers were first allocated.

The subfamily is of great economic importance, with numerous species used as food and fodder, for timber and firewood, for medicine and pesticides, as green

* Volume 3 part 3 by R.K. Brummitt, D.K. Harder, G.P. Lewis, J.M. Lock & B. Verdcourt; part 4 by B.D. Schrire; part 5 by B. Mackinder, R. Pasquet, R.M. Polhill and B. Verdcourt; part 6 by B. Verdcourt; part 7 by R.K. Brummitt, M.A. Diniz, E.S. Martins, R.M. Polhill, B. Verdcourt, M.P. Vidigal and B.-E. van Wyk. Subfamily and tribal descriptions and keys to tribes and genera by R.M. Polhill.

manures and for many local domestic products. The International Legume Database and Information Service (ILDIS) provides many useful references in electronic and published form. The more important introduced species, represented by herbarium specimens or reliable bibliographic references, are mentioned briefly under their respective tribes or genera.

*Key to Papilionoideae tribes**

1. Leaves ending in a pair of leaflets (paripinnate), often with a bristle or tendril between them (Fig. 3.3.**1**/1 & 4) · 2
   – Leaves ending in a leaflet (simple, 1–3-foliolate or imparipinnate) or digitate (Fig. 3.3.**1**/2, 3, 5 & 7) · 5
2. Leaves generally ending in a tendril, lacking a basal pulvinus (Fig. 3.3.**1**/1), generally in 2 ranks (distichous); herbs; pods generally dehiscent, with spherical seeds; introduced in the Flora area · · · · · · · **14. Fabeae [Vicieae]** (vol. 3.7: 22)
   – Leaves without tendrils, pulvinate, spiral or crowded; herbs, shrubs or trees; pods various, with more elongate or flattened seeds · · · · · · · · · · · · · · · · · · 3
3. Fruit jointed (Fig. 3.3.**3**/3 & 4) or sometimes reduced to a single 1-seeded article, or buried underground; stipules several-nerved, often cordate or spurred at the base (Fig. 3.3.**1**/9) or adnate to the petiole; calyx generally 2-lipped; plants often with glands or glandular hairs · · · · · · · · · · · **11. Aeschynomeneae** (vol. 3.6: 50)
   – Fruits neither articled nor geocarpic; stipules 1–few-nerved, not prolonged below the point of insertion, free; calyx with 3 lower lobes separate, the upper 2 lobes often joined higher · 4
4. Flowers in clusters along the rachis or in the axils; stamens 9 (upper one lacking); pods narrowly oblong, compressed; stems sometimes climbing or trailing · **5. Abreae** (vol. 3.3: 210)
   – Flowers inserted singly along an extended axis; stamens 10 (upper one free); pods linear to cylindrical, sometimes ± torulose; stems strictly erect · **6. Sesbanieae** (vol. 3.3: 217)
5. Petiole without a pulvinus at the base and attached to the stipules (except *Cicer* and *Hedysarum*) or leaves sessile, then sometimes with stipules reduced to glands and the basal leaflets resembling stipules (Loteae); leaves compound, generally distichous; mostly north temperate or Cape herbs occurring largely in the highlands or in cultivation · 6
   – Petiole generally with a distinct swelling at the base, at least when the leaves are compound, and free from the stipules if these are present; leaves 1–many-foliolate, generally spiral or crowded · 11
6. Leaves conspicuously glandular-punctate or pustulate, usually sessile; pods 1-seeded, glandular; inflorescences compound, with more than one series of bracts · **10. Psoraleeae** (vol. 3.6: 43)
   – Leaves not glandular; pods 2–many-seeded except some Trifolieae, eglandular; inflorescences with bracts only subtending flowers, sometimes with involucral bracts at the base · 7
7. Nerves looped within the margin of entire leaflets · · · · · · · · · · · · · · · · · · · 8
   – Nerves extended to ± toothed margin of leaflets · 10
8. Leaves sessile or nearly so with stipules reduced to glands (lower pair of leaflets in *Lotus* resemble stipules, Fig. 3.3.**1**/3); flowers in umbels in Flora area; filaments of all or alternate stamens dilated apically · · · **12. Loteae** (vol. 3.7: 1)

---

* An alternative artificial key to genera is given on p. 4.

  – Leaves petiolate and with conspicuous stipules generally adnate to the base of
     the petiole; flowers generally in racemes; filaments not dilated apically · · · · · 9
 9. Fruits not jointed ·**13. Galegeae** (vol. 3.7: 10)
  – Fruits composed of several 1-seeded articles (cultivated) · · · · · · · · · · · · · · · ·
     · **Hedysareae**, see under **Galegeae**
10. Stipules free; leaves pinnate in the Flora area; seeds globular to ovoid, beaked,
     with a very short radicle ·**15. Cicereae** (vol. 3.7: 32)
  – Stipules adnate to the petiole; leaves 3-foliolate in the Flora area; seeds small,
     with well-developed radicle · · · · · · · · · · · · · · · · · · ·**16. Trifolieae** (vol. 3.7: 34)
11. Stamens free or nearly so ·12
  – Stamens joined to a considerable degree ·14
12. Calyx entire in bud, opening irregularly or in (2)4–5 distinct lobes; stamens c.
     20–60 (5–many elsewhere); petals 0–1 (0–5 elsewhere), not much differentiated
     · **1. Swartzieae** (vol. 3.3: 23)
  – Calyx lobes apparent in bud, and/or corolla papilionoid; stamens (5)10(30);
     petals 5, differentiated as standard, wings and keel or 4 lower ones similar
     (elsewhere occasionally 1 or 0) ·13
13. Calyx rounded and then tapered into the hypanthium and/or pedicel · · · · · · ·
     · **2. Sophoreae** (vol. 3.3: 29)
  – Calyx of mature flowers with the hypanthium intruding at the base · · · · · · · · ·
     ·**17. Podalyrieae** (vol. 3.7: 52)
14. Anthers dimorphic (Fig. 3.3.**2**/4), alternately basifixed and shorter dorsifixed
     (the lower one sometimes shorter) or 5 aborted; leaves digitate, 3(7)-foliolate, 1-
     foliolate or simple ·15
  – Anthers all similar, or if somewhat dimorphic to partly aborted (a few
     Phaseoleae, Fig. 3.3.**2**/3) then leaves pinnately 3-foliolate · · · · · · · · · · · · · · ·16
15. Calyx without a 3-fid lower lip; upper lobes usually not joined higher up, not
     spathaceous; pods usually horizontal or nodding ·
     ·**18. Crotalarieae** (vol. 3.7: 55)
  – Calyx with a 3-fid lower lip in the Flora area; upper lobes often fused higher to
     form an upper lip or occasionally (*Spartium*) spathaceous; pods usually pointing
     upwards ·**19. Genisteae** (vol. 3.7: 246)
16. Anthers apiculate, or hairy (Fig. 3.3.**2**/5); petals usually reddish and caducous;
     2-branched hairs (Fig. 3.3.**2**/1) usually present (simple hairs sometimes also
     present) ·**7. Indigofereae** (vol. 3.4)
  – Anthers not apiculate if petals falling early; 2-branched hairs lacking · · · · · ·17
17. Fruits jointed, comprising 1–numerous 1-seeded segments, generally breaking
     transversely at the joints (not in *Pseudarthria*, Fig. 3.3.**3**/1), the segments
     generally apparent from an early stage of development, occasionally fruits
     buried underground (*Arachis*) ·18
  – Fruits not jointed ·19
18. Leaflets 1 or 3, or if up to 9 (*Uraria*) then opposite with conspicuous stipels at
     the base of the leaflets; tubercle-based glandular hairs lacking; inflorescences
     generally compound with more than one series of bracts · · · · · · · · · · · · · · · · · ·
     ·**9. Desmodieae** (vol. 3.6: 1)
  – Leaves imparipinnate with leaflets generally alternate, or digitately to pinnately
     2- or 4-foliolate, rarely 3-foliolate (*Stylosanthes*), exceptionally simple (*Brya*, rarely
     cultivated), but always without stipels; tubercle-based glandular hairs often
     conspicuous; flowers in racemes, the pedicels subtended by a single (sometimes
     3-partite) bract, congested or single · · · · · · **11. Aeschynomeneae** (vol. 3.6: 50)

19. Calyx protracted at the base into a ± conical hypanthium, containing the nectary inside; nectariferous disc lacking; stamens not forming nostrils at base of staminal tube; flowers inserted singly on extended axes; leaflets alternate or opposite, without stipels · 20
   – Calyx abruptly contracted at the base, without a hypanthium, generally with a nectariferous disc at the base inside; upper stamen often connate medially but free, arched and thickened to form opening near the base in Millettieae (Fig. 3.3.**2**/2), free or connate in Phaseoleae; flowers often clustered on knobs or short branches of the inflorescence (pseudoracemes), sometimes in extensive panicles or clustered in the axils; leaflets generally strictly opposite and stipellate, sometimes 1-foliolate or simple ·21
20. Fruits indehiscent, the 1–4 seeds generally in separate seed-chambers with hard endocarp (Fig. 3.3.**3**/2); stamens often rather shortly and irregularly joined · **3. Dalbergieae** (vol. 3.3: 50)
   – Fruits dehiscent, elongate pods, with numerous seeds; stamens ± $^3/_4$-joined in a sheath with a generally free upper stamen · · · · · · **6. Sesbanieae** (vol. 3.3: 217)
21. Leaves 1–many foliolate, if 3-foliolate the lateral leaflets only slightly asymmetrical; trees, shrubs or lianes with hard wood, less often subshrubs or herbs (*Tephrosia*, with closely parallel lateral nerves in the leaflets, Fig. 3.3.**1**/6); flowering (except herbs) often massive, the inflorescences aggregated towards the branch ends, the flowers relatively unspecialized · **4. Millettieae** (vol. 3.3: 76)
   – Leaves (1)3-foliolate, when 3-foliolate the lateral leaflets usually markedly asymmetrical and regularly stipellate, sometimes 5–9-foliolate, but then generally herbs without striately nerved leaflets; twining, prostrate or erect herbs, sometimes shrubs, lianes or trees; flowering generally protracted, the inflorescences often in many axils; flowers often as in Millettieae, but with a tendency to elaboration in style, standard appendages, abortion of anthers, resupination and bird-flowers · · · · · · · · · · · · · · · · · **8. Phaseoleae** (vol. 3.5: 1)

Artificial key to Papilionoideae genera

1. Leaves not present at time of flowering · · · · · · · · · · · · · · · · · · **Group A**, p. 7
   – Leaves present · 2
2. Leaves all simple, 1-foliolate or reduced to spine-tipped phyllodes · **Group B**, p. 8
   – Leaves, or some of them, compound · 3
3. Leaves digitately 2–11-foliolate, the leaflets all arising at the same point and without any extension of the axis beyond them (Fig. 3.3.1/2) · · **Group C**, p. 10
   – Leaves pinnate, either 3–many-foliolate with leaflets arising from separate points, or sometimes (in *Humularia, Lathyrus, Pisum, Vicia*) 2-foliolate with the rachis prolonged beyond them · 4
4. Leaves paripinnate with the rachis sometimes produced as a spine, bristle or tendril, or imparipinnate with a much reduced terminal leaflet (Fig. 3.3.1/1 & 4) · **Group D**, p.12
   – Leaves imparipinnate, terminating in a leaflet about as large or larger than the lateral leaflets (Fig. 3.3.1/5 & 7) · 5
5. Leaves 5–many-foliolate, except sometimes a few at top or bottom of plant · **Group E**, p. 13
   – Leaves all or mostly 3-foliolate, occasionally 5-foliolate with the 4 lower leaflets arising at the same point · **Group F**, p. 17

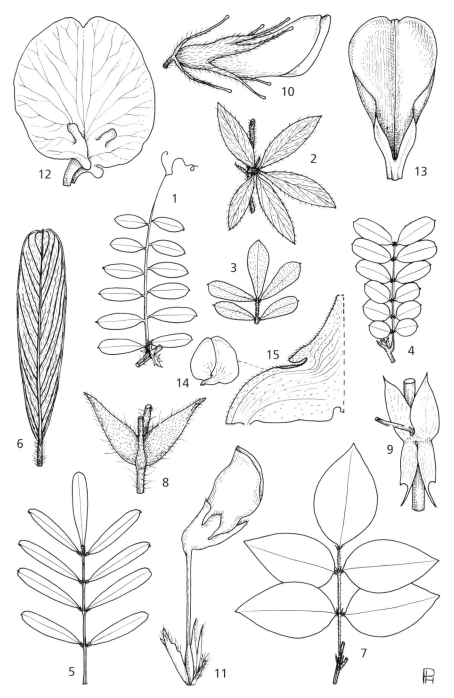

Fig. 3.3.**1**. KEY CHARACTERS. 1, paripinnate leaf of *Vicia*; 2, digitate leaf of *Ophrestia digitata*; 3, imparipinnate leaf of *Lotus*; 4, paripinnate leaf of *Abrus*; 5, imparipinnate leaf of *Tephrosia*; 6, detail of *Tephrosia* leaflet, showing venation; 7, imparipinnate leaf of *Clitoria*, showing stipels; 8, stipules of *Kotschya*; 9, stipules of *Smithia*; 10, flower of *Eminia*; 11, flower of *Stylosanthes*; 12, standard of *Dolichos*; 13, standard of *Lotus*; 14, standard of *Centrosema*; 15, detail of *Centrosema* standard, showing spur. Drawn by Pat Halliday. From F.T.E.A.

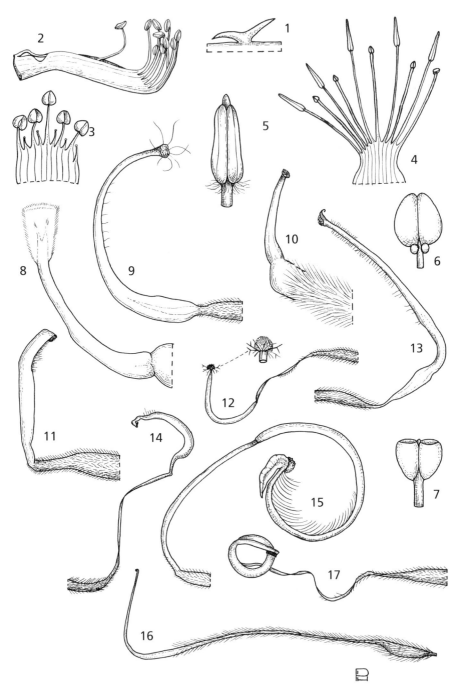

Fig. 3.3.**2**. KEY CHARACTERS. 1, biramous hair of *Indigofera*; 2, stamens of *Philenoptera*; 3, stamens of *Teramnus*, spread out; 4, stamens of *Crotalaria*, spread out; 5, anther of *Indigofera*; 6, anther of *Vigna* (subgen. *Haydonia*); 7, anther of *Dalbergia*; 8–17, styles of: 8, *Sphenostylis*; 9, *Dolichos*; 10, *Neorautanenia*; 11, *Lablab*; 12, *Decorsea*, with detail of stigma enlarged; 13, *Vigna*; 14, *Macroptilium*; 15, *Physostigma*; 16, *Pseudeminia*; 17, *Phaseolus*. Drawn by Pat Halliday. From F.T.E.A.

Group A: Leaves not present at time of flowering.

1. Standard 2–7.5 cm long, red, often much exceeding the other petals; prickles generally present; fruits rather woody, sometimes torulose, dehiscent pods; seeds red, yellow or orange with a white or black hilum · · · **29. Erythrina** (vol. 3.5: 6)
– Standard smaller; prickles lacking (branches may be spine-tipped); other features not as combined above · 2
2. Leaves falling early, leaving a distinct raised leaf-base; branches virgate, green, striate to ribbed; flowers yellow, with spathaceous or 2-lipped calyx; cultivated ornamental trees or shrubs · · · · · · · · · · · · · · · · · to **Group B**, couplet 8, p. 8
– Leaves appearing after the flowers except aphyllous species of *Aeschynomene* with conspicuous stipules; branches and flowers not as above · · · · · · · · · · · · · · · 3
3. Trees, shrubs or woody climbers · · · · · · · · · · · · · to **Group E**, couplet 30, p. 15
– Herbs or subshrubs, then usually with annual shoots from a perennial rootstock · 4
4. Stipules conspicuous, spurred, but rachis and leaflets lacking or reduced to a simple small projection · · · · · · · · · · · · · · · · · **75. Aeschynomene** (vol. 3.6: 58)
– Leaves not so modified · 5
5. Fruit of 1–5 1-seeded segments on a long plumose stipe; free part of united filaments unequal in length, 4 short and filiform, alternating with 5 longer ones much dilated and pincer-shaped beneath the anther · **68. Droogmansia** (vol. 3.6: 27)
– Fruit not jointed, nor on a long plumose stipe; filaments subequal in length, all linear · 6
6. Anthers apiculate (Fig. 3.3.**2**/5); hairs, or some of them, 2-branched (Fig. 3.3.**2**/1); corolla usually red in part, generally falling early · **28. Indigofera** (vol. 3.4)
– Anthers not apiculate, hairs simple or lacking; corolla variously coloured · · · 7
7. Calyx-teeth ending in 1–5 long-stalked gland-tipped projections (Fig. 3.3.**1**/10) · **37. Eminia** (vol. 3.5: 42)
– Calyx-teeth without glandular projections · 8
8. Calyx (and often also corolla) with small rounded sessile resinous glands; pedicels mostly arising singly from the rachis, without a swelling at the point of insertion; pods 1–2-seeded · · · · · · · · · · · · · · · to **Group F**, couplet 12, p. 18
– Calyx without glands; pedicels generally arising several together from swollen cushion-like nodes of the inflorescence; pods often more than 2-seeded · · · · 9
9. Style with a distinct dorsal appendage above the oblique stigma, the appendage reflexed (Fig. 3.3.**2**/15); style and keel markedly spirally twisted for just over 360° · **54. Physostigma** (vol. 3.5: 119)
– Style without such an appendage, style and keel not so twisted · · · · · · · · · · 10
10. Style entirely thick and hardened with no thin portion connecting it to the top of the ovary; stigma terminal · 11
– Style consisting of a thin flexible portion and a thicker hardened apical portion · 13
11. A small but distinct boss present at the junction of the style and ovary (Fig. 3.3.**2**/10); bracteoles absent or vestigial · · · · · · **49. Neorautanenia** (vol. 3.5: 74)
– No boss present at junction of style and ovary; bracteoles present · · · · · · · · 12
12. Apex of style distinctly spathulate and flattened (Fig. 3.3.**2**/8) · **47. Sphenostylis** (vol. 3.5: 68)
– Apex of style not distinctly flattened (Fig. 3.3.**2**/9) · · **52. Dolichos** (vol. 3.5: 84)

13. Stigma terminal, bearded with branched hairs (Fig. 3.3.**2**/12);* ovary (6)9–14-ovuled; thin part of style very tenuous; pollen grains finely reticulate ·········
· **56. Decorsea** (vol. 3.5: 156)
 – Stigma lateral or oblique, without branched hairs, sometimes conspicuously beaked beyond the style (Fig. 3.3.**2**/13); thin part of style less tenuous; pollen grains very openly reticulate (except *V. juncea*) · · · · · · **55. Vigna** (vol. 3.5: 121)

Group B: Leaves simple, 1-foliolate or reduced to spine-tipped phyllodes (garden plants only).

1. Stamens free; trees or shrubs · 2
 – Stamens united; herbs or small shrubs except *Philenoptera* and some cultivated ornamentals · 3
2. Leaves 1-foliolate, with 2 distinct pulvini (sometimes confluent); calyx rounded or tapered at the base; keel petals incompletely fused along lower margin, disorientated at anthesis exposing fertile parts; corolla usually mostly white · · ·
· **7. Baphia** (vol. 3.3: 41)
 – Leaves simple; calyx of mature flowers intruding at the base; keel petals joined, falcate, hiding the fertile parts; corolla mostly pinkish or purple; cultivated ornamental · *Podalyria calyptrata* (vol. 3.7: 52)
3. Leaves reduced to spine-tipped phyllodes; corolla yellow, shortly exserted from a tubular 2-lipped calyx; cultivated ornamental · · · · · · *Ulex* spp. (vol. 3.7: 247)
 – Leaves not all so reduced (primary leaves reduced to curved spines in *Brya*); flowers not as above · 4
4. Petiole with a broad foliaceous wing; alternate filaments laterally expanded beneath the anther; fruit of 1–5 1-seeded articles on a long plumose stipe · · · ·
· **68. Droogmansia** (vol. 3.6: 27)
 – Petiole not foliaceous; filaments not expanded; fruit without a long plumose stipe · 5
5. Stipules sheathing, papery, sometimes connate, persistent; fruit narrowly subcylindrical, little compressed, breaking into several 1-seeded segments; calyx scarious · **70. Alysicarpus** (vol. 3.6: 36)
 – Stipules spreading or lacking; fruit not breaking into segments, or if so (*Desmodium*) then pods usually flat and calyx herbaceous · · · · · · · · · · · · · · 6
6. Leaves in clusters in axil of a reduced leaf with its petiole modified as a small recurved spine; ornamental shrub or small tree with racemes of orange-yellow flowers and stipitate 1-seeded fruits; cultivated · · · · · · *Brya ebenus* (vol. 3.6: 51)
 – Subtending leaves not so modified; other features not as combined above · · · 7
7. Leaves falling early, leaving a distinct raised leaf-base; branches virgate, green, striate or ribbed; flowers yellow, with a spathaceous or 2-lipped calyx; ornamental trees and shrubs · 8
 – Leaves mostly persistent; branches and flowers not as above · · · · · · · · · · · · 10
8. Calyx spathaceous, deeply split on upper side, the teeth very short; corolla 2–2.5 cm long; fruit a linear dehiscent pod; cultivated · · · · *Spartium junceum* (vol. 3.7: 247)
 – Calyx shortly 2-lipped; corolla c.1 cm long; fruits as next couplet · · · · · · · · · 9
9. Fruit a shortly oblong, dehiscent pod; keel petals longer than the standard; cultivated · *Genista aetnensis* (vol. 3.7: 247)
 – Fruit drupe-like, indehiscent; keel petals about as long as the standard; cultivated · *Retama monosperma* (vol. 3.7: 247)

---

* Visible under low-powered binocular.

10. Tree with hard wood; upper filament joined above, but free, arched and thickened to form an opening near the base (Fig. 3.3.**2**/2); flowers in panicles with young leaves, mauve to purplish, 1–1.5 cm long; fruit linear-oblong, flat, indehiscent · **15. Philenoptera** (vol. 3.3: 93)

 − Herbs or small shrubs; stamens not forming nostrils near the base; other features not as combined above · 11

11. Anthers alternately long and short (Fig. 3.3.**2**/4); keel oblong-falcate (*Lebeckia*) to beaked; pod inflated or at least slightly turgid and continuous over the seeds; seeds oblique-cordiform · 12

 − Anthers uniform; other characters not as combined above · · · · · · · · · · · · · 14

12. Keel oblong-falcate, shortly hairy; pods slightly turgid · **105. Lebeckia** (vol. 3.7: 245)

 − Keel distinctly beaked, usually glabrous on the sides; pods inflated · · · · · · · 13

13. Beak of the keel not spirally coiled · · · · · · · · · · · · **102. Crotalaria** (vol. 3.7: 68)

 − Beak of the keel spirally coiled · · · · · · · · · · · · · · · **103. Bolusia** (vol. 3.7: 228)

14. Anthers apiculate (Fig. 3.3.**2**/5) or with tufts of hairs above and below; hairs, or some of them, 2-branched (Fig. 3.3.**2**/1); corolla usually red in part, falling early · 15

 − Anthers not apiculate nor with tufts of hairs above and below; hairs simple or lacking; corolla variously coloured · 17

15. Stamens free distally for c.2 mm, sheath L-shaped; anthers without an apiculate connective, tufted hairy above and below; style twisted, dilated medially or abruptly constricted at the base; keel prolonged-rostrate; endocarp without tannin deposits · **24. Rhynchotropis** (vol. 3.4)

 − Stamens free distally for less than 1 mm; anthers with distinct apical connective; style tapering evenly to stigma; keel prolonged or not; endocarp with or without tannin deposits · 16

16. Keel and back of the standard hairy, upper margin of keel with a fringe of hairs; standard mostly broad at the base, tapering abruptly to a short claw; keel with lateral spurs, not prolonged-rostrate at the apex; endocarp tannins usually present · **28. Indigofera** (vol. 3.4)

 − Keel and usually back of the standard glabrous; standard narrow at the base, tapering gradually to the claw; keel with lateral pockets (spurs absent), often prolonged-rostrate at the apex; endocarp without tannin deposits · **25. Microcharis** (vol. 3.4)

17. Anthers of the upper and the 4 short stamens with a pair of glands beneath (Fig. 3.3.**2**/6) · **55. Vigna** subgen. **Haydonia** (vol. 3.5: 121)

 − Anthers without glands beneath · 18

18. Paired bracteoles absent or vestigial (but secondary bracts may be present and the pedicel may bear a cupule) · 19

 − Paired bracteoles present on the pedicel or at the base of the calyx · · · · · · 23

19. Fruits constricted between the seeds, mostly breaking up into 1-seeded segments; no gland dots on foliage · · · · · · · · · · · · **66. Desmodium** (vol. 3.6: 1)

 − Fruits not constricted between the seeds nor breaking up · · · · · · · · · · · · · · 20

20. Undersides of leaves and generally the calyx, etc. with gland dots (may be difficult to see if indumentum is thick); venation open; standard glabrous or hairy outside; pods 1–2-seeded · · · · · · · · · · · · · to **Group F**, couplet 13, p. 18

 − Undersides of leaves, etc. without glands; leaflets with many rather steeply ascending, closely parallel, almost straight lateral nerves running to the margin (Fig. 3.3.**1**/6); standard always hairy outside · 21

21. Pods ± thin-walled, variously contorted, indehiscent · **21. Ptycholobium** (vol. 3.3: 206)

  – Pods straight to slightly curved, dehiscent · 22
22. Pods with several to numerous seeds, mostly oblong to linear, often slightly
    curved · **19. Tephrosia** (vol. 3.3: 119)
  – Pods 1-seeded, shortly falcate, up to 10 × 4 mm · · **20. Requienia** (vol. 3.3: 204)
23. Standard hairy outside; pods small, 12–18 mm long, mostly 1–2-seeded · · · · 24
  – Standard glabrous outside; pods longer and more cylindrical, 3–many-seeded
    · 25
24. Inflorescences axillary, elongated, lax · · · · · · · · · · · **31. Ophrestia** (vol. 3.5: 24)
  – Inflorescences terminal and in upper axils, dense, subcapitate · · · · · · · · · · · · ·
    · **32. Pseudoeriosema** (vol. 3.5: 28)
25. Style thick and hardened, widened at base and with no thin part separating it
    from the ovary, glabrous; stigma terminal, surrounded by a ring of hairs (Fig.
    3.3.**2**/9); flowers mostly purple · · · · · · · · · · · · · · **52. Dolichos** (vol. 3.5: 84)
  – Style thin and not particularly hardened, or if so in upper part then with a thin
    portion connecting to the ovary · 26
26. Style tenuous throughout, with a terminal stigma surrounded by a ring of hairs;
    small erect stout-rooted herb with cream-green or whitish flowers sometimes with
    a mauve spot; pollen grains finely tuberculate · · **53. Macrotyloma** (vol. 3.5: 104)
  – Style (Fig. 3.3.**2**/13) with thickened hairy apical part and a tenuous basal part
    connecting to the ovary; stigma lateral; flowers pink, purple and yellow; pollen
    grains widely reticulate · **55. Vigna*** (vol. 3.5: 121)

Group C: Leaves digitately 2–11 foliolate.

 1. Leaflets slightly to conspicuously toothed; stipule-bases encircling the stem;
    stipels absent; flowers in umbels, heads or short condensed racemes · · · · · · 2
  – Leaflets entire or (in *Otholobium*) slightly crenate (but then glandular-punctate);
    other characters not as combined above · 3
 2. Pod 3–4 times as long as calyx, 10–15-seeded; peduncle 1–4-flowered · · · · · · · · ·
    · **93. Parochetus** (vol. 3.7: 35)
  – Pod usually shorter than calyx, never as much as twice as long, 1–9-seeded;
    peduncles usually with more than 4 flowers · · · · · · **97. Trifolium** (vol. 3.7: 45)
 3. Leaflets 2 or 4 (occasionally a few leaves with 3); stipules and bracts conspicuous,
    produced below the point of insertion, often glandular-punctate; fruits
    transversely jointed, often bristly · · · · · · · · · · · · · · · · **80. Zornia** (vol. 3.6: 159)
  – Leaflets 3, 5 or more; stipules and bracts not so produced; fruits not bristly · · 4
 4. Anthers 4–5, basifixed, alternating with 5–6 shorter dorsifixed ones (Fig.
    3.3.**2**/4) or 5 only (Fig. 3.3.**2**/3) · 5
  – Anthers uniform or subuniform (6 attached a little higher in *Pearsonia*), none
    aborted · 15
 5. Calyx 2-lipped, the lower lip 3-fid to entire · 6
  – Calyx not 2-lipped, or if slightly so the lower lobes much longer than the united
    part · 11
 6. Leaflets 5–11; keel beaked · · · · · · · · · · · · · · · · · **110. Lupinus** (vol. 3.7: 263)
  – Leaflets 3; keel obtuse · 7
 7. Small cultivated tree, with pendent racemes of medium-sized flowers (standard 1.5
    –2 cm long) and leathery dehiscent pods · · *Laburnum anagyroides* (vol. 3.7: 247)
  – Herbs or small shrubs; native · 8

-----

* If features incongruent, see key to genera under Phaseoleae in vol. 3.5 (*Neorautanenia* and
perhaps some others may have occasional plants with wholly 1-foliolate leaves, though generally
3-foliolate).

8. Stipules semi-sagittate or semi-cordate at base, with a narrow point of attachment; glandular tubercles often present ··········· **107. Melolobium** (vol. 3.7: 250)
 – Stipules if present not lobed; glandular tubercles absent except in *Adenocarpus* ·· 9
9. Stipules inconspicuous or absent, less than 1 mm long if present; wings much shorter than the keel, distinctly spurred ········· **106. Dichilus** (vol. 3.7: 248)
 – Stipules conspicuous, usually with a wide point of attachment (often encircling the stem); wings as long as or longer than the keel, auriculate but not spurred ··· 10
10. Pods without glandular papillae; lip formed by upper calyx lobes more than twice as long as tube, as long as the lower lip; leaves mostly well spaced ·······
 ······································· **108. Argyrolobium** (vol. 3.7: 253)
 – Pods with conspicuous glandular papillae; lip formed by upper calyx lobes about as long as tube, shorter than the lower lip; leaves mostly crowded on short shoots
 ······························· **109. Adenocarpus** (vol. 3.7: 262)
11. Stipels present; pod linear, compressed with an upturned beak (formed by the accrescent style), with partitions of endocarp between the seeds; plants trailing or twining ································· **40. Teramnus** (vol. 3.5: 49)
 – Stipels lacking; pod variously shaped, but if linear without an upturned stylar beak, continuous within; plants never twining ····················· 12
12. Stamens 9, 5 with small anthers, the others aborted; style straight, projecting forwards; calyx lobes long, narrow, equal; plants small, annual ··············
 ································· **101. Robynsiophyton** (vol. 3.7: 66)
 – Stamens and anthers 10; style curved upwards in flower; calyx lobes as next couplet ·· 13
13. Keel obtuse to shortly beaked; lateral calyx lobes more united with the upper than the lower lobe, which is often narrower; 4 anthers large, 6 anthers small; pods flat to somewhat turgid ················· **104. Lotononis** (vol. 3.7: 233)
 – Keel distinctly beaked to coiled; calyx lobes subequal or the 3 lower slightly more united or with upper two also united to form 2 lips; 5 large anthers alternating with 5 small ones; pods markedly inflated ························· 14
14. Beak of keel not coiled ···················· **102. Crotalaria** (vol. 3.7: 68)
 – Beak of keel helically coiled through several turns ·· **103. Bolusia** (vol. 3.7: 228)
15. Leaflets and calyx (often also corolla and pods) with sessile rounded resinous glands or pustules ··· 16
 – Leaflets and calyx not glandular ··································· 19
16. Pods 1-seeded, indehiscent; leaves very shortly petiolate, with embedded pustules ························· **72. Otholobium** (vol. 3.6: 46)
 – Pods 2-seeded, dehiscent; leaves mostly petiolate (except some *Eriosema*), with glands on surface ······································· 17
17. Bracteoles present; style stiff, thickened below, bearded; pod oblong-falcate, much narrowed at base ·············· **63. Adenodolichos** (vol. 3.5: 242)
 – Bracteoles absent; style thinner; pod not narrowed so conspicuously at base ··18
18. Shrubby herbs, with fasciculate dense sessile axillary inflorescences; pods oblong-ovoid, somewhat inflated, 9–12 × 6 mm, densely but irregularly covered with dark red globules of secretion which stain yellow ······ **60. Flemingia** (vol. 3.5: 165)
 – Herbs, mostly with pedunculate inflorescences or if sessile diverging in other characters; pods oblong-elliptic in outline, flattened, not bearing red globules of secretion ···························· **62. Eriosema** (vol. 3.5: 211)
19. Leaflets with the lateral nerves numerous, closely parallel, steeply ascending and united into a horny marginal nerve (Fig. 3.3.1/6) ····················· 20
 – Leaflets without lateral nerves, not as above ·························· 21
20. Pods narrowly oblong, dehiscent; leaves 3–7-foliolate ·····················
 ····································· **19. Tephrosia** (vol. 3.3: 119)

– Pods variously contorted, ± thin-walled, indehiscent; leaves mostly 3-foliolate (some 1- or 5-foliolate) · · · · · · · · · · · · · · · · **21. Ptycholobium** (vol. 3.3: 206)

21. Anthers apiculate (Fig. 3.3.**2**/5); at least some hairs 2-branched (Fig. 3.3.**2**/1); corolla red in part, usually falling early · · · · · · · · · · · · **28. Indigofera** (vol. 3.4)

– Anthers not apiculate; hairs simple; corolla, apart from markings, usually not red · 22

22. Style straight or almost so, projecting forwards or downwards (flowers upside down in *Pearsonia*) · 23

– Style curved upwards · 24

23. Anthers elongate, 4 basifixed, 6 attached a little higher; flowers upside down; standard 8–30 mm long · **99. Pearsonia** (vol. 3.7: 56)

– Anthers all small, rounded; flower not inverted; standard c.5 mm long · **100. Rothia** (vol. 3.7: 64)

24. Pods hairy, breaking up transversely into 1-seeded segments; flowers small in dense heads, blue, reddish or white; calyx densely brown hairy · **66. Desmodium** (vol. 3.6: 1)

– Pods not breaking up into 1-seeded segments; other characters not as combined as above · 25

25. Flowers in umbelliform inflorescences, subtended by a 1–3-foliolate bract; leaves subsessile (Fig. 3.3.**1**/3), with stipules reduced to small glands and without stipels · **83. Lotus** (vol. 3.7:1)

– Flowers in axillary clusters, sometimes aggregated terminally, without an involucral bract; leaves subsessile to petiolate; stipules herbaceous, sometimes caducous; stipels present · 26

26. Style hardened, thicker towards the base (Fig. 3.3.**2**/9); standard appendages short, conic or oblong (Fig. 3.3.**1**/12); wings wide; flowers mainly purple; pollen grains widely reticulate · **52. Dolichos** (vol. 3.5: 84)

– Style tenuous, not thickened; standard appendages long, linear and lamelliform; wings very narrow; flowers mainly yellow; pollen grains finely tuberculate · **53. Macrotyloma** (vol. 3.5: 104)

Group D: Leaves paripinnate.

1. Fruits developing underground; receptacle (extended hypanthium) long and filiform and easily mistaken for a pedicel but flowers are actually sessile; anthers alternately long and short; leaves 4-foliolate; cultivated for seeds (peanuts) but sometimes found as an escape · · · · · · · · · · · · · · · · · **82. Arachis** (vol. 3.6: 169)

– Fruits not developing underground (or if so, in *Aeschynomene nematopoda*, then leaves 4–16-foliolate and leaflets very narrow); receptacle not developed; anthers mostly uniform; leaves 2–many-foliolate · 2

2. Pods coiled in a ring or spiral 4–5 mm in diameter, the articles numerous, dehiscent, separating, the suture persistent after they have fallen; annual herb · **79. Cyclocarpa** (vol. 3.6: 157)

– Pods not coiled into a ring, or if so pods very much larger and a tree · · · · · · 3

3. Calyx 2-lipped (rarely not, then tiny herbs); bracts small to large and papery; fruits (Fig. 3.3.**3**/3 & 4) with 1–28 1-seeded segments which mostly separate easily (in some *Aeschynomene* spp. only one segment is present which is straight above, strongly rounded below and flattened) · 4

– Calyx not distinctly 2-lipped; bracts usually small; fruits not jointed, 1–many seeded · 7

4. Inflorescences mostly lax, never scorpioid, often few-flowered; pods easily visible, well exserted from the calyx; bracts mostly small, but in sect. *Rubrofarinaceae* large

and papery, but then bark of branches breaking down to form a soft ferruginous floury covering; calyx never scarious; fruit 1–28-seeded · · · · · · · · · · · · · · · · ·
· **75. Aeschynomene** (vol. 3.6: 58)

– Inflorescences mostly dense and ± scorpioid, often strobilate (e.g. as in *Humulus*, the hop); fruits hidden, folded like a concertina, of 1–9 segments · · · · · · · · · 5

5. Fruits of 1–2 segments not included in the calyx; bracts membranous or scarious, much larger than the flower and entirely hiding the fruit, persistent; stipules mostly leaf-like, less often membranous or scarious; bracteoles and calyx membranous; bark not breaking down to form a soft floury covering · · · · · · ·
· **78. Humularia** (vol. 3.6: 140)

– Fruits included in the calyx, consisting of 2–9 segments; bracts scarious, much smaller than the flowers; stipules membranous or scarious; bracteoles and calyx scarious · 6

6. Stipules not spurred (Fig. 3.3.**1**/8); leaflets alternate, with 2–7 basal nerves; inflorescences in 2 rows, usually dense and strobilate; bracts persistent; lateral appendages of keel petals short or lacking · · · · · · · **76. Kotschya** (vol. 3.6: 117)

– Stipules spurred (Fig. 3.3.**1**/9); leaflets opposite, with only one main nerve; inflorescences subumbellate (in Flora area); bracts deciduous; lateral appendages of keel petals nearly as long as claws · · · **77. Smithia** (vol. 3.6: 138)

7. Stiff erect herbs, softly woody shrubs or small trees or rather woody twiners; tendrils absent · 8

– Weak herbs; leaves usually ending in a tendril · 9

8. Flowers in clusters along the rachis; pods oblong or narrowly oblong, compressed; stems sometimes twining or trailing; stamens 9 (upper one absent)
· **22. Abrus** (vol. 3.3: 210)

– Flowers inserted singly along the rachis; pods linear, narrowly cylindrical or winged, often somewhat torulose; stems erect; stamens 10, upper one free · · · ·
· **23. Sesbania** (vol. 3.3: 218)

9. Style winged at the base, folded longitudinally above with hairs on the inner side; stipules large and leaf-like; garden pea · · · · · · · · · · · · · **91. Pisum** (vol. 3.7: 30)

– Style not winged at base and folded above; stipules smaller · · · · · · · · · · · · · 10

10. Style pubescent on leading edge, all round or glabrous; leaflets usually in rather numerous pairs (one native subspecies with only 2 leaflets), folded flat lengthwise in bud; seeds globular · · · · · · · · · · · · · · · · · · **88. Vicia** (vol. 3.7: 22)

– Style pubescent on the inner side; leaflets usually 1–few pairs; other characters not as combined above · 11

11. Leaflets inrolled in bud; seeds globular · · · · · · · · · · · ·**90. Lathyrus** (vol. 3.7: 26)

– Leaflets folded flat lengthwise; seeds lenticular · · · · · · · · **89. Lens** (vol. 3.7: 24)

Group E: Leaves imparipinnate, 5–many-foliolate.

1. Stamens numerous; petals 0–1 · 2
– Stamens 10 or rarely fewer; petals 5 · 3
2. Petal 1; fruit elongate cylindrical · · · · · · · · · · · · · · · · · · **1. Swartzia** (vol. 3.3: 23)
– Petals 0; fruit ellipsoid to subglobose · · · · · · · · · · · · · **2. Cordyla** (vol. 3.3: 26)
3. Corolla with 4 lower petals all similar · 4
– Corolla papilionoid with 4 lower petals differentiated as wings and keel · · · · 5
4. Large bird-pollinated flower, with stamens long-exserted; fruit turgid, woody, dehiscent (cultivated) · · · · · · · · · · · · · · · · *Castanospermum australe* (vol. 3.3: 29)
– Small insect-pollinated flowers, with stamens about as long as lower petals; fruit like a drupe · **3. Xanthocercis** (vol. 3.3: 30)
5. Stamens free or nearly so · 6

- Stamens joined to a considerable extent in bundles, a sheath or tube, the upper one sometimes free ·· 11
6. Calyx 2-lipped; keel beaked; pods dehiscent, with rather thick valves (cultivated) ································· *Virgilia oroboides* (vol. 3.7: 53)
- Calyx not 2-lipped, the lower 3 teeth subequal, the upper 2 more united; keel not beaked; pods various, mostly indehiscent ························ 7
7. Fruits moniliform, sinuate between the seeds; foliage tomentose or silky pubescent in African species (less hairy in cultivated species); flowers yellow in native species, white or blue in some cultivated plants ················· 8
- Fruits flattened; foliage not conspicuously hairy; flowers yellow only in *Calpurnia* ··· 9
8. Bracteoles and stipels lacking; herbs, shrubs or small trees ················ ·· **6. Sophora** (vol. 3.3: 38)
- Bracteoles present on pedicel; stipels present; large ornamental tree (cultivated) ···························· *Styphnolobium japonicum* (vol. 3.3: 29)
9. Leaflets falcate-lanceolate, markedly unequal-sided; fruit not winged; corolla blue ····································· **5. Bolusanthus** (vol. 3.3: 35)
- Leaflets not markedly unequal-sided; fruit winged ·················· 10
10. Fruits woody, winged along both margins; flowers greenish-white to pink or violet with purple veins ···················· **4. Pericopsis** (vol. 3.3: 32)
- Fruits papery, winged on one margin only; flowers bright yellow ··········· ·· **98. Calpurnia** (vol. 3.7: 53)
11. Hairs or some of them usually 2-branched (Fig. 3.3.**2**/1); anthers apiculate (Fig. 3.3.**2**/5); corolla usually red in part, often falling early ················ 12
- Hairs simple; anthers not markedly apiculate; corolla variously coloured, but rarely red and faling early ·································· 15
12. Pods 3–5 mm wide, erect, straw-coloured, longitudinally ridged; leaflets often toothed; upper filament lightly attached to sheath; corolla darkly veined when dry ···································· **27. Cyamopsis** (vol. 3.4)
- Pods less than 3 mm wide, erect, spreading or deflexed, not as above; leaflets entire; upper filament free; corolla not darkly veined when dry ········· 13
13. Keel and back of standard hairy; upper margin of keel with a fringe of hairs; standard mostly broad at the base, tapering abruptly to a short claw; keel with lateral spurs, not prolonged-rostrate at the apex; endocarp tannins usually present ································· **28. Indigofera** (vol. 3.4)
- Keel and back of standard glabrous; standard narrow at the base, tapering gradually to the claw; keel with lateral pockets (spurs absent), often prolonged-rostrate at the apex; endocarp without tannin deposits ··············· 14
14. Style short, thick; stigma discoid; fruiting pedicels mostly over 2 mm long, spreading, pods often held ± at a right-angle; anthers, at least those of the 4 shorter stamens, with hyaline scales at the base; bracts persistent; inflorescences lax in bud ································· **25. Microcharis** (vol. 3.4)
- Style narrow, tapering upwards to an oblique or capitate stigma; fruiting pedicels rarely over 2 mm long, deflexed (as are pods); anthers of the shorter stamens mostly with scales reduced or absent; bracts falling early; inflorescences dense in bud ···································· **26. Indigastrum** (vol. 3.4)
15. Fruits folded like a concertina, constricted between the seeds, largely enclosed within the persistent calyx; flowers crowded in mostly terminal spike-like inflorescences, the pedicels (in African species) covered with long deciduous hairs and short persistent hook-tipped hairs ········· **69. Uraria** (vol. 3.6: 34)
- Fruits not so folded; pedicels without hooked hairs ·················· 16
16. Standard 2.5–5 cm long, much exceeding other petals ··············· 17

– Standard smaller or not much exceeding other petals ················· 18

17. Standard whitish to blue, held lowermost, straight; bracteoles ovate or round, persistent; pods linear-oblong, flattened; climbing or trailing perennial with solitary or paired flowers ······················· **34. Clitoria** (vol. 3.5: 34)

– Standard red, held uppermost, reflexed towards apex; bracteoles small; pods bladdery; small shrub with flowers in racemes ·· **85. Sutherlandia** (vol. 3.7: 12)

18. Leaves and calyx conspicuously glandular-pustulate; flower-stalks with a conspicuous cupule formed by 2 fused bracts; pods 1-seeded, indehiscent, chaffy ··· **73. Psoralea** (vol. 3.6: 48)

– Leaves and calyx without embedded glands, but sometimes with surface glands (see next couplet) or glandular-based hairs; flowers not subtended by a cupula; pods not chaffy ··· 19

19. Leaflet lower surface and calyx with discrete, rounded, sessile glands, visible wherever indumentum not dense; leaflets mostly 5, the 2 lower pairs often conjugate ······························· **62. Eriosema** (vol. 3.5: 211)

– Leaflet lower surface and calyx without glands; leaves more strictly imparipinnate ··· 20

20. Standard pubescent to silky tomentose outside ····················· 21

– Standard glabrous or with small scattered hairs mostly towards apex and margins ··· 24

21. Pods leathery, linear, not exceeding 1 cm wide, velvety, ultimately breaking up irregularly or only tardily dehiscent; filament tips somewhat widened; small tree or shrub ······································· **18. Mundulea** (vol. 3.3: 117)

– Pods dehiscent, sometimes tardily so, or if indehiscent then much broader; filament tips narrowed ··· 22

22. Fruits indehiscent, sinuate between the few seeds, narrowly winged along the upper margin; cultivated tree ············ *Lonchocarpus sericeus* (vol. 3.3: 77)

– Fruits dehiscent, not winged ······································· 23

23. Lateral nerves of leaflets not very closely parallel and steeply ascending; stipels usually present; pod somewhat woody; seeds with a generally thin testa and well developed aril; trees, shrubs or lianes ············ **17. Millettia** (vol. 3.3: 103)

– Lateral nerves of leaflets numerous, closely parallel, steeply ascending, extending to margin and often united into a horny marginal nerve (Fig. 3.3.**1**/6); stipels absent; pod not woody; seeds with hard testa and fairly small to negligible aril; herbs or shrubs ·············· **19. Tephrosia** (vol. 3.3: 119)

24. Trees, shrubs or climbers with hard wood, rarely (*Millettia makondensis*) semi-woody stems from a woody rootstock ····························· 25

– Annual or short-lived perennial herbs, not climbing ················· 40

25. Fruit breaking transversely into 1–6 1-seeded segments; branchlets with swollen-based glandular hairs, which persist as small conical tubercles; calyx lobes usually longer than the tube; stipules striate, persistent ·· **74. Ormocarpum** (vol. 3.6: 51)

– Fruits not jointed; glandular hairs absent or if present (in *Pterocarpus, Robinia*) then calyx lobes short and stipules deciduous ······················· 26

26. Anthers basifixed with apical dehiscence (Fig. 3.3.**2**/7); flowers on one side of rachis only; fruits flat, often papery and venose, with 1–few seeds in hardened seed-chambers (Fig. 3.3.**3**/2); some branches modified as spines or climbing aids ··· **9. Dalbergia** (vol. 3.3: 62)

– Anthers dorsifixed, with longitudinal dehiscence; flowers not along only one side of the rachis; fruits diverse, rarely as above; branches not modified ···· 27

27. Corolla yellow or orange; wing petals broad and ± crimped around the edges; fruits with a relatively large wing-like expansion either around or distal to the seed cavity ··· 28

– Corolla white, pink, blue or purplish, at least in part; wing petals not crinkly around the edges; fruits without such a large wing-like expansion ········· 29

28. Fruit obovate to circular, the seed ± median; leaflets mostly broadly oblong to rounded, more than 2 cm wide ················ **8. Pterocarpus** (vol. 3.3: 51)

– Fruit with a basal seed-bearing portion and a distal semielliptic-ovate wing; leaflets oblong, 1.5–2 cm wide (cultivated) ········· *Tipuana tipu* (vol. 3.3: 51)

29. Upper calyx lobes greatly enlarged to form a lip as large as the standard ······
··· **13. Platysepalum** (vol. 3.3: 86)

– Upper calyx lobes not so enlarged ··· 30

30. Panicles lateral, precocious, several together from young shoots which grow out into leafy branches; fruit winged along both margins ··························
··· **12. Xeroderris** (vol. 3.3: 83)

– Panicles or racemes terminal and/or in axils of fallen leaves; fruits not winged along both margins ··· 31

31. Fruit a dehiscent pod; seeds (in native species) ± quadrate, flat, often with thin testa and prominent aril; corolla 1–3 cm long ························· 32

– Fruit indehiscent; seeds (in native species at least) oblong-reniform, with negligible aril; corolla 0.7–1.8 cm long ··························· 36

32. Vigorous woody climber, with racemes 15–30 cm long of pendent flowers on pedicels 1.5–2.5 cm long, without evident bracts or bracteoles; cultivated ·····
··································· *Wisteria sinensis* (vol. 3.3: 77)

– Characters not as combined above ··································· 33

33. Flowers in pseudoracemes, the flowers 2 or more at the nodes or at the ends of the ultimate inflorescence-branches; bracteoles present, sometimes falling early; ovary hairy; often woody climbers ············· **17. Millettia** (vol. 3.3: 103)

– Flowers in racemes or panicles with the pedicels inserted singly; bracteoles absent except *Craibia*; ovary glabrous or with long hairs, falling early; erect trees or shrubs ··· 34

34. Pod tapering to base, with 1–2(3) seeds usually in the distal half; leaflets usually alternate; bracteoles present ··················· **14. Craibia** (vol. 3.3: 88)

– Pod usually ± oblong, often with more than 2 seeds distributed evenly along its length; leaflets opposite, rarely subalternate; bracteoles absent; introduced trees ··· 35

35. Style glabrous; calyx truncate, the lobes very reduced; standard reflexed 180°; plants cultivated ························ *Gliricidia sepium* (vol. 3.3: 217)

– Style hairy on upper part; calyx lobes ± as long as the tube; standard reflexed 90°; plants cultivated ··························· *Robinia* (vol. 3.3: 217)

36. Panicles well branched, the pedicels inserted singly; fruit flat, ± strap-shaped, with or without a narrow wing along the upper margin; trees ················
··· **15. Philenoptera** (vol. 3.3: 93)

– Panicles contracted, the pedicels paired, clustered or densely crowded on short or ± obsolete ultimate inflorescence-branches; if fruit winged then generally climbers (unless introduced) ··· 37

37. Margins of fruit fringed with long densely matted plumose hairs, not winged; tree, with leaves mostly crowded on spur-shoots, usually flowering precociously
··· **11. Dalbergiella** (vol. 3.3: 81)

– Margins of fruit without plumose hairs, often winged along the upper margin; woody climbers or semi-scandent trees, also some cultivated trees ········ 38

38. Fruits turgid, somewhat compressed laterally, thick-walled, obliquely oblong-ellipsoid in outline; cultivated tree with dark glossy foliage ················
··· *Millettia pinnata* (vol. 3.3: 104)

– Fruits flat, winged along the upper margin; climbers or cultivated trees ···· 39

39. Standard subcircular; inflorescences not obviously branched, the flowers clustered severally on very short axes lateral from the otherwise undivided rachis ·· **16. Derris** (vol. 3.3: 101)
 − Standard narrowly elliptic-oblong; inflorescences paniculate, well branched, with flowers densely crowded on abbreviated ultimate branches, subsidiary axillary inflorescences less branched ········· **10. Leptoderris** (vol. 3.3: 78)
40. Leaflets toothed ································ **92. Cicer** (vol. 3.7: 32)
 − Leaflets entire ·· 41
41. Leaves subsessile, the lowermost pair of leaflets sometimes resembling leafy stipules; stipules reduced to glands or absent; flowers white, yellow or pink, often marked with red, in pedunculate axillary umbels ···················· 42
 − Leaves petiolate and/or stipulate; other features not as combined above ··· 43
42. Pod straight, narrowly cylindrical, dehiscent as a whole; leaflets 5 ············ ·· **83. Lotus** (vol. 3.7: 1)
 − Pod curved, composed of spherical 1-seeded individually dehiscent segments; leaflets 5–11 ······························ **84. Antopetitia** (vol. 3.7: 10)
43. Fruits jointed, with several spinulose segments apparent from an early stage of development; introduced fodder plant ··········· **Hedysarum** (vol. 3.7: 12)
 − Fruits not jointed; native plants ································· 44
44. Style with tuft of hairs below the stigma; pods oblong, thinly membranous, not longitudinally septate ····················· **86. Lessertia** (vol. 3.7: 15)
 − Style glabrous; pod divided lengthwise by a vertical septum ··············· ·································· **87. Astragalus** (vol. 3.7: 19)

Group F. Leaves imparipinnate, all or mostly 3-foliolate, rarely 5-foliolate with lower 4 leaflets conjugate.

1. Leaflets toothed ··· 2
 − Leaflets entire or 3–5-lobed ····································· 6
2. Leaflets glandular-pustulate, the glands somewhat obscured by silky hairs but evident at least with glands terminating the teeth; fruit glandular-warty, 1-seeded, indehiscent ····························· **71. Cullen** (vol. 3.6: 43)
 − Leaflets and fruits not glandular ································· 3
3. Pods enclosed within the calyx and persistent corolla, 1-seeded ············· ····························· *Trifolium campestre* (vol. 3.7: 47)
 − Pods exserted, coiled or straight ································· 4
4. Pod spirally coiled; inflorescence-rachis short, the flowers few or crowded above the peduncle ······························· **96. Medicago** (vol. 3.7: 42)
 − Pods not coiled ··· 5
5. Pods ovoid to globose, 1–2-seeded, indehiscent; flowers in extended racemes ··· **94. Melilotus** (vol. 3.7: 37)
 − Pods oblong to linear, 2–20-seeded, indehiscent or opening down one side; flowers solitary or in axillary clusters ··········· **95. Trigonella** (vol. 3.7: 39)
6. Indumentum mainly of 2-branched hairs (Fig. 3.3.**2**/1); anthers apiculate (Fig. 3.3.**2**/5); corolla usually red ···································· 7
 − Indumentum of simple hairs; anthers not apiculate ··············· 8
7. Pods 3–5 mm wide, erect, stramineous, longitudinally ridged; stigma generally oblique; corolla darkly veined when dry; standard glabrous ··············· ······································· **27. Cyamopsis** (vol. 3.4)
 − Pods less than 3 mm wide, erect, spreading or deflexed, not as above; stigma capitate; standard generally hairy outside; corolla not darkly veined when dry ·· ······································· **28. Indigofera** (vol. 3.4)

8. Calyx lobes, bracts and bracteoles with clavate gland-tipped apices (Fig. 3.3.**1**/10); erect or climbing herb or subshrub; pod 2–3 × 0.8–0.9 cm, compressed, densely appressed bristly hairy ········ **37. Eminia** (vol. 3.5: 42)
– Calyx lobes, bracts and bracteoles not gland-tipped ···················9
9. Calyx obliquely truncate with no trace of teeth or scarcely any; seeds subglobose, black, remaining attached to the placenta after the pod dehisces ············
··**42. Dumasia** (vol. 3.5: 58)
– Calyx-teeth developed ··10
10. Undersides of leaflets, calyces and sometimes the petals and other organs covered with gland dots (difficult to see if indumentum dense) ···········11
– Undersides of leaflets, etc. without conspicuous gland dots ············14
11. Ovary 3–8-ovuled; pod grooved between the seeds, linear-oblong, inflated, 4.5–10 cm long; cultivated shrub ················**59. Cajanus** (vol. 3.5: 163)
– Ovary 2-ovuled; pod not grooved between the seeds, mostly smaller or more tapered to the base ··12
12. Bracteoles present at the base of the calyx, falling early; style stiff, thickened below, bearded above; standard white, mauve or crimson-purple; pod oblong-falcate, much narrowed at the base ········ **63. Adenodolichos** (vol. 3.5: 242)
– Bracteoles absent; style thinner, glabrous on upper part; standard usually yellowish, often with extensive reddish-purple veining; pod not conspicuously narrowed at the base ··13
13. Funicle inserted at middle of circular hilum (Fig. 3.3.**3**/6 & 7); flowers spreading; pod broadened upwards, 3–4 times as long as broad, pubescent, sometimes with longer hairs interspersed ······ **61. Rhynchosia** (vol. 3.5: 168)
– Funicle attached at one end of linear hilum (Fig. 3.3.**3**/8 & 9); flowers deflexed; pod ovate or elliptic-oblong, abruptly contracted to the stipe, up to twice as long as broad, covered with long hairs (or in one section velvety) ·············
····························· **62. Eriosema** (vol. 3.5: 211)
14. Style distinctly flattened and spathulate at the apex (Fig. 3.3.**2**/8); standard glabrous ···15
– Style various but not apically flattened and spathulate or if slightly so then standard hairy outside ·······································16
15. Bracteoles shorter than the calyx tube; upper calyx lobes completely or not completely connate; standard without appendages; filaments less dilated above, the upper one basally dilated but without a tooth at the base; seeds not arillate
···**47. Sphenostylis** (vol. 3.5: 68)
– Bracteoles as long as the calyx tube and sometimes broad; upper calyx lobes completely joined to form a truncate lip; standard with appendages; filaments dilated at the apex, the upper one with a tooth at base; seeds arillate ········
···································· **48. Nesphostylis** (vol. 3.5: 74)
16. Flowers jade-green in long pendent racemes, the standard 4.5–6 cm long; ornamental woody climber ··········· *Strongylodon macrobotrys* (vol. 3.5: 1)
– Flowers not jade-green ··17
17. Pods covered with bristly irritant hairs (except in some cultivated variants of *M. pruriens*); standard much shorter than other petals; keel petals mostly horny at apex; 5 larger subbasifixed anthers alternating with 5 shorter versatile or basifixed anthers; seeds large, either globose or oblong, with short hilum and conspicuous rim aril, or discoid with hilum occupying over half the circumference and no rim aril ················· **30. Mucuna** (vol. 3.5: 14)
– Pods not covered with bristly irritant hairs and without other characters combined (Note: many species have bristly hairs on their pods) ·········18
18. Trees, shrubs or woody climbers, rarely annual stems from a woody rootstock,

but then with prickles · 19
– Herbs, subshrubs or herbaceous climbers, without prickles · · · · · · · · · · · · 20
19. Mostly trees armed with strong prickles, stems sometimes annual from a woody rootstock; corolla mostly large, orange, scarlet or crimson, the standard usually larger than other petals; pod often constricted between the seeds; seeds mostly red, orange or yellow, with a white or black hilum · · **29. Erythrina** (vol. 3.5: 6)
– Unarmed tree without prickles; corolla, pods and seeds not as above · · · · · · · ·
· return to **Group E**, couplet 30, p. 16
20. Standard hairy outside (sometimes only finely so, but then standard much larger than the other petals); upper calyx lobes not completely joined · · · · · · · · · 21
– Standard glabrous outside, or if with a few marginal hairs then upper calyx lobes completely joined · 26
21. Standard with a small spur 1–2 mm long on the outside above the claw (Fig. 3.3.**1**/14 & 15); stipules and bracteoles with close parallel raised nerves; pods with longitudinal raised ribs on either side of the margins · · · · · · · · · · · · · · · · · ·
· **33. Centrosema** (vol. 3.5: 32)
– Standard without such a spur and without other characters combined · · · · · 22
22. Standard much larger than the other petals, 3–4.5 cm long, very finely puberulous · **34. Clitoria** (vol. 3.5: 34)
– Standard not much larger than the other petals and often less than 3 cm long
· 23

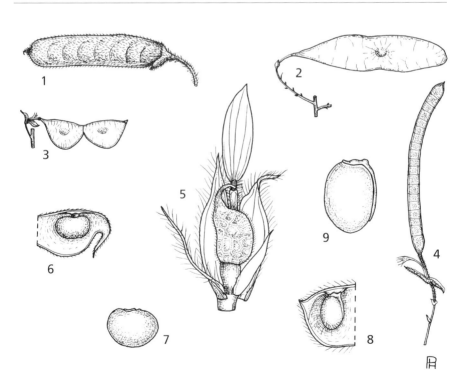

Fig. 3.3.**3**. KEY CHARACTERS. 1–5, fruits of: 1, *Pseudarthria*; 2, *Dalbergia*; 3 & 4, *Aeschynomene*; 5, *Stylosanthes*; 6, valve of *Rhynchosia* pod showing attachment of seed; 7, seed of *Rhynchosia*; 8, valve of *Eriosema* pod showing attachment of seed; 9, seed of *Eriosema*. Drawn by Pat Halliday. From F.T.E.A.

23. Leaflets with characteristic close parallel venation of rather steeply ascending almost straight lateral nerves running up into the margin (Fig. 3.3.**1**/6); ovary and pods more than 2-seeded · · · · · · · · · · · · · · · **19. Tephrosia** (vol. 3.3: 119)
   – Leaflets with more open venation, with fewer lateral nerves not reaching the margin · 24
24. Corolla 10–20 mm long; inflorescences with long peduncles; style long, with tenuous basal part and curved bearded thickened upper part beaked beyond the lateral stigma (Fig. 3.3.**2**/13); pods linear-cylindric · **55. Vigna** (in small part) (vol. 3.5: 121)
   – Corolla 4–11 mm long, if over 10 mm then inflorescences sessile clusters; style short, not as above; pods linear-oblong to oblong · · · · · · · · · · · · · · · · · · · 25
25. Flowers in elongated lax inflorescences or if in subsessile axillary clusters then pods over 2 cm long; ovules 2–8 and pods mostly more than 2-seeded; pods 1.7–6.5 cm × 4.5–8 mm · **31. Ophrestia** (vol. 3.5: 24)
   – Flowers in sessile or mostly stalked condensed inflorescences; ovules 2 and pods 1–2-seeded; pods 1.2–2 cm × 6–7 mm · · · · · · **32. Pseudoeriosema** (vol. 3.5: 28)
26. Alternate stamens sterile, lacking anthers (Fig. 3.3.**2**/3); pods elongate, linear, distinctly upturned at the apex · · · · · · · · · · · · · · · · **40. Teramnus** (vol. 3.5: 49)
   – All stamens with anthers; pods without a distinct upturned beak, or if beaked then very short and never linear · 27
27. Fruit dividing transversely into distinct segments or if only one segment then the calyx extended basally as a pedicel-like receptacle (Figs. 3.3.**1**/11 and 3.3.**3**/5) · 28
   – Fruit not divided into distinct segments, but in some few cases transversely furrowed and appearing as if it might be so (particularly *Pseudarthria*, Fig. 3.3.**3**/1, and *Calopogonium*) · 30
28. Calyx extended basally as a pedicel-like receptacle (Fig. 3.3.**1**/11); upper filament joined with the others; fruits with 1–2 segments, the apical one narrowed into a distinct hooked beak (Fig. 3.3.**3**/5); stipels absent · **81. Stylosanthes** (vol. 3.6: 165)
   – Calyx not so extended; upper filament usually free; fruit not so conspicuously beaked; stipels present · 29
29. Fruit ± straight and exserted from the calyx (if scarcely exserted from a scarious calyx see *Alysicarpus rugosus*) · · · · · · · · · · · · · · · · · · · **66. Desmodium** (vol. 3.6:1)
   – Fruit folded up like a concertina, included within the calyx; leaves mostly more than 3-foliolate · **69. Uraria** (vol. 3.6: 34)
30. Pods longitudinally 4-winged; upper stamen connate above, free below; stigma terminal or internal, penicillate, or style with a conspicuous ring of hairs beneath it; bracteoles large; corolla blue or purplish · · · **45. Psophocarpus** (vol. 3.5: 63)
   – Pods not longitudinally 4-winged and without other characters combined · ·31
31. Style with a reflexed appendage beyond the stigma (Fig. 3.3.**2**/15), curved through 360–450°; keel curved inwards through 250–280°; basal corner of keel produced into a long erect spur · · · · · · · · · · · · · **54. Physostigma** (vol. 3.5: 119)
   – Style sometimes produced beyond the stigma as a beak but never reflexed back as an appendage; keel, if coiled, without such a prominent spur · · · · · · · · · 32
32. Fruits large, linear-oblong to oblong, usually with longitudinal ribs close to the upper margin, 8–40 × 2.5–5 cm; standard large, reflexed, 2.5–3.5 cm long; seeds 1.4–3.5 cm long; upper filament free below but connate above with the rest; calyx lobes unequal, the upper 2 rounded and larger than the lower 3; ± woody climbers or lianes · **35. Canavalia** (vol. 3.5: 36)
   – Fruits and seeds never so large or if nearly so then pods without ribs; if upper stamen connate then corolla smaller · 33

33. Style with a distinct bulging eccentric callus at its junction with the ovary (Fig. 3.3.**2**/10); style less than half the length of ovary; bracteoles absent; usually an extensive climber but sometimes erect and flowering when leafless or almost so · **49. Neorautanenia** (vol. 3.5: 74)
– Style without a callus at its junction with the ovary · · · · · · · · · · · · · · · · · · 34
34. Upper stamen and other short stamens with a pair of fused glands below the anthers (Fig. 3.3.**2**/6); division of style into tenuous and thickened portions obscure, but upper part densely bearded; pollen grains without evident sculpture · · · · · · · · · · · · · · · · · · **55. Vigna** (subgen. **Haydonia**) (vol. 3.5: 121)
– Upper and other short stamens without a pair of glands below the anthers · · · 35
35. Style divided into a thin basal part and a thick hardened upper part · · · · · · 36
– Style either uniformly thick or uniformly thin or tapering upwards or downwards, but not distinctly of two parts · 39
36. Style with thickened part (together with the keel) curved through more than 360°, with the stigma terminal but not penicillate (Fig. 3.3.**2**/17); stipules not produced below point of attachment; standard mostly without appendages; pods oblong to linear-lanceolate, 4.5–20 × 0.7–2.5 cm; pollen grains with no or very fine sculpture · **58. Phaseolus** (vol. 3.5: 160)
– Style with thickened part curved through no more than 360°, usually up to 180° (but in some non-African species the keel itself can be curved much more) · · 37
37. Thickened part of the style (Fig. 3.3.**2**/14) characteristically abruptly curved through 90° just above junction with the tenuous part and narrowed and slightly curved towards the apex, resembling a squarish hook; stipules not produced below the point of attachment; wings round, large, longer than the standard and keel; claws of wings and keel long, partly adnate to the staminal tube; pods long, cylindrical; pollen grains finely reticulate (introduced) · **57. Macroptilium** (vol. 3.5: 159)
– Thickened part of the style not with this characteristic hook-like shape and without the other characters combined · 38
38. Thickened part of the style (Fig. 3.3.**2**/12) curved through 180–360°; stigma terminal, penicillate with a ring of simple or branched hairs, but rest of thick part of the style glabrous; thin part of the style extremely slender, filiform; stipules not produced below the point of insertion; pod long, linear; pollen grains strongly but not widely reticulate; twiner with usually precocious flowers in axillary fascicles · **56. Decorsea** (vol. 3.5: 156)
– Thickened part of the style curved through 90–360°; stigma lateral or nearly so, not penicillate and style often produced as a beak beyond it; thick part of style usually densely bearded; thin part often ribbon-like (Fig. 3.3.**2**/13); stipules often but not always produced below the point of insertion; pollen grains with a very open raised reticulation, except in a few species · · · · · · **55. Vigna** (vol. 3.5: 121)
39. Style distinctly thickened; standard mostly purple and with distinct appendages inside near the base (Fig. 3.3.**1**/12) · 40
– Style not distinctly thickened, often short and inconspicuous; standard mostly without appendages (not to be confused with auricles where claw and blade meet) · 43
40. Style conspicuously flattened laterally, straight and blade-like throughout its length, forming an angle of just less than 90° with the ovary · · · · · · · · · · · · 41
– Style subterete at least in upper part (sometimes expanded apically) · · · · · · 42
41. Style not winged (Fig. 3.3.**2**/11), with a line of hairs near top of inner margin; pods oblong-falcate, broadened distally, to linear-oblong, more than 1 cm wide; seeds with a long arillate hilum · · · · · · · · · · · · · · · · · **50. Lablab** (vol. 3.5: 80)

- Style thin along each margin so as to appear winged on both margins, glabrous (but penicillate beneath the stigma); pods falcate, less than 1 cm wide; seeds unknown · **51. Alistilus** (vol. 3.5: 84)

42. Style expanded at tip into a horizontal spoon-like cover, from which is suspended the spherical stigma, glabrous; upper wing spurs greatly developed · · · · · · · · · ·
· **46. Otoptera** (vol. 3.5: 66)

- Style tapered (Fig. 3.3.**2**/9), with a terminal penicillate stigma; wings not markedly auriculate · **52. Dolichos** (vol. 3.5: 84)

43. Pods rather distinctly transversely grooved with the grooves corresponding to strong internal septa, densely covered with spreading somewhat ferruginous hairs as on the stems; climber with blue flowers in sessile or stalked clusters; standard 0.7–1 cm long; plants introduced · · · **44. Calopogonium** (vol. 3.5: 62)

- Pods not distinctly transversely grooved or, if slightly so, then not so conspicuously septate and inflorescences lax and elongate · · · · · · · · · · · · · 44

44. Pods oblong, 6–9 mm wide, densely covered with stiff spreading bristly hairs; leaflets usually lobed and sericeous beneath with long appressed hairs; standard purple, violet, pink or white, 1.3–2.6 cm long, without appendages; seeds with a reticulate coating · 45

- Pods glabrous to densely hairy but if covered with spreading hairs then pods longer, linear and usually under 5 mm wide (except the cultivated soya bean, *Glycine max*) or inflated · 46

45. Stigma not penicillate; inflorescence long, long-stalked, terminal; style filiform and hairy at the base (Fig. 3.3.**2**/16), upper third virtually glabrous · · · · · · · · · ·
· **38. Pseudeminia** (vol. 3.5: 45)

- Stigma penicillate; inflorescence short, subsessile, axillary; style tenuous and glabrous, the upper part more cylindrical but not really thickened · · · · · · · · · ·
· **39. Pseudovigna** (vol. 3.5: 47)

46. Upper stamen free below but connate above with the rest; inflorescence elongate, with the groups of flowers arising from very conspicuous nodes; corolla mauve, with standard 1–2 cm long; pods linear, compressed, 4–11 cm × 3–4.5 mm; climber (introduced) · · · · · · · · · · · · · · · *Pueraria lobata* (vol. 3.5: 1)

- Upper stamen free, or if connate then free above; other characters not as combined above · 47

47. Erect herbs or subshrubs with somewhat lax to very dense terminal and axillary many-flowered panicles of small reddish-purple or white flowers; standard 6–8 mm long; leaves somewhat undulate; pods narrowly linear-oblong, very flattened, 0.7–3.8 cm × 3–4(5) mm, with fine raised venation (Fig. 3.3.3/1) · · ·
· **67. Pseudarthria** (vol. 3.6: 25)

- Climbing or trailing herbs with less dense inflorescences* · · · · · · · · · · · · · 48

48. Bracteoles absent; inflorescence slender and lax, the many flowers in groups of 2–3 without noticeable swellings of the rachis; standard mauve or violet (sometimes marked yellow), 8–16 mm long; pods glabrous except sometimes for densely ciliate margins · 49

- Bracteoles present; other features not as combined above · · · · · · · · · · · · · 50

49. Bracts conspicuous, striate, persistent; calyx tubular, the standard thus reflexing only towards the apex, without a yellow-green central marking; pods flattened; seeds not arillate · **43. Amphicarpaea** (vol. 3.5: 60)

- Bracts not apparent except at the base of the inflorescence; calyx cupular, with an erect broad standard brilliant-blue to purple with a yellow-green centre; pods

---

* If small creeping perennial herb with sessile 3-foliolate leaves and 2–5-flowered umbels, see *Lotus robsonii* (vol. 3.7: 3).

inflated; seeds with a conspicuous aril; leaves sometimes 5-foliolate with the lower 4 conjugate; cultivated ornamental · · · · · · *Hardenbergia comptoniana* (vol. 3.5: 1)

50. Corolla greenish-yellow, cream or yellow, sometimes with a pink or purple mark; standard 0.6–2.6 cm long, provided inside with 2 long linear lamelliform appendages; stigma terminal, usually ± distinctly penicillate; pollen grains covered with short blunt tubercles · · · · · · · · · · **53. Macrotyloma** (vol. 3.5: 104)

– Corolla blue, mauve, violet or white; standard 0.35–1.4 cm long, devoid of appendages · 51

51. Leaflets elliptic-oblong, obtuse or emarginate; pods (1.5)3–6 cm × 5–9 mm; upper calyx lobes united to form an entire lip; upper filament usually free · **36. Galactia** (vol. 3.5: 40)

– Leaflets elliptic or ovate, acute or acuminate; pods (0.7)1.5–3.6 cm × 2.5–5 mm; upper calyx lobes almost completely united but lip narrowly bifid at the apex · 52

52. Flowers in clusters along the extended rachis; climbing or trailing · **41. Neonotonia** (vol. 3.5: 54)

– Flowers inserted singly along the short axes; erect cultivated annual · *Glycine max* (vol. 3.5: 1)

Tribe 1. **SWARTZIEAE**

by R.K. Brummitt

Swartzieae DC., Prodr. **2**: 422 (1825).

Trees, shrubs or rarely woody climbers. Leaves generally imparipinnate, the leaflets opposite or, less commonly, alternate, sometimes 1-foliolate or simple; stipels sometimes evident; stipules usually present. Inflorescences racemose or paniculate, axillary, terminal or borne on the stems, flowers rarely solitary or fasciculate. Calyx entire in bud, opening irregularly in several segments or in (2)4–5 regular lobes. Petals 0–5. Stamens free (except in *Cordyla* and *Mildbraediodendron*), 5–many, uniform or dimorphic; anthers basifixed or dorsifixed. Ovary sessile to stipitate. Fruits various, drupaceous, moniliform, dehiscent or not, 1–few-seeded. Seeds sometimes arillate, without a hard testa (in the Flora Zambesiaca area, see note under *Swartzia madagascariensis*); radicle short, straight or longer and curved.

A tribe of 16 genera, principally in tropical parts of Central and South America, with four genera in tropical Africa. Formerly included in the Caesalpinioideae or considered intermediate between the two subfamilies, but the wood anatomy, the tendency to form root nodules, the chemistry, cytology, palynology and chromosome numbers all accord with a position at the base of Papilionoideae. This is supported by molecular evidence.

Petals 1; fruit elongate cylindrical · **1. Swartzia**
Petals 0; fruit ellipsoid to subglobose · **2. Cordyla**

1. **SWARTZIA** Schreb.

Swartzia Schreb. in Linnaeus, Gen. Pl., ed.8, **2**: 518 (1791) nom. conserv.
Bobgunnia J.H. Kirkbr. & Wiersema in Brittonia **49**: 1 (1997).

Unarmed trees or rarely shrubs. Leaves imparipinnate or (in South America only) 1–3-foliolate; leaflets opposite or alternate; stipules small. Inflorescence of lateral racemes or sometimes the flowers in panicles or fascicles, or apparently solitary in leaf axils. Flowers hermaphrodite. Receptacle inconspicuous; calyx globose or ellipsoid and entire before dehiscence. Petals 1, or (not in the Flora area) 0 or 3. Stamens many (more than 30), in several series; filaments free or almost so; anthers affixed near the base, dehiscing by longitudinal slits; connective not glandular. Ovary on a long gynophore, with several to many ovules; stigma small, rarely capitate. Pods coriaceous or woody, generally elongate, indehiscent or dehiscing into two valves, 1–several-seeded. Seeds arillate or not, with or without endosperm, the embryo with the radicle curved or straight.

A genus of c.150 species of which two are in tropical Africa (one as below and one from Nigeria to Angola); the others all in tropical America.

Kirkbride & Wiersema in Brittonia **49**: 1–23 (1997) segregated the African species as a separate genus, *Bobgunnia*. The species of *Swartzia*, in their sense, differ from *Bobgunnia* by having 'overgrown' seeds (as in *Cordyla* and *Mildbraediodendron* Harms) with a thin testa and lacking several characteristic features of hard seeds in Papilionoideae. Neotropical species of *Swartzia* also lack resin canals in the mesocarp of the fruit, and the leaflets are generally opposite. The structure of the flowers and inflorescences are, however, very similar and recent molecular analyses group the Old World and New World species of *Swartzia* sensu lato together and apart from other genera in the tribe.

Swartzia madagascariensis Desv. in Ann. Sci. Nat. (Paris) **9**: 424 (1826). —Baker in F.T.A. **2**: 257 (1871). —Sim, For. Fl. Port. E. Africa: 46 (1909). —Baker f., Legum. Trop. Africa: 605 (1929).—Burtt Davy & Hoyle, Check-list For. Trees Shrubs, Nyasaland: 61 (1936).—Gomes e Sousa, Dendrol. Moçamb. **1**: 228 (1950). —Miller in J.S. Afr. Bot. **18**: 36 (1952). —Gilbert & Boutique in F.C.B. **3**: 551 (1952). —Pardy in Rhod. Agric. J. **51**: 274 (1954). —Torre & Hillcoat in C.F.A. **2**: 167 (1956). —White, F.F.N.R.: 128, fig.21K (1962). —Brenan in F.T.E.A., Legum., Caesalp.: 219, fig. 50 (1967). —Schreiber in Merxmüller, Prodr. Fl. SW Afrika, fam. 59: 19 (1970). —Palmer & Pitman, Trees Sthn. Africa: 891 (1972). —K. Coates Palgrave, Trees Sthn. Africa, ed.2: 296 (1988). —M. Coates Palgrave, Trees Sthn. Africa: 351, fig.92 (2002). Origin of type unknown, originally but incorrectly thought to be from Madagascar, *Herb. Desvaux* (P holotype). FIGURE 3.3.4.

Bobgunnia madagascariensis (Desv.) J.H. Kirkbr. & Wiersema in Brittonia **49**: 7 (1997).

Small or medium deciduous tree up to 18 m or occasionally a shrub; bark grey or brown, deeply fissured or reticulate. Young branches, leaf rachides and inflorescence axes and pedicels rather conspicuously covered with a rusty-brown or greyish dense pubescence or tomentum, but fairly quickly somewhat glabrescent, the indumentum inconspicuous at time of fruiting. Leaves: petiole and rachis together (3)6–15(18) cm long; leaflets (3)5–9(13), alternate or occasionally opposite, each (1)2–6(10) × (0.8)1.2–3(5.7) cm, usually elliptic or broadly elliptic, occasionally ovate- or obovate- or oblong-elliptic, rounded to emarginate at the apex, rounded at the base, usually markedly darker above than beneath (particularly when young), the lower surface densely appressed-pubescent and the upper surface sparsely so when young, tending to be glabrescent with age; petiolules 1–3(4) mm; stipels absent or present at the base of the terminal leaflet only. Racemes up to 5(8) cm long, sometimes appearing branched, with up to 10 flowers but sometimes much reduced so that the flowers appear to be in sessile fascicles of 2 or 3 or solitary in the leaf axil; bracts subtending pedicels triangular, up to 3 mm long, rapidly falling and often leaving two short stipules; pedicels 1–3(4) cm long. Calyx globose and 6–7 mm in diameter before anthesis, then splitting for about half its length into 2–5 lobes and reflexing. Petal 1, 2–3.6 × (1.8)2.2–3 cm including the short claw, white with a yellow patch at the base inside, crinkled, densely covered with appressed brown or greyish hairs outside, glabrous inside. Stamens probably 50–60, exceeded by the petal, somewhat unequal, yellow or orange. Gynophore 7–12 mm long, bearing the ovary usually largely clear of the stamens. Pods (6)8–18(30) × 1–1.7(2.3) cm, roughly cylindrical, usually somewhat bumpy and twisted, hard, dark brown to black, pendent,

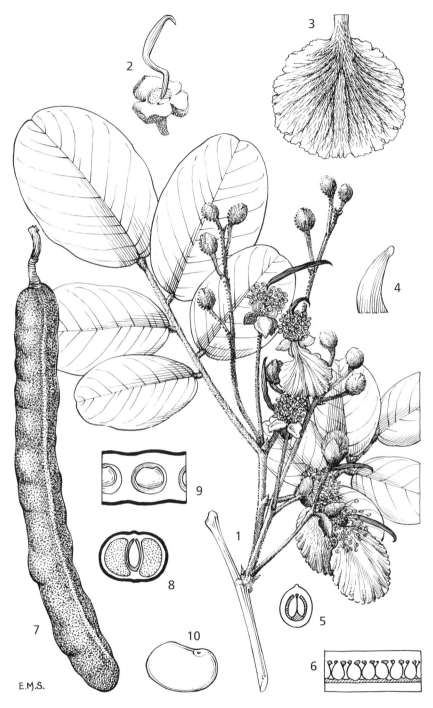

Fig. 3.3.**4**. SWARTZIA MADAGASCARIENSIS. 1, part of flowering branch (× 1); 2, flower, with petal and stamens removed (× 1½); 3, petal, outer side (× 1½); 4, apex of style with stigma (× 6); 5, ovary in transverse section, diagrammatic; 6, ovary in longitudinal section, diagrammatic, 1–6 from *B.D. Burtt* 3417; 7, pod (× ⅔); 8, pod in transverse section, diagrammatic; 9, portion of pod in longitudinal section, diagrammatic; 10, seed (× 3), 7–10 from *B.D. Burtt* 3382. Drawn by Margaret Stones. From F.T.E.A.

indehiscent, with several to many seeds. Seeds 7–8 × 5–7 × 3 mm, oblong-reniform, compressed, with a small hilum below the radicular lobe, pale brown or greyish, shiny.

Caprivi Strip. Andara Mission Station, imm.fr. 16.i.1956, *de Winter* 4279 (K, PRE). **Botswana**. N: Ngamiland Dist., Lake Ngami, fr., vi.1886, *Schinz* s.n. (K). **Zambia**. B: Sesheke Dist., N of Katima Mulilo, st. 18.vii.1952, *Codd* 7110 (K, PRE). N: Mbala Dist., Kawimbe, fl. 19.ix.1959, *Richards* 11481 (K). W: Mwinilunga Dist., c.6 km N of Kalene Hill Mission, fl. 22.ix.1952, *Angus* 516 (FHO, K). C: Lusaka Dist., Chakwenga headwaters, 100–129 km E of Lusaka, fl. 16.xi.1963, *Robinson* 5846 (K). E: Chadiza, fl. 1.xii.1958, *Robson* 790 (K). S: Choma Dist., Choma Nat. Forest (Siamambo Forest Res.), fr. 8.vi.1952, *White* 2926 (FHO, K). **Zimbabwe**. N: Murehwa Dist., Chitowa Native Purchase Area, imm.fr. 25.xii.1966, *Mavi* 101 (K, SRGH). W: Hwange Dist., Victoria Falls, fl.& fr. 12.xi.1919, *Shantz* 409 (K). C: Marondera, Grasslands (Marandellas) Res. Station, Driffield Farm, fl. 27.x.1966, *Corby* 1662 (K, SRGH). E: Mutasa Dist., Honde Valley, fl. 17.xi.1948, *Chase* 1265 (BM, COI, K). S: Bikita Dist., between Bikita and Save R., fl. 21.x.1930, *Fries, Norlindh & Weimarck* 2165 (K). **Malawi**. N: Rumphi Dist., 2 km in, Nyika road, fr. 26.iv.1973, *Pawek* 6569 (K). C: Nkhotakota Dist., Chia area, fr. 1.ix.1946, *Brass* 17479 (K). S: Zomba, fr. 1933, *Clements* 347 (K). **Mozambique**. N: Malema Dist., 10 km along road from Mutuali to Lioma, fr. 9.iii.1953, *Gomes e Sousa* 4057 (COI, K). Z: Mopeia Dist., 26.6 km from road crossing for Morrumbala and Mopeia, st. 30.vii.1949, *Barbosa & Carvalho* 3795 (K). T: between Chicoa and Fíngoè, 19.4 km from Chicoa, fr. 26.vi.1949, *Barbosa & Carvalho* 3290 (K). MS: Manica Dist., near Chimoio (Vila de Manica), by road to Mina André, 18°55'S, 32°55'E, c.850 m, fl.& fr. 25.xi.1961, *Gomes e Sousa* 4736 (COI, K).

Also in West Africa, Congo and Tanzania southwards to N Namibia and the Flora area. (Despite the specific epithet, which was given in error, the species does not occur in Madagascar.) In deciduous woodland and wooded grassland, mainly in sandy soils; 150–1740 m.

The vegetative parts show a remarkable similarity to those of *Pericopsis angolensis* in the Sophoreae, and sterile or immature specimens may easily be confused. *Pericopsis angolensis* usually has a small stipel at the base of each lateral petiolule, but these often fall early and are not an infallible means of distinction. The buds of the *Pericopsis* are more cuneate at the base than in the *Swartzia*, and have two bracteoles at their base, at least when young.

The seeds are used for fish poison by the Shona people in Zimbabwe.

2. CORDYLA Lour.

Cordyla Lour., Fl. Cochinch. **2**: 412 (1790). —Milne-Redhead in Repert. Spec. Nov. Regni Veg. **41**: 227–235 (1937).

Calycandra A. Rich. in Guillemin, Perrottet & Richard, Fl. Seneg. Tent.: 30 (1831).

Unarmed deciduous trees. Leaves imparipinnate; leaflets alternate to rarely subopposite, with minute pellucid dots or dashes between the smaller veins; stipules small, caducous. Inflorescences of rather short and usually clustered racemes. Flowers hermaphrodite or male, the most conspicuous part being the many stamens. Hypanthium (receptacle) well developed, campanulate; calyx subglobose and entire before dehiscence, splitting into (3)5 ± reflexed lobes. Petals 0. Stamens many (23–126), usually crowded into several series round the rim of the hypanthium; filaments connate for up to 3 mm at base; anthers very small (up to 0.5 mm long), dorsifixed, dehiscing by longitudinal slits; connective glandular at the top. Ovary (in hermaphrodite flowers) on a long gynophore exceeding the hypanthium, with several ovules, tapering to a short style with inconspicuous stigma. Fruits ellipsoid to subglobose with a ± oblique beak, fleshy, indehiscent, with 1–6 seeds embedded in pulp. Seeds large, with a thin testa; embryo with a straight radicle.

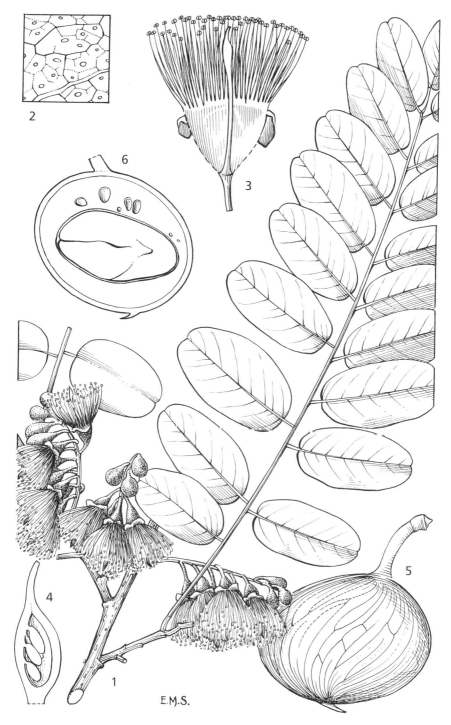

Fig. 3.3.5. CORDYLA AFRICANA. 1, part of flowering branchlet (× 1), from *McCoy-Hill* 19; 2, detail of leaflet surface, showing venation and gland dots (× 6); 3, flower, longitudinal section (× 1½); 4, ovary, longitudinal section (× 6), 2–4 from *Lewis* 38; 5, fruit (× 1); 6, fruit, longitudinal section (× 1), 5 & 6 from *Wild* 2408. Drawn by Margaret Stones. From F.T.E.A.

A genus of 7 species, of which one is from West Africa, two from Madagascar and three from E and NE tropical Africa in addition to the species described below.

Cordyla africana Lour., Fl. Cochinch. **2**: 412 (1790). —Bolle in Peters, Naturw. Reise Mossamb. **6**(1): 17 (1861). —Baker in F.T.A. **2**: 257 (1871) in part. —Sim, For. Fl. Port. E. Africa: 46 (1909). —Baker f., Legum. Trop. Africa: 606 (1929) in part. —Burtt Davy & Hoyle, Check-list For. Trees Shrubs Brit. Emp., Nyasaland: 58 (1936). —Gomes e Sousa, Dendrol. Moçamb., Estudo Geral **1**: 276 (1966). — Pardy in Rhod. Agric. J. **51**: 110 (1954). —White, F.F.N.R.: 121, fig.21F (1962). —Brenan in F.T.E.A., Legum., Caesalp.: 221, fig. 51 (1967). —Palmer & Pitman, Trees Sthn. Africa: 889 (1972). —K. Coates Palgrave, Trees Sthn. Africa, ed.2: 295 (1988). —Pooley, Trees Natal: 156 (1993). —White, Dowsett-Lemaire & Chapman, Evergr. For. Fl. Malawi: 307 (2001). —M. Coates Palgrave, Trees Sthn. Africa: 351, fig.91 (2002). Type: East African coast, *Loureiro* s.n. (P holotype, BM fragm.). FIGURE 3.3.**5**.

Usually a large deciduous tree up to 20(25) m high with a spreading crown; bark brown or grey, longitudinally fissured. Young branches glabrous or minutely pubescent. Leaves: petiole and rachis together (5)9–24 cm long; leaflets 11–30, each (1)2–4(5) × (0.6)1–1.8(2.4) cm, oblong or oblong-elliptic or oblong-ovate, rounded to emarginate at the apex, rounded to truncate at the base, glabrous above, minutely appressed-pubescent beneath; petiolules up to 4 mm long. Racemes usually borne in clusters on shoots of current season below the leaves, 1.5–6(11) cm long, appressed-pubescent; pedicels 3–4 mm long. Hypanthium 5–6 mm long, green. Stamens 23–45, 13–21 mm long, yellow or orange-yellow. Fruits up to 8 × 6 cm, the stipe c.2 cm long, ripening yellow. Seeds embedded in yellow sticky flesh, 1–4, c.3 × 1.5–1.8 cm, oblong.

Zambia. C: Luangwa Dist., Zambezi R. at Luangwa (Feira), fl.& imm.fr. 27.ix.1962, *Angus* 3344 (K, LISC); Mpika Dist., Luangwa Game Res., Mfuwe pontoon, st. 28.iv.1965, *Mitchell* 2707 (K). E: Chipata Dist., foothills E of Machinje Hills, fl.& fr. 12.x.1958, *Robson & Angus* 66 (K; LISC). S: Gwembe, fl. 1.ix.1955, *Bainbridge* 100/55 (FHO, K, LISC). **Zimbabwe**. N: Hurungwe Dist., Mana Pools, fl. ix.1969, *Gordon* 44 (K, SRGH). E: Chipinge Dist., c.6 km E of Birchenough Bridge, fl. 25.xi.1965, *Leach & Müller* 13136 (K, LISC, SRGH). **Malawi**. C: Dedza Dist., c.98 km N of Balaka on lakeshore road, fr. 14.xi.1973, *Pawek* 7504 (K, MAL, MO, SRGH). S: road between Blantyre and Chikwawa, 600 m, fl. 1.x.1946, *Brass* 17884 (K). **Mozambique**. N: Macomia Dist., mouth of Messalo (Msalu) R., fl. 11.ix.1909, *Allen* 55 (K). Z: between Maganja da Costa and Régulo Ingive, 7 km from Maganja, fl. 26.ix.1949, *Barbosa & Carvalho* 4189 (K). T: Tete, fr. *Kirk*, fruit No.68 (K). MS: Chibabava Dist., Madanda Forest, fl. ix.1911, *Dawe* 424 (K). GI: Guijá Dist., between Javanhane and Caniçado, near R. Limpopo, fl. 9.x.1958, *Barbosa* 8348 (COI, K). M: Namaacha Dist., Fonte de Goba, imm.fr. 11.xi.1971, *Marques* 2342 (COI).

Also in the coastal districts of Kenya and Tanzania, and in Swaziland and South Africa (Mpumalanga and N KwaZulu-Natal Province). In the Flora area occurring at low altitudes in hot areas, apparently confined to the coastal plain and larger river valleys, growing in riverine forest or in miombo woodland on escarpment slopes or alluvial plains; sea level to 1000 m.

Fruits reported to be eaten by animals, and sometimes by humans.

Tribe 2. **SOPHOREAE**

by R.K. Brummitt

Sophoreae DC., Anleit. Kennt. Gewäsche **2**: 741 (1818).

Trees, shrubs, woody climbers or rarely herbs. Leaves pinnately 1–many-foliolate or rarely digitately 3-foliolate, sometimes with stipels. Flowers in racemes or panicles or rarely solitary, regular to papilionoid. Calyx valvate to imbricate in bud, rarely closed. Petals (1)5. Stamens (6)10(30), usually free or joined only at the base, outside the Flora Zambesiaca area sometimes joined higher; anthers ± similar. Ovary 1–many-ovulate, sessile to shortly stipitate; style glabrous at least above. Fruits various, dehiscent or not. Seeds oblong-reniform to oblong-ellipsoid or globose, with a small or rarely long apical or lateral hilum, with the hilar groove apparent if hard but often overgrown (i.e. with a thin testa and lacking a hilar groove), sometimes arillate; radicle short, straight or incurved.

A tribe of about 45 genera, widely distributed in tropical and subtropical regions, extending more thinly into temperate areas.

A tribe including primitive genera with free stamens. See also *Podalyrieae*, where the woody genera have a calyx with an intrusive base.

Cultivated species

Castanospermum australe Hook., the Moreton Bay Chestnut, native to NE Australia, New Caledonia and New Hebrides, is cultivated in the Flora Zambesiaca area. Tree with large imparipinnate leaves; leaflets 7–19, alternate, 8–20 × 2–6 cm, elliptic-oblong, coriaceous; flowers borne in racemes on older wood along the branches, yellow to orange-red or red, the standard spathulate, the wings and keel petals little differentiated, the stamens red exserted and 3–4 cm long; pod woody, with 2–5 globose shiny brown seeds up to 4 cm long. It was recorded as early as 1912 in Harare (*H.G. Mundy* (K)).

Styphnolobium japonicum (L.) Schott (*Sophora japonica* L.), Japanese Pagoda Tree, native of China, is grown in Harare as a garden ornamental. Tree to 25 m; leaves with 7–17 ovate-oblong leaflets, stipellate; flowers 1.2–1.5 cm long in terminal panicles, white to pink; fruits 3–12 × 0.7–1 cm, moniliform, indehiscent. **Zimbabwe** C: Highlands, Pevensey Road, fl. 12.xii.1968, *Biegel* 2726 (K, SRGH); Greendale, fr. 9.iv.1969, *Biegel* 2920 (K, SRGH).

1. Leaves 1-foliolate; calyx splitting into 2 or spathaceous · · · · · · · · · · · · **7. Baphia**
– Leaves compound; calyx subentire to 5-lobed · 2
2. Corolla with 4 lower petals all similar · 3
– Corolla papilionoid, with 4 lower petals differentiated as wings and keel · · · · 4
3. Flowers c.1 cm long, petals white; fruits indehiscent, drupaceous · · · · · · · · · · · ·
· **3. Xanthocercis**
– Flowers 3–4 cm long, petals yellow to red; fruit a dehiscent pod; ornamental tree (cultivated) · **Castanospermum**
4. Calyx lobes nearly as long as the tube or longer; fruits flattened · · · · · · · · · · 5
– Calyx teeth usually very short, less than ⅓ length of calyx; fruits moniliform, constricted between the seeds · 6
5. Leaflets not markedly unequal-sided; petals greenish-white to pink or violet with purple veins; fruit winged along both edges* · · · · · · · · · · · · · · · · · **4. Pericopsis**

---

* If a cultivated tree with falcate keel and velvety pods, see also *Virgilia* vol. 3.7: 53.

 – Leaflets falcate-lanceolate, markedly unequal-sided; petals blue (rarely white); fruit not winged · **5. Bolusanthus**
6. Bracteoles and stipels lacking; herbs, shrubs or small trees · · · · · · · · **6. Sophora**
 – Bracteoles present on pedicel; stipels present; large ornamental tree (cultivated) · **Styphnolobium**

3. XANTHOCERCIS Baill.

Xanthocercis Baill. in Adansonia **9**: 293 (1870).
Pseudocadia Harms in Bot. Jahrb. Syst. **33**: 162 (1902).

Trees up to 30 m high. Leaves imparipinnate; leaflets (7)9–11(13), alternate or in subopposite pairs, coriaceous, without stipels. Inflorescence of terminal and axillary simple racemes; bracts small, falling early; bracteoles usually near the base of the pedicel, insignificant. Flowers with a small perigynous disc up to 1 mm broad. Calyx campanulate, ± truncate above, apparently entire or with 5 very short teeth. Petals (at least in *X. zambesiaca*) white, the median part somewhat thickened and darker; each with a claw ± as long as the limb, the standard broader than the others with an oblong limb, the other 4 petals similar to each other, ± entire and lanceolate, free. Ovary subsessile, pubescent with a short glabrous style. Fruit brown or greenish-brown and shiny, an indehiscent oblongoid to subspherical drupe, stipitate, 1–3-seeded. Seeds brown or black, ± rectangular or reniform, attached in the centre of their larger side, with a small hilum.

A genus of 3 species, the others in Madagascar and Gabon.
Root nodules have been confirmed as absent from this genus, see Sprent, Nodulation in Legumes: 63 (2001).

Xanthocercis zambesiaca (Baker) Dumaz-le-Grand in Bull. Soc. Bot. France **99**: 314 (1953). —Palmer & Pitman, Trees Sthn. Africa: 951–952 (1972). —K. Coates Palgrave, Trees Sthn. Africa, ed.2: 297 (1988). —M. Coates Palgrave, Trees Sthn. Africa: 352, fig. 93 (2002). Type: Zambia S, highlands of Batoka Country, x.1860, *Kirk* s.n. (K holotype). FIGURE 3.3.**6**.
    Sophora ?zambesiaca Baker in F.T.A. **2**: 253 (1871).
    Pseudocadia zambesiaca (Baker) Harms in Engler, Pflanzenw. Afr. **3**(1): 524 (1915). —Baker f., Legum. Trop. Africa: 604 (1929). —Miller in J.S. Afr. Bot. **18**: 35 (1952). —Codd, Trees & Shrubs Kruger Nat. Park: 75, fig.71 (1951). —Pardy in Rhod. Agric. J. **51**: 172 (1954). —White, F.F.N.R: 161 (1962).

Usually a large spreading round-topped tree up to 20(30) m high, with a stout trunk branching low down, evergreen, or ± deciduous in drier places. Leaves with (7)9–11(13) leaflets, each (1.5)2–5(7) × (0.8)1–2(2.4) cm, elliptic or the proximal ones ovate and the distal ones obovate, proximal ones usually ± strongly asymmetrical at the base and rounded or emarginate at the apex, the distal ones subsymmetrical and cuneate at the base and rounded or acute at the apex, all coriaceous, dark green and shiny above, pubescent but glabrescent beneath, glabrous above; rachis and petiole up to 12(15) cm long, pubescent. Racemes many-flowered, 4–8(12) cm long, softly pubescent. Calyx c.3 mm long, greyish appressed-pubescent, with 5 very short teeth. Petals white, (8)9–10 mm long. Fruit up to 2.5(3) cm long on a stipe up to 1 cm long, 1- or rarely 2-seeded.

Botswana. SE: Central Dist., Mangwato Reserve, under Sefare (Sofala) Hill, fl.& fr. x.1940, *Miller* B249 (K, PRE, SRGH). **Zambia**. C: Luangwa Dist., Luangwa (Feira), on anthill, fr. 16.ii.1963, *Grout* 290 (K). S: Gwembe Dist., Zongwe R., c.10 km in from Zambezi R., fr. 14.vi.1957, *Scudder* 69 (K; SRGH). **Zimbabwe**. N: Hurungwe Dist., Zambezi Valley, Chirundu road, fl. 29.x.1952, *Lovemore* 282 (K, PRE, SRGH). E: Chipinge Dist., Birchenough Bridge, fl. 22.x.1948, *Chase* 1204 (K, LISC, SRGH). S:

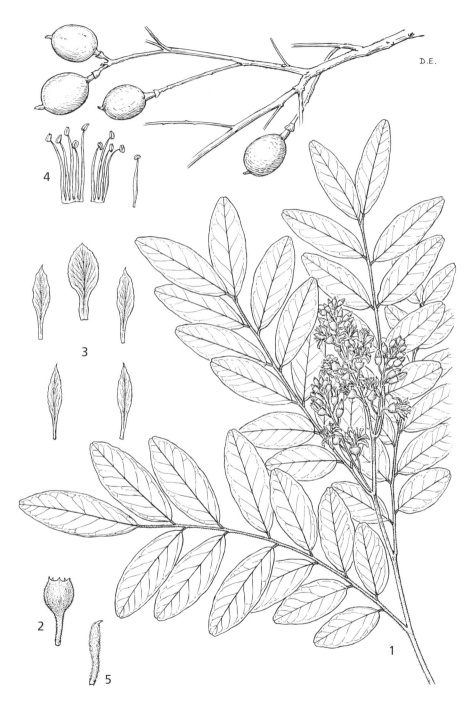

Fig. 3.3.**6**. XANTHOCERCIS ZAMBESIACA. 1, flowering branchlet (× ²/₃); 2, calyx (× 2); 3, petals (× 2); 4, stamens (× 2); 5, gynoecium (× 2), 1–5 from *Davies* 1572; 6, infructescence (× ²/₃), from *Torre* 7961. Drawn by Derrick Erasmus.

Chiredzi Dist., N side of Runde (Lundi) R. near Chipinda Pools, fl. 7.xi.1959, *Goodier* 616 (K, PRE, SRGH). **Malawi**. C: Dedza Dist., Chipaka near DC's Resthouse, fl.& fr. 22.xi.1954, *Adlard* 202 (K, PRE). S: Mangochi Dist., Lundwe, Mangochi (Fort Johnston), fr. 12.vii.1954, *G. Jackson* 1366 (K). **Mozambique**. T: Changara Dist., at side of road from Tete to Chipembere (Chioco), 30 km from Chipembere (Chioco), fr. 6.vi.1962, *Gomes e Sousa* 4769 (K, LMA, PRE). MS: Marromeu Dist., near Chupanga, fr. 13.vii.1941, *Torre* 3091 (LISC). GI: Guijá, road towards Pafuri, c.3 km from bridge (barragem), fl. 16.xi.1957, *Barbosa & Lemos* 8169 (COI, K, LISC, LMA). M: Moamba Dist., R. Incomati, between Moamba and Sábiè, fl. 28.xi.1944, *Mendonça* 3103 (LISC).

Also in South Africa (Limpopo and Mpumalanga Provinces). At low altitudes in hot areas, in rich alluvial soils of the major river valleys, on river banks and beside lakes, sometimes in drier places, often associated with termite mounds; up to 1500 m.

Reported as grown in Zimbabwe gardens by Biegel, Check-list Ornam. Pl. Rhod. Parks & Gard.: 108 (1977).

The fruits are reported to be eaten by birds and animals, and sometimes by man.

4. PERICOPSIS Thwaites

Pericopsis Thwaites, Enum. Pl. Zeyl.: 413 (1864). —Meeuwen in Bull. Jard. Bot. État **32**: 213–219 (1962).

Afrormosia Harms in Engler & Prantl, Nat. Pflanzenfam., Nachtr. **3**: 158 (1906).

Large shrubs or more usually trees. Leaves with (5)7–12(13) leaflets, which usually appear alternate but are occasionally in opposite pairs, lateral leaflets with a single stipel at the base, terminal one with 2 usually unequal stipels. Inflorescence a terminal panicle or rarely a simple raceme, sparsely to densely brown- or greyish-pubescent or subglabrous; bracts small, ± linear-oblong, caducous; bracteoles similar, caducous. Flowers with a small hypanthium (receptacular disc). Calyx campanulate below, with 5 teeth 2–4 times as long as the campanulate part, the upper 2 remaining connate for most of their length, the whole calyx eventually falling as one piece. Petals white, greenish-white or violet, with dark purple veins, the standard with a purple or yellowish blotch near the base; standard suborbicular with a short claw, the limb usually reflexed; wings with a short claw up and a well-developed auricle at the base of the limb; keel petals with a claw up and a well-developed auricle. Ovary subsessile, pubescent with a long glabrous stipe. Pods indehiscent, variable in size and shape, ± flat, broadly oblong to linear-oblong, ± winged along the upper margin, slightly woody, glabrous and smooth or rarely pubescent, the proximal part often constricted to form a false stipe, the longer pods often also constricted about the middle. Seeds flat, oblong to suborbicular, reddish, with a small hilum; radicle short, straight.

A genus of 4 or 5 species, one occurring in Sri Lanka and the Malay Archipelago to Micronesia, the others restricted to Africa.

Sprent (Nodulation in Legumes: 85, 2001) recorded root nodules for 3 species including *P. angolensis*.

Pericopsis angolensis (Baker) Meeuwen in Bull. Jard. Bot. État **32**: 216 (1962). —Polhill in F.T.E.A., Legum., Pap.: 41, fig.7 (1971). —Gonçalves in Garcia de Orta, sér.Bot. **5**: 100 (1982). —K. Coates Palgrave, Trees Sthn. Africa, ed.2: 297 (1988). —M. Coates Palgrave, Trees Sthn. Africa: 353, fig. 92 (2002). Type: Angola, Huila, Humpata–Mumpula, *Welwitsch* 615 (BM, K).

    Ormosia angolensis Baker in F.T.A. **2**: 255 (1871). —Steedman, Trees, Shrubs & Lianes S. Rhod.: 26 (1933).

    Afrormosia angolensis (Baker) De Wild. in Repert. Spec. Nov. Regni Veg. **11**: 507 (Jan. 1913). —Harms in Bot. Jahrb. Syst. **49**: 431 (Mar. 1913); in Engler, Pflanzenw. Afrikas **3**(1): 527 (1915). —Baker f., Legum. Trop. Africa: 600 (1929). —Gomes e Sousa, Dendrol. Moçam. **1**:

80 (1950). —Pardy in Rhod. Agric. J. **49**: 257 (1952). —Toussaint in F.C.B. **4**: 41 (1953). —
O. Coates Palgrave, Trees Central Africa: 309 (1957). —White, F.F.N.R.: 143 (1962).

 Afrormosia bequaertii De Wild. in Repert. Spec. Nov. Regni Veg. **11**: 506 (1913). —Baker f.,
Legum. Trop. Africa: 600 (1929). Type from Congo (Katanga).

 Afrormosia brasseuriana var. *subtomentosa* De Wild., Pl. Bequaert. **3**: 246 (1925). —Baker f.,
Legum. Trop. Africa: 601 (1929). Type from Congo.

 Afrormosia angolensis var. *subtomentosa* (De Wild.) Louis in Louis & Fouarge, Ess. For. et Bois
Cong. **2**: 4 (1943); in Bull. Jard. Bot. État **17**: 115 (1943). —Toussaint in F.C.B. **4**: 42 (1953).

 Pericopsis angolensis var. *subtomentosa* (De Wild.) Meeuwen in Bull. Jard. Bot. État **32**: 216
(1962).

Tree up to 17 m high, or sometimes a large shrub, deciduous, flowering with young leaves.
Leaves with 7–10(13) leaflets arranged alternately or rarely in opposite pairs; leaflets
(2)3.5–6.5(9.5) × (1.5)2–3.5(4) cm, ovate to elliptic or the distal ones obovate, rounded to
cuneate and ± asymmetrical at the base, rounded to emarginate at the apex; terminal leaflet
usually rather broader than the laterals; stipels usually inconspicuous and caducous, 1–2(4) mm
long; leaflets glabrous except on the underside of the midrib, or sometimes with varying
development of hairs on the lower or both surfaces (see notes below); petiole, rachis and
petiolules rusty-brown or rarely greyish puberulous to tomentose, sometimes ± glabrescent.
Inflorescence (including calyces) usually rusty-brown or rarely greyish tomentose. Calyx
(excluding hypanthium) 7–10(11) mm long, including lobes (5)6–8 mm long. Petals white,
greenish-white or violet with dark purple veins; standard 13–15 mm long, wings slightly longer.
Ovary, pods and seeds as in generic description.

Forma **angolensis**. FIGURE 3.3.**7**.

 Petiole, leaf rachis and midrib of leaflets beneath conspicuously hairy.

 Zambia. N: Mbala Dist., Uningi pans near Mbala, fl. 1.x.1956, *Richards* 6333 (K,
SRGH). W: Copperbelt, Mufulira Dist., fl. 31.x.1948, *Cruse* 410 (K). C: Lusaka Dist.,
Mt. Makulu, 19 km S of Lusaka, fl. 31.x.1956, *Angus* 1432 (K, PRE). E: Chipata, fl.
1.xi.1950, *Gilges* 3597A (SRGH). S: Choma Dist., Mapanza, fl. 13.xi.1958, *E.A.
Robinson* 2928 (K, PRE, SRGH). **Zimbabwe**. N: Hurungwe Dist., Zvipani (Zwipani), fl.
12.x.1957, *Phipps* 786 (K, LISC, PRE, SRGH). E: Mutare (Umtali) Golf Course, fl.
11.xi.1948, *Chase* 1335 (BM, K, LISC, SRGH). **Malawi**. N: Njakwa Gorge, 2 km from
Rumphi Boma in gorge beside Rumphi R., fl. xi.1953, *Chapman* 175 (K). C:
Nkhotakota Dist., Chia area, fr. 6.ix.1946, *Brass* 17555 (K, SRGH). S: Zomba, fl.& fr.
i.1916, *Purves* 235 (K). **Mozambique**. N: Malema Dist., Mutuáli–Malema road, 12 km
from Mutuáli, fr. 25.ii.1954, *Gomes e Sousa* 4211 (COI, K, LISC, LMA, PRE, SRGH). Z:
near Milange, fl. 16.xi.1942, *Mendonça* 1434 (LISC). T: between Kazula (Casula) and
Furancungo, 32.3 km from Kazula (Casula), fr. 9.vii.1949, *Barbosa & Carvalho* 3516
(K, LMA, SRGH). MS: Gondola Dist., Chimoio, near Tembe, fr. 10.v.1948, *Andrada*
1242 (LISC).

Also in Tanzania, Congo and Angola. Usually in dry open woodland or wooded
grassland at low to medium altitudes; up to 1700 m.

This species shows considerable variation in pubescence of leaflets and leaf rachis.
In the south and east of the Flora area the leaflets are consistently glabrous except
on the midrib on the lower surface, but to the north and west there is a tendency
towards development of sparse hairs over the lower surface and occasionally on the
upper surface as well. A single specimen from the extreme north of Zambia (N:
Chiengi, fl. 12.x.1949, *Bullock* 1248 (K)) differs conspicuously from all others in
having the leaflets densely appressed-pubescent and almost sericeous, the
pubescence of the whole plant being greyish rather than the usual rusty-brown. It
may be referable to var. *subtomentosa* (De Wild.) Meeuwen (see synonymy above)
described from neighbouring Katanga. This variety is, however, distinguished by

Fig. 3.3.7. PERICOPSIS ANGOLENSIS forma ANGOLENSIS. 1, flowering branchlet (× ²/₃), from *Richards* 6333; 2, young leaf (× ²/₃); 3, detail of leaflet surface (× 6), 2 & 3 from *Bullock* 1248; 4, calyx (× 1); 5, standard (× 1); 6, wings (× 1); 7, keel petals (× 1); 8, stamen (× 1); 9, gynoecium (× 1), 4–9 from *Richards* 6333; 10, infructescence (× ²/₃), from *Robinson* 213. Drawn by Derrick Erasmus.

Louis and by Toussaint (1953) mainly on the basis of its pubescent pod rather than pubescent leaflets, but our specimen is in flower and its pod is unknown. Another specimen (Kaputa Dist., Msanka Plain, fl. 25.ix.1956, *Richards* 6292 (K)) is intermediate between *Bullock* 1248 and the typical plant. Further investigation of this variety is required. Other specimens differ from the typical plant in having the petiole, rachis and leaflet midrib completely glabrous and so appearing conspicuously dark, even when young. They occur sporadically throughout the range of the species and are distinguished only as a form.

The wood is fine-grained, hard and termite resistant; it is valued particularly for fence posts, panelling and flooring blocks.

Forma **brasseuriana** (De Wild.) Brummitt in Kew Bull. **21**: 244 (1967). —Gonçalves in Garcia de Orta, sér. Bot. **5**: 101 (1982). Type: Congo, Katanga, Lukafu, *Verdick* 38, 180 & 197 (BR syntypes).

    Ormosia brasseuriana De Wild. in Ann. Mus. Congo, Bot. **1**: 183 (1903).

    Afrormosia brasseuriana (De Wild.) De Wild. in Repert. Spec. Nov. Regni Veg. **11**: 507 (Jan. 1913). —Harms in Bot. Jahrb. Syst. **49**: 431 (Mar. 1913). —Baker f., Legum. Trop. Africa: 601 (1929). —Delevoy, Quest. For. Katanga **2**: 21 (1929).

    Afrormosia angolensis var. *brasseuriana* (De Wild.) Louis in Louis & Fouarge, Ess. For. et Bois Cong. **2**: 4 (1943); in Bull. Jard. Bot. État **17**: 114 (1943). —Topham, Check List For. Trees Shrubs Nyasaland Prot.: 74 (1958).

Petiole, leaf rachis and leaflet midrib glabrous or subglabrous, even when young, and often conspicuously dark in colour.

Zambia. N: Chungu, fl. 19.x.1961, *Astle* 970 (SRGH). S: between Livingstone and Kalomo, fr. 10.vii.1930, *Pole Evans* 2789 (46) (SRGH). **Zimbabwe**. N: Hurungwe Dist., between Ranger's camp and main Chirundu–Harare road, fr. 10.v.1951, *Lovemore* 45 (SRGH). **Mozambique**. T: Moatize Dist., Montes de Zóbuè, fl. 20.x.1941, *Torre* 3687 (LISC). MS: Sussundenga Dist., between Mavita and Rotanda, st. 7.ii.1948, *Barbosa* 990 (LISC).

Also in Congo and Tanzania; 400–1250 m.

5. BOLUSANTHUS Harms

Bolusanthus Harms in Repert. Spec. Nov. Regni Veg. **2**: 14 (1906).

Small tree. Leaves imparipinnate; leaflets alternate to subopposite proximally, more strictly opposite distally, well spaced, slightly falcate, unequal-sided, on well-developed petiolules, without stipels; stipules small. Racemes terminal and from the upper axils; bracts fused with the lower part of the pedicels and so apparently arising a short distance above their base; bracteoles on the pedicel small. Calyx campanulate above a short hypanthium (receptacle); lateral lobes about as long as the tube, upper ones joined higher. Petals blue to violet or purple or rarely white; standard elliptic-obovate, reflexed above the claw; wings sculptured between the upper veins; keel oblong-falcate. Stamens 10, free or almost so; anthers dorsifixed. Ovary very shortly stalked; style curved, tapered, glabrous above, with a small stigma. Pod shortly stipitate, narrowly oblong, flattened, indehiscent, compressed between the few slightly hardened seed chambers. Seed oblong-reniform, with a small radicular lobe and small hilum; embryo with a short inflexed radicle.

A monotypic genus, similar to *Platycelyphium* from the drier parts of E and NE Africa and several small genera in Madagascar. It is much like *Calpurnia*, currently referred to the *Podalyrieae*, a contiguous tribe (see Vol. 3.7: 53).

Root nodules are recorded by Sprent, Nodulation in Legumes: 84 (2001).

Bolusanthus speciosus (Bolus) Harms in Repert. Spec. Nov. Regni Veg. **2**: 15 (1906).
—Sim, For. Fl. Port. E. Africa: 44 (1909). —Eyles in Trans. Roy. Soc. S. Afr. **5**: 369
(1916). —Phillips in Fl. Pl. S. Afr. **1**: t.23 (1921). —Baker f., Legum. Trop. Africa:
594 (1929); in Bol. Soc. Brot., sér.2, **8**: 114 (1933). —Steedman, Trees, Shrubs &
Lianes S. Rhod.: 22 (1933). —Gomes e Sousa, Dendrol. Moçam. **2**: 62 (1949). —
Pardy in Rhod. Agric. J. **48**: 318 (1951). —Miller in J.S. Afr. Bot. **18**: 29 (1952). —
White, F.F.N.R.: 144 (1962).—Palmer & Pitman, Trees Sthn. Africa: 901 (1972).
—K. Coates Palgrave, Trees Sthn. Africa, ed.2: 300 (1988). —M. Coates Palgrave,
Trees Sthn. Africa: 356, fig. 95 (2002). Types from South Africa (Mpumalanga).
FIGURE 3.3.8.

Lonchocarpus speciosus Bolus in J. Linn. Soc., Bot. **25**: 161 (1889).

Bolusanthus spp. sensu Eyles in Trans. Roy. Soc. S. Afr. **5**: 369 (1916) with respect to *Eyles*
176 & 1079.

Smallish deciduous tree, up to 8(12) m tall, flowering with the young leaves. Young branches
and inflorescences and very young leaves ± densely appressed-pubescent and often sericeous.
Leaves up to 18(22) cm long, (2)3–6(7)-jugate; lateral leaflets up to 7(9) × 2.3(3.5) cm, falcate-
lanceolate, asymmetrical at the base, acute at the apex; terminal leaflet narrowly elliptic,
symmetrical; all leaflets glabrescent above, usually persistently appressed-pubescent beneath.
Racemes 10–20 cm long, 16–40-flowered; pedicels (7)10–20(22) mm long; bracteoles c.1 mm,
usually not opposite, quickly falling. Standard 13–18 × 13–18 mm, claw shallowly channelled;
wing and keel petals ± as long as the standard. Ovary with a gynophore up to 1 mm long. Pod
6–9 × 0.7–1 cm, narrowly oblong, grey-brown, on a stipe 3–5 mm long at maturity, 3–5-seeded.
Seeds 9–10 × 4.5–5 mm, grey-brown.

Botswana. N: Central Dist., 16 km S of Tsigara (Sigara) Pan, st. 25.iv.1957,
Drummond & Seagrief 5232 (K, SRGH). SE: Central Dist., Tshimoyapula
(Tshimoapula), fr. 8.iv.1959, *de Beer* 878a (K, SRGH). **Zambia**. S: Livingstone Dist.,
near Victoria Falls, fl. 13.ix.1957, *Angus* 1709 (K). **Zimbabwe**. N: Mount Darwin Dist.,
Machikachika's, foot of Zambezi Valley escarpment, st. 5.vi.1923, *Swynnerton* s.n. (K).
W: Umguza Dist., Nyamandhlovu Pasture Research Station, fl. 6.x.1953, *Plowes* 1639
(K, PRE, SRGH). C: Chegutu Dist., Poole Farm, fl. 2.x.1954, *R.M. Hornby* 3363
(SRGH). E: Mutare Dist., no locality, fl. ix.1948, *Chase* 1215 (BM, K, SRGH). S:
Masvingo Dist., Mushandike (Umshandige), fr. 6.x.1949, *Wild* 2983 (K, SRGH).
Malawi. S: Blantyre Dist., 48 km N of Blantyre, Matope, 2 km from Shire R., fl.
5.ix.1960, *Willan* 59 (K, SRGH). **Mozambique**. MS: Cheringoma Dist., Inhamitanga,
fr. 31.x.1945, *Simão* 613 (LISC). GI: Vilankulo Dist., Cabo Xelim, near Vilanculos, fl.
viii.1937, *Gomes e Sousa* 2000 (K, LISC). M: Namaacha Dist., between Catuane and
Goba, fr. 21.xi.1944, *Mendonça* 2996 (BM, K, LISC).

Also in South Africa (Limpopo, Mpumalanga and KwaZulu-Natal Provinces) and
Swaziland. Widespread at low to medium altitudes, usually in open woodland, often
with mopane; up to 1600 m.

Steedman (1933) noted that a white-flowered 'variety' grows in the Mbalabala area
S of Bulawayo in Zimbabwe. Two white-flowered collections have been seen, both
from Zimbabwe: Tokwani R., 3.x.1951, *Pole Evans* 4709 (K, PRE); and Esigodini
(Essexvale), s.n., *Cheeseman* 39 (BM).

Eyles (1916) gave four Zimbabwe specimens as *Bolusanthus* spp., apparently
excluding them from *B. speciosus*. Two of these (*Eyles* 176 and 1079) have been seen
and are referable to *B. speciosus*, but the other two (*Monro* 512 and *Chubb* 300) have
not been traced.

Also grown as a street tree, e.g. Zambia, Kitwe, fl. 5.ix.1968, *Fanshawe* 10316 (K,
NDO); Zimbabwe, Harare, fl.& fr. 12.x.1983, *Müller* s.n. (K, SRGH).

J.C.W.

Fig. 3.3.8. BOLUSANTHUS SPECIOSUS. 1, flowering branchlet (× 1), from *Lemos &*
Balsinhas 185 and *H. Bolus* 1144; 2, calyx, opened out (× 1¹/₂); 3, standard (× 1¹/₂); 4, wing petal
(× 1¹/₂); 5, keel petals (× 1¹/₂); 6, stamens (× 3); 7, gynoecium (× 3), 2–7 from *Lemos & Balsinhas*
185; 8, infructescence (× 1), from *White* 2231. Drawn by Joanna Webb.

6. SOPHORA L.

Sophora L., Sp. Pl.: 373 (1753); Gen. Pl., ed.5: 175 (1754).

Trees or shrubs or rarely perennial herbs. Leaves imparipinnate, with 4–18(32) leaflets per side, glabrous to densely tomentose; stipels absent. Inflorescences terminal or axillary, consisting of few- to many-flowered racemes; bracts often fairly large; bracteoles small when present but usually apparently absent. Flowers markedly perigynous, with a distinct hypanthium. Calyx campanulate to tubular, with very shallow to prominent and acute lobes, the upper 2 often fused. Petals yellow, white, blue or purple, small to rather large; standard usually gradually narrowed below into a short claw, the limb ± reflexed; keel petals overlapping or joined on the lower side. Stamens free to shortly joined at the base; anthers dorsifixed. Ovary shortly stalked. Pod moniliform, often winged, with 1–14 seeds, dehiscent or tardily breaking up irregularly. Seeds ovoid, ellipsoid or globose, usually without a distinct radicular lobe and with a small hilum; radicle short, ± straight or incurved.

A genus of about 50 species, widely distributed in most of the warmer parts of the world, but only three species in Africa.

Root nodules have been recorded for all three native species in the Flora area, see Sprent, Nodulation in Legumes: 85, 103 (2001).

Sophora tetraptera J.S. Mill., Kowhai, native of New Zealand, is mentioned by Biegel, Check-list Ornam. Pl. Rhod. Parks & Gard.: 99 (1977). Shrub or small tree to 12 m; leaves with 20–40 small, hairy, ovate or oblong-elliptic leaflets; flowers 4–5 cm long, yellow, in pendent racemes, with keel much exceeding the standard; fruits 5–20 cm long, 4-winged, with constrictions between the seeds.

1. Leaflets glabrous on upper surface; calyx with conspicuous teeth c.2 mm long; inland · **3.** *velutina*
   – Leaflets pubescent to sericeous on upper surface, at least when young; calyx with shallow teeth up to 1 mm long; coastal · 2
2. Leaflets (except terminal one) broadly elliptic to suborbicular, rounded at base and apex, tomentose beneath, densely pubescent but glabrescent above, not sericeous, or slightly so on lower surface only · · · · · · · · · · · · · · · · **1.** *tomentosa*
   – Leaflets elliptic to obovate, cuneate at base, obtuse at apex, conspicuously sericeous on both surfaces · **2.** *inhambanensis*

1. **Sophora tomentosa** L., Sp. Pl.: 373 (1753). —Baker in F.T.A. **2**: 254 (1871) in part excl. syn. *S. inhambanensis*. —Sim, For. Fl. Port. E. Africa: 46 (1909) in part excl. syn. *S. inhambanensis*. —Baker f., Legum. Trop. Africa: 595 (1929) in part excl. syn. *S. inhambanensis*. —Brummitt & Gillett in Kirkia **5**: 264 (1966). —Polhill in F.T.E.A., Legum., Pap.: 44 (1971). Type: Sri Lanka, *Hermann* 1: 61 & 3: 13 (BM syntypes).

Subsp. **tomentosa**

Shrub 1–3 m high. Young branches, leaf petioles and rachides, undersurface of leaflets, inflorescences and calyces densely white- or grey-tomentose, upper surface of leaflets densely pubescent when young but glabrescent later. Leaves (rachis and petiole) 7–25 cm long, with (4)5–7(9) pairs of leaflets; lateral leaflets up to 4 × 3 cm, broadly elliptic to suborbicular, asymmetrical at the base with the distal side rounded and the proximal side ± cuneate, rounded at the apex; terminal leaflet slightly larger than the lateral ones, equal and ± cuneate at the base. Flowers in terminal many-flowered racemes up to 25 cm long; bracts 3–5 mm long, linear, caducous; pedicels 5–10 mm long; bracteoles apparently absent. Calyx 4–6 mm long, campanulate, with 5 shallow teeth up to 1 mm long. Petals yellow; standard 14–20 × 11–14 mm, the limb cuneate and gradually narrowed below to a short broad claw; wings and keel petals about equalling or slightly shorter than the standard. Ovary densely ± appressed-pubescent. Pods moniliform, mostly with 4–8 seeds developing. Seeds c.6–7 mm long, subglobose, brownish.

Mozambique. N: Palma Dist., Tecomaji Is. (Tecomaze), edge of beach, fl.& fr. 29.iii.1961, *Gomes e Sousa* 4678 (COI, K, LMA, PRE, SRGH). GI: Vilanculos, fr. viii.1928, *Earthy* 174 (PRE).

This subspecies is widespread on tropical and subtropical coasts from Kenya, Tanzania, Madagascar, Seychelles, S India, Sri Lanka, Malay Peninsula, Indo-China, Taiwan, Indonesia, Philippines, New Guinea, E Australia and Polynesia. Sandy foreshores, often as a pioneer colonist.

Subsp. *occidentalis* (L.) Brummitt occurs in W Africa, and in E South America and West Indies, and differs in having a much finer indumentum.

2. **Sophora inhambanensis** Klotzsch in Peters, Naturw. Reise Mossamb. **6**(1): 26 (1861). —Brummitt & Gillett in Kirkia **5**: 264 (1966). —Polhill in F.T.E.A., Legum., Pap.: 46, fig.8 (1971). —Palmer & Pitman, Trees Sthn. Africa: 897 (1972). —K. Coates Palgrave, Trees Sthn. Africa, ed.2: 298 (1988). —M. Coates Palgrave, Trees Sthn. Africa: 353 (2002). Types: Mozambique, Inhambane, *Peters* (B† syntype); Maputo (Delagoa Bay), *Peters* (B† syntype); Mozambique, Zambezi R. mouth (coast), viii.1862, *Kirk* (K neotype, see Brummitt & Gillett 1966). FIGURE 3.3.**9**.

Sophora nitens Harv. in F.C. **2**: 266 (1862), non Schumach. & Thonn. (1827). Type: South Africa, KwaZulu-Natal, near Durban (Port Natal), *T. Williamson* (TCD holotype, K).

Sophora tomentosa sensu Baker in F.T.A. **2**: 254 (1871) in part as regards syn. *S. inhambanensis*. —sensu Sim, For. Fl. Port. E. Africa: 46 (1909) in part for syn. *S. inhambanensis*, excl. fig.51. —sensu Baker f., Legum. Trop. Africa: 595 (1929) in part for syn. *S. inhambanensis*. —sensu Henkel, Woody Pl. Natal & Zululand: 210 (1934). —sensu Mogg in Macnae & Kalk, Nat. Hist. Inhaca Is., ed.2: 13, 146 (1969).

Shrub 1–2 m high. Young branches, leaves, inflorescences and calyces densely whitish to silvery or golden pubescent, the calyces and both surfaces of the leaflets strongly sericeous. Leaves (rachis and petiole) 7–12(16) cm long, with (3)5–7(8) pairs of leaflets; lateral leaflets 1.8–4(5) × 0.7–1.5(2) cm, elliptic to obovate or oblanceolate, cuneate and slightly asymmetrical at the base, obtuse to subacute at the apex; terminal leaflet 1.5–2 times as long as the lateral ones, cuneate and equal at the base. Flowers in terminal many-flowered racemes up to 25 cm long; bracts up to 14 mm long, narrowly triangular to linear, often persistent, sometimes fused with base of pedicel; pedicels 5–12 mm long; bracteoles apparently absent. Calyx 4–6 mm long, campanulate, with 5 shallow teeth up to 1 mm long. Petals yellow; standard 17–22 × 15–18 mm, the limb suborbicular, cordate, the claw ± distinct; wings and keel petals slightly shorter than the standard. Ovary densely appressed-pubescent. Pods moniliform, with 1–3(5) seeds developing. Seeds 8–9 mm long, subglobose, brownish.

Mozambique. N: Mogincual Dist., Quinga beach, fl.& fr. 28.iii.1964, *Torre & Paiva* 11437 (LISC). Z: Pebane, maritime sands, fr. 4.x.1949, *Barbosa & Carvalho* 4290 (K, LMA, SRGH). MS: Beira Dist., Macúti, sand dunes above high water mark, fr. c.1921, *Honey* 656 (K). GI: Inhassoro Dist., Bazaruto Is., by sea, fl. viii.1936, *Gomes e Sousa* 1886 (COI, K). M: Maputo Dist., Ilha da Inhaca, Ponta Torres, sand dune pioneer, fr. vii.1957, *Mogg* 27298 (K, SRGH).

Also in Kenya, Tanzania, South Africa (KwaZulu-Natal) and on the W coast of Madagascar. Apparently occupying the same ecological position on sandy foreshores as *S. tomentosa*; 0–20 m.

Fig. 3.3.**9**. SOPHORA INHAMBANENSIS. 1, flowering branch ($\times$ ^{2}/$_3$); 2, standard ($\times$ 1); 3, wings ($\times$ 1); 4, keel petals ($\times$ 1); 5, stamen ($\times$ 1^1/$_2$); 6, gynoecium ($\times$ 1^1/$_2$), 1–6 from *Elliot* in *Kenya Forest Dept.* 1375; 7, infructescence ($\times$ ^{2}/$_3$), from *Honey* 656. Drawn by Derrick Erasmus. From F.T.E.A.

3. **Sophora velutina** Lindl. in Bot. Reg. **14**: t.485 (1828). —Brummitt & Gillett in Kirkia **5**: 261 (1966). Type cultivated in England (Whitley), said to have been introduced from Nepal, *Herb. Lindley* s.n. (CGE lectotype, selected by Brummitt & Gillett 1966).

> *Sophora glauca* DC. in Ann. Sci. Nat. (Paris) **4**: 98 (Jan. 1825) non Salisb. (1796); in Prodr. **2**: 95 (Nov. 1825). Type from S India.

Shrub 1–3 m high. Young branches, leaf petioles and rachides, inflorescence and calyces subglabrous to densely whitish or brownish pubescent or tomentose; lower surface of leaflets sparsely to densely appressed-pubescent, upper surface glabrous or appressed-pubescent. Leaves (rachis and petiole) 6–20 cm long, with 6–13 pairs of leaflets; largest lateral leaflets on each leaf 1.2–4(6) × 0.6–1.7(2) cm, ovate to elliptic or rarely slightly obovate, usually very slightly asymmetrical at the base, the terminal leaflet from slightly smaller to slightly larger than the lateral ones and symmetrical, leaflet margins often inrolled beneath. Flowers in terminal few- to many-flowered racemes up to 14(20) cm long; bracts 3–16 mm long, linear, caducous or persistent; pedicels 3–7 mm long; bracteoles apparently absent. Calyx campanulate, 5–9 mm long including the prominent teeth c.2 mm long, the two upper teeth pointed or rounded, slightly to almost completely connate. Petals purple, pink or whitish; standard 9–15 × 3–6 mm, gradually narrowed towards the base but sometimes with 2 lateral cusps demarcating the base of the limb; wings ± equalling the standard; keel petals slightly shorter than the standard. Ovary ± densely pubescent. Young pods either sparsely to densely closely appressed-pubescent or densely tomentose; mature pods moniliform, with 1–4(6) seeds developed. Seeds c.8 mm long, ellipsoid, tumid, light brown.

Subsp. **zimbabweensis** J.B. Gillett & Brummitt in Kirkia **5**: 261 (1966). Type: Zimbabwe, 7–8 km NW of Great Zimbabwe, fl. 6.xii.1960, *Leach* 10566 (K holotype, BM, K, LISC, PRE, SRGH).

Upper surface of leaflets glabrous except for occasional hairs on or near the midrib. Bracts c.3 mm long, shorter than the pedicel, quickly caducous. Calyx c.5 mm long. Petals creamy-white; standard 9–12 × 4.5–6 mm with 2 lateral cusps, the claw 1.5–2 mm broad; keel petals with 2 cusps. Young pods densely appressed-pubescent.

Zimbabwe. S: Masvingo Dist., 8 km NW of Great Zimbabwe, fr. 7.v.1963, *Leach* 11649 (K, SRGH).

The typical subspecies occurs in SW China, NE and W India, and Indonesia (Sumba) and Timor. Subsp. *zimbabweensis*, first collected by L.C. Leach in 1960, is known only from two colonies 1 km apart near Great Zimbabwe, where it forms an important constituent of undergrowth of indigenous woodland on a hillside on stony red soil. For discussion of the question of whether this plant has been introduced into Africa by man or is an ancient relict of a species once widespread in the forests of Africa and Asia, see Brummitt & Gillett in Kirkia **5**: 259–264 (1966).

7. BAPHIA Lodd.

Baphia Lodd., Bot. Cab.: t.367 (1825). —Soladoye in Kew Bull. **40**: 291–386 (1985).

Shrubs, sometimes scrambling or climbing, or trees up to 20(25) m high. Leaves unifoliolate; leaflet with entire margin, rarely cordate at base; petiole with upper and lower pulvini, these occasionally contiguous and so appearing as one; stipules caducous. Inflorescence a lax terminal or axillary raceme, or this often condensed to an axillary fascicle which may be reduced to a solitary, apparently axillary, flower; bracts small (up to c.5 mm) but distinct in lax inflorescences, or sometimes clearly replaced by 2 separate or fused stipules, or not obvious in fasciculate inflorescences; bracteoles linear to suborbicular, immediately beneath the flower or distant from it, sometimes both on one side of the pedicel and fused at the base or not. Calyx at anthesis splitting to the base either down one side only (spathaceous) or down both sides (i.e. splitting into 2) and usually splitting at the apex into 2–5 teeth, the whole persistent or caducous, sometimes shed by means of a horizontal split just above the base so that a small ring remains. Petals white or purplish-pink, with a yellow blotch at the base of the standard; standard broadly ovate to suborbicular or reniform, with the claw very short or absent, apex emarginate; wings with a claw, oblong to narrowly obovate; keel petals with a claw, incompletely fused along their dorsal margins. Stamens 10, free; filaments glabrous or occasionally hairy; anthers dorsifixed. Ovary subsessile, glabrous to velutinous, with a long curved style. Pods linear-oblong to oblanceolate, often strongly curved near the apex, laterally compressed, greyish or straw-coloured to purplish or black, sometimes pubescent. Seeds reddish or blackish, ± laterally compressed, with a small hilum.

A genus of around 45 species confined to tropical Africa apart from one species and one subspecies in Madagascar, one reaching South Africa (Limpopo Province) and Namibia, and one restricted to South Africa (KwaZulu-Natal).

Root nodules have been recorded for three species, including *B. massaiensis* (see Sprent, Nodulation in Legumes: 85, 2001).

1. Flowers in axillary fascicles (sometimes reduced to 1 or 2 flowers) or ± lax terminal racemes or panicles; calyx splitting to base on one side only; pods brown pubescent, or glabrous and purplish to black · · · · · · · · · · · · · · · · · 2
– Flowers in axillary racemes (these sometimes laxly grouped towards ends of main branches); calyx splitting to base down both sides, so dividing into two; mature pods glabrous and ± straw-coloured · 7
2. Flowers in ± lax terminal racemes or panicles · 3
– Flowers in axillary fascicles (rarely apparently solitary in the axil of a foliage leaf) · 4
3. Bracteoles broadly elliptic to suborbicular; petals usually pinkish; ovary ± villous and pod tomentose · **1.** *macrocalyx*
– Bracteoles narrowly triangular, caducous; petals white; ovary and pod glabrous · **2.** *speciosa*
4. Petioles to more than 4 cm long; younger vegetative parts and inflorescence densely brown-tomentose · **3.** *bequaertii*
– Petioles up to 2.7 cm long; younger vegetative parts and inflorescence glabrous to softly pubescent · 5
5. Bracteoles reniform or semicircular, persisting after flowering · · · **6.** *punctulata*
– Bracteoles longer than broad, caducous · 6
6. Leaves coriaceous, with fine reticulate venation; petals usually purplish-pink · **4.** *whitei*
– Leaves not coriaceous, without conspicuously reticulate venation; petals white · **5.** *massaiensis*
7. Large shrub or tree to 10 m or more high; bracts and bracteoles quickly caducous, glabrous; pedicels and calyces appressed-pubescent; pods 7–10 × 1.7–3 cm · **7.** *ovata*
– Scrambling or climbing shrub; bracts and bracteoles ± persistent, pubescent; pedicels and calyces densely spreading pubescent to tomentose; pods 4–7 × 0.8–1.4 cm · **8.** *capparidifolia*

1. **Baphia macrocalyx** Harms in Bot. Jahrb. Syst. **40**: 33, t.3 (1907); in Engler, Pflanzenw. Afrikas **3**(1): 536, fig.282 (1915). —Lester-Garland in J. Linn. Soc., Bot. **45**: 237 (1921). —Baker f., Legum. Trop. Africa: 589 (1929). —Brummitt in Kew Bull. **22**: 515 (1968); in F.T.E.A., Legum., Pap.: 51 (1971). —Soladoye in Kew Bull. **40**: 316, fig.4 (1985). Syntypes: Tanzania, Lindi Dist., Rondo Plateau, *Busse* 2557 (B lectotype, BM, EA).

Baphia mocimboensis Pires de Lima in Brotéria, sér.Bot. **19**: 120 (1921). Type: Mozambique, Mocímboa da Praia, 7.ix.1917, *Pires de Lima* 265 (PO holotype).

Small tree up to 10 m high. Young branches, inflorescence, bracteoles and calyces covered with a dense dark brown tomentum. Leaflet coriaceous, 4–16 × 2–7 cm, elliptic to obovate, the base rounded to cuneate, apex obtuse to emarginate, glabrous; petiole 0.8–5.5 cm long. Flowers in lax terminal or axillary racemes often aggregated to form pseudopanicles; pedicels 4–9 mm long, with 2 large stipules or a ± concave fused stipular 'bract' at the base; bracteoles 4–6 mm long, broadly elliptic to suborbicular, both inserted on upper side of pedicel and usually partially fused near the base. Calyx 13–17 mm long, splitting to the base on one side and almost to halfway on the opposite side, all 5 teeth usually becoming free. Petals white to pink- or violet-tinged; standard (16)18–26 mm long; wings 16–21 mm long; keel petals 16–18 mm long, appressed brown pubescent outside towards their fused margin. Stamen filaments hairy. Ovary 5–8 mm long, brown-villous. Pod up to 17 × 2.5 cm, brown-tomentose.

Mozambique. N: Palma Dist., near the road from Mocímboa da Praia to Palma, 7 km S of Palma, fl.& imm.fr. 23.x.1960, *Gomes e Sousa* 4568A (COI, K, LMA, PRE, SRGH).

Also in SE Tanzania. Open woodland on sandy soil, sometimes forming small communities; up to 250 m.

2. **Baphia speciosa** J.B. Gillett & Brummitt in Bol. Soc. Brot., sér.2 **39**: 161 (1965). — Soladoye in Kew Bull. **40**: 331 (1985). Type: Zambia, Kaputa Dist., Bulaya to Sumbu road, 5.iv.1957, *Richards* 9041 (K holotype). FIGURE 3.3.**10**.

Baphia sp. 2 of White, F.F.N.R.: 144 (1962).

Bush or small tree, sometimes scrambling, up to 7 m. Young branches, petioles, inflorescences, bracteoles and calyces with a short dark brown pubescence. Leaflet up to 8 × 4.5 cm, elliptic to narrowly elliptic or slightly obovate, rounded or cuneate at the base, obtuse or rounded at the apex, glabrous on upper surface, pubescent on lower; petiole 9–16 mm long. Flowers in lax terminal panicles; pedicels 7–14 mm long, with small glandular hairs at the base; bracts up to 6 mm long, narrowly triangular to lanceolate, quickly caducous; bracteoles opposite, up to 6–10 × 1.5–4 mm, triangular, acute, quickly caducous. Calyx 14–19 mm long, spathaceous, remaining ± entire at the apex. Petals white, with a yellow mark near base of the standard; standard 18–24 mm long, wing and keel petals 16–24 mm long. Ovary 8–10 mm long, glabrous. Pods up to 13 × 1.5 cm, greyish-black, glabrous.

Zambia. N: Mbala Dist., lower Lufubu (Lufu) R. valley, fl. iii.1937, *Trapnell* 1746 (K). Known only from N Zambia. In grassland, scrub and thickets on sandy alluvial soil; 780–990 m.

3. **Baphia bequaertii** De Wild. in Repert. Spec. Nov. Regni Veg. **13**: 116 (1914); Notes Fl. Katanga **4**: 13 (1914); in Ann. Sci. Nat. Bot., sér.10 **1**: 206 (1919); Contrib. Fl. Katanga: 75 (1921); Pl. Bequaert. **3**: 266 (1925). —R.E. Fries, Wiss. Ergebn. Schwed. Rhod.-Kongo-Exped. 1: 73 (1914). —Lester-Garland in J. Linn. Soc., Bot. **45**: 241 (1921). —Baker f., Legum. Trop. Africa: 585 (1929). —Toussaint in F.C.B. **4**: 21 (1953). —White, F.F.N.R.: 143 (1962). —Soladoye in Kew Bull. **40**: 347, fig. 8D–K (1985). Type from Congo (Katanga).

Baphia ringoetii De Wild. in Repert. Spec. Nov. Regni Veg. **13**: 116 (1914); Notes Fl. Katanga **4**: 14 (1914); Pl. Bequaert. **3**: 300 (1925). Type from Congo (Katanga).

Shrub or small tree up to 10 m high. Young branches, petioles and surfaces of young leaves, stipules, pedicels, bracteoles and calyces covered with dense brown tomentum, upper surface of leaves soon glabrescent. Leaflet up to 15(17.5) × 12(15.5) cm, broadly elliptic to suborbicular, rounded to slightly cordate at the base, obtuse to rounded or emarginate at the

Fig. 3.3.**10**. BAPHIA SPECIOSA. 1, flowering branchlet (× ²⁄₃), from *Richards* 9041; 2, detail of underside of leaflet (× 1); 3, extrafloral nectary at base of pedicel (× 6), 2 & 3 from *Phipps & Vesey-FitzGerald* 3202; 4, flower bud (× 2), from *Richards* 9041; 5, standard (× 1); 6, gynoecium (× 1), 5 & 6 from *Phipps & Vesey-FitzGerald* 3202. Drawn by Eleanor Catherine. From Kew Bulletin.

apex; petiole up to 10(12.5) cm long. Flowers in fairly dense axillary fascicles; pedicels 20–40 mm long; bracteoles inserted immediately beneath the calyx, often slightly to one side, 4–6 × 1.5–2 mm, triangular to oblong, caducous. Calyx 10–18 mm long, spathaceous, usually falling before the petals. Petals white with yellow markings at the base of the standard; standard 16–21 mm long; wings slightly shorter; keel petals about as long as the standard. Ovary c.7 mm long, densely brown- or greyish villous. Pod up to 18 × 3.5 cm, brown-tomentose.

Zambia. N: Kasama Dist., without locality, fl. 10.x.1960, *E.A. Robinson* 3966 (K, SRGH). W: Solwezi Boma, near guesthouse, fl. 22.ix.1930, *Milne-Redhead* 1170 (K). C: Mkushi Dist., 6 km W of Chiwefwe, fr. 25.vii.1930, *Hutchinson & Gillett* 4092 (K). S: Mazabuka Dist., Mazabuka, fr. 1931, *Stevenson* 307 (K).

Also in Angola and Congo. In miombo and chipya woodland (fire-degraded dry evergreen forest); 1200–1500 m.

4. **Baphia whitei** Brummitt in Bol. Soc. Brot., sér.2 **39**: 159 (1965). Type: Zambia, Western Province, 56 km NE of Kabompo, 13 km from Old Manyinga, 1.6 km from Chikonkwelo R., *Angus* 2251 (SRGH holotype, K).

   Baphia sp. 1 of White, F.F.N.R.: 144 (1962).

   Baphia massaiensis subsp. *obovata* var. *whitei* (Brummitt) Soladoye in Kew Bull. **40**: 339 (1985).

Shrub or small tree up to 10 m high. Leaflet coriaceous, up to 10 × 7 cm, ovate or elliptic or ± obovate, acute to acuminate or broadly rounded at the apex, with a very fine reticulate venation on both surfaces; the upper surface densely puberulous when young but glabrescent, the lower surface appressed-pubescent when young but slowly glabrescent and somewhat glaucous, usually with minute scattered dark purplish hairs with persistent swollen bases; petiole ± short and stout, 6–14 mm long, with the upper and lower pulvini often contiguous. Flowers in dense or somewhat lax axillary fascicles; pedicels up to 14 mm long; pedicels and buds covered with spreading to appressed brownish hairs; bracteoles 2–3 mm long, narrowly ovate to linear-oblong, pubescent. Calyx 8–11 mm long, spathaceous, splitting into at least 3 at the apex, persistent. Petals usually purplish-pink, sometimes white, with a yellow blotch near base of the standard; standard 10–18 × 12–15 mm; wings 12–18 × 4–7 mm; keel petals 12–16 mm long. Ovary ± densely brown-pubescent to villous. Young fruits dark purplish and densely pubescent; mature fruits c.9 × 12 cm, light purplish-brown, glabrous.

Zambia. B: Zambezi (Balovale), fl. vi.1952, *Gilges* 81 (K, PRE, SRGH). W: Mwinilunga Dist., fl. ix.1934, *Trapnell* 1608 (K).

Also in Angola. In dry *Cryptosepalum* woodland on Kalahari sands; 900–1500 m.

Soladoye (1985) has reduced *B. whitei* to a variety of *B. massaiensis* subsp. *obovata* on the grounds that the pattern of leaf venation was found sporadically elsewhere in the *B. massaiensis* complex, and that both pink and white flowers can be found. As the two taxa grow in the same ecogeographical region, varietal status seemed to him most appropriate. In fact the differences in leaf shape and indumentum are quite consistent in the area of geographical overlap. The prominence of the leaf venation is of a different order – halfway up the leaflet and halfway from the midrib to the margin there are about 10 areoles between one lateral vein and the next in *B. massaiensis* subsp. *obovata* and about 50 in *B. whitei* when the leaf has hardened (Fig. 7R in Soladoye (1985) is drawn from a soft immature leaflet). Since *B. whitei* was described in 1965 the number of specimens from Zambia has doubled and the distinctions are not as blurred as would be expected between varieties, but field observations might provide further information.

5. **Baphia massaiensis** Taub. in Engler, Pflanzenw. Ost-Afrikas **C**: 203 (1895). —Lester-Garland in J. Linn. Soc., Bot. **45**: 240 (1921). —Baker f., Legum. Trop. Africa: 584

(1929). —Toussaint in F.C.B. **4**: 20 (1953). —Brummitt in Bol. Soc. Brot. **39**: 172–183 (1965); in Kew Bull. **22**: 528 (1968); in F.T.E.A., Legum., Pap.: 58 (1971). —Soladoye in Kew Bull. **40**: 334, fig.7 (1985). —K. Coates Palgrave, Trees Sthn. Africa, ed.2: 302 (1988). —M. Coates Palgrave, Trees Sthn. Africa: 359, fig. 97 (2002). Type: Tanzania, Dodoma Dist., Saranda (Salanda), *Fischer* 195 (BM).

Shrub or small tree up to 8 m high. Leaflet up to 9 × 5.5 cm, elliptic to obovate or suborbicular, glabrous to pubescent; petiole 4–27 mm long. Flowers in fascicles of up to 7(12), sometimes reduced to a single flower apparently solitary in the axil of a foliage leaf; pedicels 6–16(24) mm long; bracteoles inserted immediately beneath the flower or up to 5 mm distant, 1.5–5 × 0.3–0.8 mm, linear-oblong to linear. Calyx 6–12 mm long, sparsely to densely pubescent. Petals white, the standard with a yellow mark near the base; standard 8–22 × 8–21 mm; wings 8–22 × 3.5–8 mm; keel petals 8–19 mm long. Ovary sparsely to densely covered with longish white hairs. Pod 7–11(13) × 1–1.5 cm, blackish- or purplish-brown. Seeds 8–11 × 7–9 mm, lenticular, reddish-black.

The species (including 5 subspecies) is found in Tanzania, Congo (Katanga), Angola, Botswana, Zambia, Zimbabwe, ?Malawi, Mozambique, South Africa (Limpopo Province) and Namibia. Usually in open woodland or thickets, often locally abundant.

1. Buds and pedicels with ± closely appressed hairs; standard and wings 15–22 mm long · i) subsp. *floribunda*
– Buds and pedicels with spreading hairs; standard and wings up to 16 mm long · 2
2. Leaf surfaces usually glabrous except on veins beneath; leaves ± elliptic to rarely suborbicular; petioles 11–25 mm long; standard and wings 8–13 mm long · ii) subsp. *gomesii*
– Leaf surfaces pubescent; leaves ± obovate; petioles 5–12 mm long; standard and wings 12–16 mm long · iii) subsp. *obovata*

i) Subsp. **floribunda** Brummitt in Bol. Soc. Brot. **39**: 179 (1965). —Soladoye in Kew Bull. **40**: 337, fig.7A–C (1985). Type: Zambia, Kaputa Dist., Nsama–Bulaya road, *Richards* 6240 (K holotype).

 Baphia massaiensis sensu Toussaint in F.C.B. **4**: 20 (1953). —sensu White, F.F.N.R.: 144 (1962).

Leaflet broadly elliptic to suborbicular or sometimes obovate, obtuse to retuse at the apex, with 4–6(7) major lateral veins on each side of the midrib, the surfaces pubescent when young, the upper becoming glabrous; petiole (8)12–20(27) mm long. Pedicels and buds covered with closely appressed hairs. Calyx 9–12 mm long, usually splitting into at least 3 at the apex. Standard 15–22 × 13–21 mm; wings 15–21 × 5.5–8 mm; keel petals 12–18 mm long. Ovary densely ± appressed-hairy.

 Zambia. N: Mbala Dist., Mpulungu–Mbala road, fl. 26.ix.1959, *Richards* 11474 (K, SRGH).

Also in Congo (Katanga). In woodland and thickets, often in clearings, also by lakes and streams; 780–1500 m.

ii) Subsp. **gomesii** (Baker f.) Brummitt in Bol. Soc. Brot. **39**: 181 (1965). —Soladoye in Kew Bull. **40**: 337, fig.7G–J (1985). Type: Mozambique, Serra de Ribáuè, *Gomes e Sousa* 828 (BM holotype, COI, K).

 Baphia gomesii Baker f. in Bol. Soc. Brot. **8**: 113 (1933).

Leaflet elliptic to broadly elliptic or rarely suborbicular, with 4–7 major lateral veins on each side of the midrib, usually glabrous except on the major veins beneath; petiole (12)14–25(27) mm long. Pedicels and buds with ± spreading hairs. Calyx 6–9 mm long, usually splitting into at least 3 at the apex. Standard (6)8–13 × 9–12 mm; wings (6)8–13 × 3.5–5 mm; keel petals (6)8–11 mm long. Ovary ± sparsely spreading-hairy, or very rarely glabrous.

Mozambique. N: Nampula, Campo Experimental do C.I.C.A., fl.& fr. 10.iv.1961, *Balsinhas & Marrime* 367 (COI, K, LISC, LMA, PRE).

Apparently restricted to N Mozambique where it occurs in dry evergreen forest and woodland; up to c.800 m. However, a specimen from SE Tanzania is very similar, but in having hairs regularly distributed over the leaf undersurface it tends towards subsp. *massaiensis.*

Two specimens from S Mozambique with the same collecting number but different dates – GI: between Massinga and Vilanculos, Urrongas, st. x.1947, *Pedro & Pedrógão* 2311 (PRE); between Massinga and Mapinhane, fr. 3.x.1948, *Pedro & Pedrógão* 2311 (LISC) – have leaflets broadly obovate to suborbicular with petioles up to 17 mm long, so resembling those of subsp. *gomesii,* but pubescent at least on the lower surface and with a rather close venation, so resembling those of subsp. *obovata* to which they are nearest geographically. They appear to be an isolated location for this species and might prove to represent a distinct subspecies.

iii) Subsp. **obovata** (Schinz) Brummitt in Bol. Soc. Brot. **39**: 176 (1965). —Schreiber in Merxmüller, Prodr. Fl. SW Afrika, fam. 60: 14 (1970). —Palmer & Pitman, Trees Sthn. Africa: 907 (1972). —Soladoye in Kew Bull. **40**: 339, fig.7D–F (1985) in part excl. var. *whitei.* —M. Coates Palgrave, Trees Sthn. Africa: 359, fig. 97 (2002). Type from Namibia.

> *Baphia henriquesiana* Taub. in Bot. Jahrb. Syst. **23**: 176 (Sept. 1896). —Lester-Garland in J. Linn. Soc., Bot. **45**: 241 (1921). —Baker, Legum. Trop. Africa: 587 (1929). Type from Angola.
>
> *Baphia obovata* Schinz in Bull. Herb. Boiss. **4**: 815 (Dec. 1896). —Lester-Garland in J. Linn. Soc., Bot. **45**: 241 (1921). —Baker f., Legum. Trop. Africa: 586 (1929). —Bremekamp & Obermeyer in Ann. Transvaal Mus. **16**: 418 (1935). —Miller in J. S. African Bot. **18**: 28 (1952). —Pardy in Rhod. Agric. J. **53**: 954 (1956). —White, F.F.N.R.: 144 (1962).

Leaflet usually obovate, sometimes elliptic, with (5)6–10 major lateral veins on each side of the midrib, both surfaces pubescent when young, the upper usually glabrescent; petiole 3–13 mm long. Pedicels and buds with dense spreading hairs. Calyx 8–11 mm long, usually splitting into only 2 at apex, occasionally into 3. Standard 12–16 × 12–15 mm; wings 12–15 × 5–7 mm; keel petals 10–13 mm long. Ovary densely spreading-hairy.

Caprivi Strip. Katima Mulilo area, fl. 24.xii.1958, *Killick & Leistner* 3051 (K, SRGH). **Botswana**. N: Ngamiland Dist., Savuti Channel, area of drift above Gubatsaa Hills, map sq. 1824, fl. 24.x.1972, *Biegel, Pope & Russell* 4055 (K, SRGH). **Zambia**. B: Sesheke Dist., near edge of Kazu Forest, near Machili, fl. 20.xii.1952, *Angus* 979A (K). N: Chinsali Dist., Mbesuma Ranch, fr. 17.vii.1961, *Astle* 813 (K, SRGH). W: Mwinilunga Dist., 6 km N of Mayowa Plains, fl. 4.x.1952, *White* 3450 (K). C: South Luangwa Nat. Park, S of Katete R., st. 4.v.1965, *Mitchell* 2809 (K). E: Lundazi Dist., Lukusuzi Nat. Park (Game Res.), Chikomeni to Changachanga road, 35 km W of Chikomeni, fl. 2.ii.1961, *Feely* 114 (K, SRGH). S: Livingstone Dist., Dambwa Forest Res., 6 km N of Livingstone, fl. 14.i.1952, *White* 1879 (K). **Zimbabwe**. N: Hurungwe Dist., Mhenza (Mensa) Pan area, fl. 27.i.1957, *Phelps* 195 (K, SRGH). W: c.104 km NW of Bulawayo, fl. 7.xi.1956, *Turner & Shantz* 4139 (K, SRGH). S: Mwenezi Dist., 3 km inside border from Vila Eduardo Mondlane (Malvérnia), fl. 3.xi.1955, *Wild* 4684 (K, LISC, PRE, SRGH). **Mozambique**. GI: 27 km SW of Banamana Salima on Mashaila road, Shinguengue, fl. x.1973, *Tinley* 2966 (K, SRGH).

Also in Angola, Namibia and South Africa (Limpopo Province); 130–1200 m.

Subsp. *obovata* occurs mainly in the Kalahari sand areas in the west-central part of the Flora area in *Baikiaea* and *Erythrophleum* woodlands; also in mopane and mixed deciduous woodlands on sandy soils, sometimes forming small thickets; 900–1350 m. Although records are given for Zambia N and E, these appear to be isolated localities outside the main distribution area.

Eyles 7159, wrongly given as from Nyanga on the Kew specimen, is from the Bulawayo area in Zimbabwe.

Subsp. *massaiensis* occurs in Tanzania. It differs from subsp. *obovata* mainly in having appressed hairs on the ovary, generally sparser pubescence on leaves and inflorescence, and the calyx splitting into 3 or more lobes; it differs from subsp. *floribunda* in having smaller flowers and ± spreading hairs on pedicels and buds; and from subsp. *gomesii* in having larger flowers, denser hairs on leaf surfaces and ovary, and shorter petioles. Subsp. *cornifolia* (Harms) Brummitt from Huíla Province, Angola, has leaves similar to subsp. *obovata*, but appressed pubescence on the pedicels and calyx and on the ovary.

Baphia busseana Harms in Bot. Jahrb. Syst. **33**: 166 (1902), described from SE Tanzania and recorded from Malawi by Topham (Check List For. Trees Shrubs Nyasaland Prot.: 74 (1958)), is apparently not specifically distinct from *B. massaiensis* and requires further investigation. The genus is apparently absent from Malawi.

A syntype from Tanzania (Songea or Tunduru Dist.), *Busse* 1001 (BM, K), is perhaps intermediate between subsp. *obovata* and subsp. *gomesii*, but it is more pubescent than either and may perhaps represent a further subspecies.

On his journey to western Zambia in 1937, Milne-Redhead took the boat to Lobito on the coast of Angola, where he made 15 collections, and then took the train presumably to Elizabethville, now Lubumbashi. His field notebook at Kew shows that he collected one specimen at each of three stations in Angola – Capeio, Silva Porto and Casai? – on the way. The last of these is a specimen of *Baphia aurivellera* Taub., a much-branched shrub to 1 m high with white petals with an orange blotch at the base of the standard (Casai Station, sandy ground with *Copaifera baumii*, 30.ix.1937, *Milne-Redhead* 2518 (K)). Cassai, some km north of the railway station at Luau, is a short distance west of the Congo border. The question mark after Casai in the notebook was not transferred to the specimen label, but it seems certain that this specimen was not collected in the Flora Zambesiaca area, despite the original label. *B. aurivellera* is known from several localities in E Angola and from 500 km NW in Congo (see map by Soladoye in Kew Bull. **40**: 321,1985). Although Cassai is some 200 km from the nearest point in Zambia, it is not impossible that it could also occur in W Zambia. The general appearance is similar to that of *B. massaiensis*, but it differs markedly in that its calyx splits to the base into two halves rather than being spathaceous.

6. **Baphia punctulata** Harms in Bot. Jahrb. Syst. **40**: 32 (1907). —Brummitt in Bol. Soc. Brot., Sér. 2, **39**: 184 (1965); in Kew Bull. **22**: 531 (1968); in F.T.E.A., Legum., Pap.: 59 (1971). Type: Tanzania, Lindi Dist., L. Tandangongoro, 13.v.1903, *Busse* 2486 (BM, EA).

Subsp. **palmensis** Soladoye in Kew Bull. **40**: 361 (1985). Type: Mozambique, Palma, 14.iv.1917, *Pires de Lima* 203 (PO holotype, BM).
     Baphia sp. of Pires de Lima in Bol. Soc. Brot. **2**: 136 (1924).

Small tree c.5 m high. Young branches shortly brown pubescent or glabrous. Leaflet ± coriaceous, up to 8 × 5 cm, ovate- or elliptic-acuminate, rounded at the base, glabrous, or pubescent beneath when young and glabrescent; petiole 6–9 mm long, with a short constricted portion between the 2 pulvini. Flowers in axillary fascicles reduced to usually 2 flowers, rarely solitary; pedicels c.10 mm long, somewhat thickened distally, brown pubescent; bracteoles small, reniform or semicircular, forming a 'collar' immediately below the flower, persistent after flowering. Calyx 7–8 mm long, glabrous. Petals white with yellow markings, 8–9 mm long. Ovary with hairs restricted to one margin. Pod unknown.

Mozambique. N: Palma, 14.iv.1917, *Pires de Lima* 203 (BM, PO).

Known only from the type specimen. Habitat not recorded.

Subsp. *punctulata* occurs in SE Tanzania and has a hairy calyx and ovary. Soladoye also included *B. descampsii* De Wild. from E Congo and W Tanzania as a further subspecies.

7. **Baphia ovata** Sim, For. Fl. Port. E. Africa: 42, t.49 (1909). —Brummitt in Bol. Soc. Brot. **39**: 183 (1965). Type: Mozambique, Zavala Dist. (M'Chopes), Quissico, passim, *Sim* 5279 (not traced).

   Baphia kirkii sensu Lester-Garland in J. Linn. Soc., Bot. **45**: 240 (1921). —sensu Baker f., Legum. Trop. Africa: 588 (1929), in part as regard syn. *B. ovata* Sim.
   Baphia kirkii subsp. *ovata* (Sim) Soladoye in Kew Bull. **40**: 328 (1985).

Tree 4–10 m high with pendulous branches. Leaflets 4–7 × 2.5–4 cm, ± ovate, acute to bluntly acuminate or obtuse at the apex, rounded or rarely subcordate or cuneate at the base, sparsely pubescent on both surfaces when young but glabrescent; petiole 8–25 mm long, rather slender and often curved. Flowers in simple axillary racemes, these usually occurring towards the ends of young main branches to form a leafy pseudo-panicle; pedicels 5–10 mm long; inflorescence axis, pedicels and buds closely appressed-pubescent; bracteoles up to 2.5 mm long, concave, ovate to narrowly elliptic or oblong, glabrous, quickly caducous. Calyx splitting to the base down each side and so into 2 halves, c.8 mm long, appressed-pubescent. Petals white; standard 10–13 × 10–12 mm; wings 10–12 mm long; keel petals 10–11 mm long. Ovary glabrous. Pods 7–10 × 1.7–3 cm, pale straw-coloured. Seeds up to 17 × 12 mm, flattened, purplish-black.

Mozambique. GI: Inhambane Dist., near Inharrime R., fl.& fr. ix.1939, *Gomes e Sousa* 2249 (K, PRE).

Not known outside the Gaza and Inhambane Provinces, Mozambique. Recorded from lakeshores, but further details of habitat are required; up to 30 m.

It is known to have been planted outside its native range at Maputo (Observatório Campos Rodrigues) and may be grown elsewhere.

Soladoye (1985) reduced this species to a subspecies of *B. kirkii*, which occurs in the coastal districts of Tanzania and Kenya from the vicinity of the Rufiji northwards, with an interval of some 1500 km between the two subspecies. *B. kirkii* has larger flowers, 15–22 mm long, and the calyx is generally glabrous except for a tuft of hairs at the apex. The species is also related to *B. racemosa* from KwaZulu-Natal.

8. **Baphia capparidifolia** Baker in J. Linn. Soc., Bot. **25**: 311 (1890). —Brummitt in Bol. Soc. Brot. **39**: 163–172 (1965); in Kew Bull. **22**: 521 (1968); in F.T.E.A., Legum., Pap.: 53 (1971). —Soladoye in Kew Bull. **40**: 323, fig.5 (1985). Type: NW Madagascar, *Baron* 5358 (K holotype, BM).

Subsp. **bangweolensis** (R.E. Fr.) Brummitt in Bol. Soc. Brot. **39**: 167 (1965); in Kew Bull. **22**: 523 (1968). —Soladoye in Kew Bull. **40**: 324, fig.5A (1985). Type: Zambia, Samfya Dist., L. Bangweulu, Mpanta (Panta), *R.E.Fries* 826 (UPS holotype).

Baphia bangweolensis R.E. Fr. in Repert. Spec. Nov. Regni Veg. **12**: 541 (1913); Wiss. Ergebn. Schwed. Rhod.-Kongo-Exped. 1: 73 (1914). —Lester-Garland in J. Linn. Soc., Bot. **45**: 229 (1921). —Baker f., Legum. Trop. Africa: 574 (1929). —Toussaint in F.C.B. **4**: 11 (1953). —White, F.F.N.R.: 143 (1962).

Baphia giorgii De Wild., Pl. Bequaert. **3**: 279 (1925). Type from Congo (Katanga).

Weak scrambling or climbing shrub up to 4(6) m high. Young branches, inflorescences, bracteoles and calyces softly spreading-pubescent to tomentose, yellowish-brown to greyish. Leaflets up to 12.5 × 8 cm, ovate to broadly ovate, obtuse or emarginate or ± acute at apex, rounded or subcordate at base, margins often undulate, glabrescent on upper surface, finely pubescent on the lower; petiole 10–30 mm long. Flowers in axillary, sometimes branched, racemes, 2–4(5.5) cm long, ± pyramidal with ascending pedicels; bracts (1)1.5–3 mm long, ovate to triangular; bracteoles up to 4.5 mm long, triangular-lanceolate, spreading or reflexed. Calyx c.7 mm long, splitting to base down both sides. Petals 7–11 mm long, white or cream, the standard with a yellow or orange mark towards the base. Ovary 4–5 mm long, densely villous. Pod 4–7 × 0.8–1.4 cm, glabrous, straw-coloured. Seeds 7–12 × 6–8 mm, oblong-compressed to ovoid-lenticular, dark red.

Zambia. N: Mbala Dist., grassy top of Kambole escarpment, fl. 1.ii.1959, *Richards* 10836 (K, PRE). W: Chingola, fl. 26.viii.1954, *Fanshawe* 1491 (BR, K, SRGH). C: Kabwe, fl. 3.x.1963, *Fanshawe* 8004 (K).

Also in SW Tanzania and Congo (Katanga). Usually in thickets (often an important component), sometimes scrambling on rocky ground; 700–1650 m.

The species includes 4 subspecies: subsp. *capparidifolia* from W and NW Madagascar; subsp. *polygalacea* Brummitt (*Baphia polygalacea* (Hook.f.) Baker nom. illegit.) from West Africa; subsp. *multiflora* (Harms) Brummitt (*B. multiflora* Harms; *B. polygalacea* sensu Toussaint in F.C.B. **4**: 12 (1953)) from Cameroon, Gabon, Congo, Rwanda, Burundi and Uganda to W Tanzania and NE Angola; and subsp. *bangweolensis*.

Tribe 3. **DALBERGIEAE**

by J.M. Lock*

Dalbergieae DC., Prodr. **2**: 415 (1825).

Trees, shrubs or climbers with hard wood. Leaves imparipinnate, with opposite or alternate leaflets, rarely 1–3-foliolate, sometimes with stipels (not in the Flora area). Flowers in racemes or panicles; bracts and bracteoles mostly small and caducous. Hypanthium (receptacle) often well developed; disk generally lacking. Calyx generally with the upper lobes joined higher, sometimes truncate. Standard generally without calluses or appendages; keel petals overlapping or shortly adnate on the lower side, obtuse, not interlocked with the wings and sometimes much smaller. Stamens all joined or the upper one free or stamens joined in bundles, the free part often at least half as long as the joined part; anthers dorsifixed, generally ± uniform, sometimes with pores or short slits. Ovary sessile to long-stipitate, generally soon showing signs of specialised seed chambers, 1–few-ovulate; style tapered, glabrous above, with a small stigma. Fruit with 1–few specialized indehiscent seed chambers, flattened, drupaceous, fibrous or winged. Seeds oblong-reniform to globose, often with a thin testa; radicle short, straight or incurved.

A tribe of nearly 20 genera in tropical and subtropical regions based on the above circumscription. Only two native genera in the Flora Zambesiaca area which belong to different clades of the tribe. A number of other genera formerly included in this

---

* F.A. Mendonça prepared draft Flora Zambesiaca accounts in the 1960s for most of this tribe and part of what is now Millettieae. These accounts were used as the basis for most generic treatments in these two tribes, although the accounts have been fully revised and the taxonomy updated.

tribe are now referred to the Millettieae. Lavin, Pennington, Klitgaard, Sprent, de Lima & Gasson (Amer. J. Bot. **88**: 503–533, 2001) redefine Dalbergieae in a broader sense to include the Aeschynomeneae (treated here in F.Z. Vol. 3.5) and two other small South American tribes, nearly 50 genera in all.

Cultivated species

   Tipuana tipu (Benth.) Kuntze, a native of South America, is grown as an ornamental. It is a small tree with imparipinnate leaves bearing c.7–11 pairs of oblong leaflets, yellow-flowered racemes resembling those of *Pterocarpus* and samaroid fruits resembling those of *Acer, Securidaca* or *Pterolobium*. **Malawi** S: Blantyre, Independence Avenue, fr. 12.iii.1970, *Brummitt* 9030 (K); Zomba Botanic Garden, fl. 3.x.1981, *Chapman* 5901 (K); fr. 20.ii.1979, *Brummitt* 15409 (K). Biegel (Check-list Ornam. Pl. Rhod. Parks & Gard.: 105, 1977)) has recorded in it cultivation in Zimbabwe where it is grown in gardens and as a street tree.

1. Flowers 3–12 mm long, white to violet-purple, with wings not much broader than the keel, arranged on one side of the inflorescence; fruits flat, often papery and venose, with 1–few hardened seed chambers; trees, shrubs or lianes, some branches sometimes modified as spines or climbing aids · · · · · · · · **9. Dalbergia**
–   Flowers 11–18 mm long, cream to yellow, with broadly expanded wings crinkly at the edges, spirally arranged on the axis; fruits with a wing-like expansion either around or distal to the single seed cavity; trees · 2
2. Fruit obovate to circular, the seed ± median; leaflets mostly broadly oblong to rounded and more than 2 cm wide · **8. Pterocarpus**
–   Fruit with a basal seed-bearing portion and a distal semielliptic-ovate wing; leaflets oblong, c.1.5–2 cm wide (cultivated) · · · · · · · · · · · · · · · · · **Tipuana**

8. PTEROCARPUS Jacq.

Pterocarpus Jacq., Select. Stirp. Amer. Hist.: 283, t.183, fig. 93 (1763), nom. conserv.

   Trees. Leaves imparipinnate, stipulate; leaflets alternate, opposite or irregularly subopposite. Inflorescence racemose or paniculate. Flowers yellow. Calyx turbinate at the base, incurved; upper teeth partially connate. Standard circular or broadly obovate; wings obliquely obovate; keel petals shorter than the wings, dorsally slightly connate or free. Stamens either all 10 connate in a dorsally-slit sheath, or 9 fused with the upper one free, or in 2 groups (phalanges) of 5; anthers dorsifixed, longitudinally dehiscent. Ovary few-ovulate; style somewhat incurved; stigma small, terminal. Pod circular, broadly winged or obovate with obsolete wings. Seed reniform or oblong-reniform; hilum small.

   A pantropical genus with c.25 species. Rojo (Phan. Monogr. **5**, 1972) recognized only 20 species, but his synonymy is here regarded as being somewhat too sweeping.

   Species 2–4, *Pterocarpus tinctorius* sensu lato, have been variously treated in monographs and African floras. Rojo (1972), treated them all as a single variable species, *P. tinctorius* Welw., but this conceals very considerable variation, at least some of it geographically correlated. Polhill in F.T.E.A. (1971), originally intended to treat three entities as good species but then followed Rojo, although he retained three informal groups as 'races', one of these with two 'variants'.

   Welwitsch's description of *Pterocarpus tinctorius* does not specifically include a type. However, the locality information fits only one of his collections, *Welwitsch* 1870, which Polhill correctly cited as the type. Hiern, in his Catalogue of Welwitsch's collections, included several other specimens (*Welwitsch* 1866, 1867 & 1872) under this name, thus (at least informally) widening its circumscription. It is certainly true that there appears to be considerable variation within Welwitsch's collections from Angola, but there is no *a priori* reason why these should all belong to a single species. There is considerable ecological variation even within the parts of Angola where Welwitsch collected.

There are also difficulties with fruit pubescence. Some authors claim that the dense yellow hairs, some of them glandular, which clothe the central part of the fruits, are later lost or fall off. This requires observation in the field. I have attempted to remove the indumentum from fruits but have been unable to leave the smooth surface that is seen in glabrous-fruited material. I regard fruit pubescence as something which is reasonably constant from early development to maturity.

It would seem that the best way to deal with this is not to use the name *P. tinctorius* in the Flora Zambesiaca area, at least for the present. This leaves three clear taxa which can be regarded as species in the area, although only one is adequately represented in herbaria.

Cultivated species

Pterocarpus indicus Willd., an Asian species, has been grown as a shade tree in Mozambique and perhaps elsewhere. The fruits are 6–7 cm long, suborbicular and pubescent, with the style-base lateral on the wing. **Mozambique** M: Maputo, Jardim Tunduru (Vasco da Gama), fr. 13.ix.1972, *Balsinhas* 2436 (LISC, SRGH); Maputo, Matola arboretum, fr. 9.vi.1949, *Cardoso* 729 (SRGH).

1. Pod circular or subcircular with a broad undulate wing, bent until the remains of the style approach or overlap the stipe · 2
 – Pod subelliptic or obovate with a narrow or obsolete wing, or if wing broad then plane (not undulate) and with the remains of the style at the apex or nearly so
 · 5
2. Inflorescence of lateral racemes borne on new shoots before the leaves, rarely also axillary; leaflets 5–8(9) on each side of the rachis; pod with a patch of dense rigid bristles over the seed · **1.** *angolensis*
 – Inflorescence of axillary and/or terminal panicles; leaflets 2–5 on each side of the rachis; pod glabrous or softly hairy (sometimes glandular) over the seed when ripe [see note above] · 3
3. Panicles axillary and terminal; pod (5.5)6–10 cm in diameter, with golden glandular indumentum · **4.** *chrysothrix*
 – Panicles axillary or in the axils of fallen leaves; pods glabrous or sparsely appressed-pubescent, sometimes with scattered bristly hairs · · · · · · · · · · · · 4
4. Pods 14.5 cm or more in diameter, sparsely appressed-pubescent · · · · · · · · · · ·
 · **2.** *megalocarpus*
 – Pods smaller, almost glabrous or minutely appressed-puberulous · · · · · **3.** *stolzii*
5. Stipules leaf-like, persistent; pod with a broad plane wing, smooth all over; style near the apex · **5.** *brenanii*
 – Stipules linear, falling early; pod with a narrow or obsolete wing · · · · · · · · · · 6
6. Inflorescence a terminal panicle; pod subelliptic or obovate, raised-venulose over the seed; remains of the style at the apex · · · · · · · · · · · · · · · **6.** *rotundifolius*
 – Inflorescence of axillary racemes; pod oblong-obovate; remains of the style at the apex or nearly so · **7.** *lucens*

1. **Pterocarpus angolensis** DC., Prodr. **2**: 419 (1825). —Gibbs in J. Linn. Soc., Bot. **37**: 426 (1906). —Baker f., Legum. Trop. Africa: 544 (1929). —Mendonça & Torre in Mem. Junta Invest. Ultramar, sér.Bot. **1**: 29 (1950). —Gomes e Sousa, Dendrol. Moçamb. **1**: 201 (1950). —Pardy in Rhod. Agric. J. **48**: 322 (1951). —Miller in J. S.

Afr. Bot. **18**: 35 (1952). —Hauman in F.C.B. **6**: 25 (1954). —Barbosa & Torre in Garcia de Orta **5**: 126 (1957). —White, F.F.N.R.: 162 (1962). —Sousa in C.F.A. **3**: 357 (1966). —Schreiber in Merxmüller, Prodr. Fl. SW Afrika, fam. 60: 97 (1970). —Polhill in F.T.E.A., Legum., Pap.: 89, fig.17 (1971). —Palmer & Pitman, Trees Sthn. Africa: 937–941 (1972). —M. Coates Palgrave, Trees Sthn. Africa: 389 (2002). Type: Angola, probably from Huíla, *J.J. Silva* s.n. (P holotype). FIGURE 3.3.**11**.

Pterocarpus bussei Harms in Bot. Jahrb. Syst. **33**: 171 (1902); in Engler, Pflanzenw. Afrikas **3**(1): 637 (1915). Type: Tanzania, Uluguru Mts., Kikundi, *Busse* 144 (BM, K).

Pterocarpus erinaceus sensu Passarge, Die Kalahari: 686 (1904). —sensu Sim, For. Fl. Port. E. Africa: 44, t.59 (1909). —sensu Eyles in Trans. Roy. Soc. S. Africa **5**: 379 (1916) as *P. angolensis* in syn.

Deciduous tree up to 18 m tall; crown rounded or spreading; bark greyish-brown, subquadrangular-fissured, exuding reddish sap when slashed. Young parts greyish or brownish subsericeous. Leaves 18–35 cm long; petiole 2.5–5 cm long, rachis 12–25 cm long, puberulous, glabrescent; petiolules 3–5 mm long; leaflets (4)5–8(9) on each side of the rachis, 3–7 × 2.5–4 cm, ovate to elliptic or subelliptic, obtuse to acuminate-acute at the apex, broadly obtuse or rounded or slightly cordate at the base, light green, glabrous on upper surface, paler appressed-pubescent and glabrescent beneath; lateral primary nerves in 12–18 pairs, secondary ones similar, rather close to each other; stipules 10–18 mm long, linear, caducous. Inflorescence of lateral racemes borne on new shoots before the leaves, rarely also axillary, fulvous-pubescent, glabrescent; bracts and bracteoles 4–7 mm long, narrowly lanceolate to linear, caducous. Flowers 15–18 mm long; pedicels 8–18 mm long. Calyx 7–10 mm long, brownish subsericeous; upper teeth c.1.5 mm long, deltoid, others narrowly triangular, acute, 1.5–3.5 mm long. Standard petal 11–15 mm in diameter, subcircular, emarginate, reflexed-patent, base narrowly decurrent into the c.3.5 mm long claw; wings 11–14 × 5.5–6 mm, broadly obovate, auriculate, claw 3.5–4.5 mm long; keel petals 10.5–12.5 × 3.5 mm, dorsally slightly overlapping and often apparently fused at the overlap, auricle obtuse, claw 5 mm long. Stamens 10, fused, the tube split to the base on the upper side and variously split on the opposite side to give two groups of 5 stamens, with the upper stamen fused or not with the rest. Ovary 3–4-ovulate, septate between the ovules, stipitate, densely appressed-pilose; style 4–5 mm long, glabrous towards the tip, stigma small. Pod 1(2)-seeded, circular or subcircular, 6.5–13 cm in diameter, wing thin and stiff, undulate, usually asymmetric at the base, densely setose in centre; style-remains bent down to stipe level or overlapping the stipe; pods sometimes replaced by subspherical bristly galls 10–15 mm in diameter. Seeds oblong-subreniform. Seedling (from *Fanshawe* 1009 (K)): germination epigeal, hypocotyl short; cotyledons 2.5–3.5 × 1.4–1.8 cm. First true leaves simple; stipules 4–5 mm long, puberulous; petiole 8–12 mm long, puberulous; lamina 1.5–3.5 × 1–3.5 cm, elliptic or circular, puberulous on the margin, otherwise glabrous, lateral nerves and reticulation visible on both surfaces.

Caprivi Strip. Between Katima Mulilo and Singalamwe, fr. 30.xii.1958, *Killick & Leistner* 3206 (K, SRGH). **Botswana**. N: Chobe Nat. Park, Gubatsa Hills, imm.fr. 24.x.1972, *Pope, Biegel & Russell* 858 (SRGH). **Zambia**. N: Mbala (Abercorn), fl. 29.ix.1949, *Bullock* 1102 (EA, K, SRGH). W: Ndola, fl. 28.ix.1947, *Brenan & Greenway* 7981 (EA, K, SRGH). C: Mt. Makulu Research Station, fr. 2.i.1957, *Angus* 1477 (BM, FHO, K). S: Mazabuka, fl. 8.x.1930, *Milne-Redhead* 1231 (K). **Zimbabwe**. N: Mount Darwin, Nyarandi (Nyatandi) R., fr. 29.i.1960, *Phipps* 2487 (K, SRGH). W: Bulawayo, fr. 18.xi.1941, *Hopkins* in *GHS* 8253 (SRGH). C: Chirumanzu Dist., Mvuma (Umvuma), fl. 20.x.1947, *McGregor* 67/47 (FHO, SRGH). E: Chipinge Dist., Chirinda, fl. 24.x.1905, *Swynnerton* 41 (BM, K, SRGH). S: Masvingo (Fort Victoria), fr. 12.vii.1929, *Rendle* 411 (BM). **Malawi**. N: Karonga Dist., Chaminade Sec. School grounds, fr. 1.i.1974, *Pawek* 7722 (MAL, MO, SRGH). C: Dedza Dist., Mua Escarpment, fr. 9.iii.1966, *Banda* 823 (SRGH). S: Mangochi Dist., Namwera Escarpment, Jalasi, fr. 15.iii.1955, *Exell, Mendonça & Wild* 909 (BM, LISC, SRGH). **Mozambique**. N: Namapa Dist., between Namapa and Nacaroa, fr. 30.x.1942, *Mendonça* 1131 (B, COI, K, LISC, LUA, M, P, Z). Z: Mopeia Dist., road to Campo, fl.&

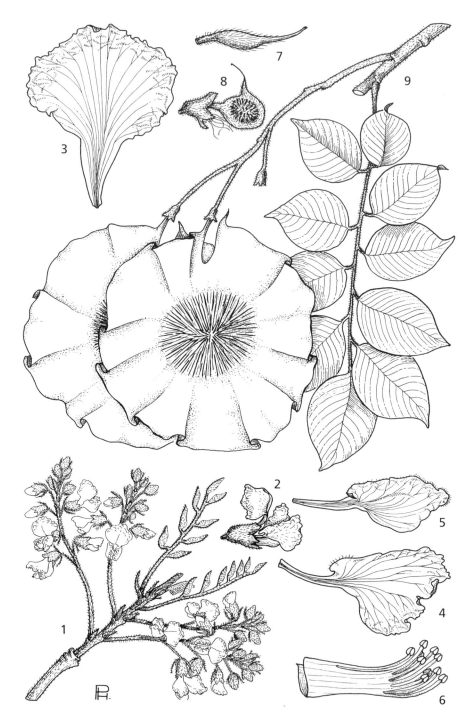

Fig. 3.3.**11**. PTEROCARPUS ANGOLENSIS. 1, flowering branchlet (× ²/₃); 2, flower (× 1); 3, standard (× 3); 4, wing (× 3); 5, keel (× 3); 6, stamens (× 3); 7, gynoecium (× 3), 1–7 from *Mendonça* 319; 8, developing fruit (× 1), from *Andrada* 1394; 9, fruiting branchlet (× ¹/₂), from *Garcia* 565. Drawn by Pat Halliday.

old fr. 25.ix.1949, *Andrada* 1933 (COI, FHO, LISC). T: Moatize Dist., railway line, 48 km from Tete, fr. 18.v.1948, *Mendonça* 4319 (COI, FHO, LISC, LMA, SRGH). MS: Manica Dist., Chimoio, Bandula, fr. 27.ii.1948, *Garcia* 418 (COI, LISC, LMA). GI: Vilankulo Dist., between Mapinhane and Muabsa, fr. x.1938, *Gomes e Sousa* 2176 (COI, EA, FHO, K, LISC). M: Maputo (Lourenço Marques), Maotas (Mahotas), fr. i.1946, *Pimenta* 17312 (LISC, PRE, SRGH).

Also in Tanzania, Congo, Angola, Namibia, South Africa (Mpumalanga and KwaZulu-Natal Provinces) and Swaziland. Widespread in miombo and other deciduous woodlands, often co-dominant, and in wooded grassland, on sandy soils and granite outcrops; 0–1600 m.

Conservation notes: A widespread taxon, sometimes exploited for timber but no evidence of scarcity or decline. Lower Risk, Least Concern.

P. angolensis occurs from south of the evergreen Guinean forest to the edge of the Kalahari, and eastwards to the coast. Pod diameter varies between 6.5 and 13 cm, apparently without any discontinuity throughout the range. In the type specimen (P) the pod is 10 cm in diameter.

An important and valuable timber tree over much of its range. The autecology of the species is treated in detail by Boaler, Ecology of *Pterocarpus angolensis* DC. in Tanzania, Overseas Research Publication 12 (H.M.S.O. 1966).

2. **Pterocarpus megalocarpus** Harms in Bot. Jahrb. Syst. **53**: 474 (1915). —Baker, Legum. Trop. Africa: 543 (1929). —Brenan, Check-list For. Trees Shrubs Tang. Terr.: 437 (1949). Type: Tanzania, Kilwa Dist., Kibata, Nandembo, *Holtz* 2532 (B† holotype).

   Pterocarpus holtzii Harms in Bot. Jahrb. Syst. **53**: 474 (1915); in Engler, Pflanzenw. Afrikas **3**(1): 635 (1915). —Baker f., Legum. Trop. Africa: 543 (1929). —Brenan, Check-list For. Trees Shrubs Tang. Terr.: 437 (1949). Types: Tanzania, Bagamoyo Dist., Kihoka-Kwa-Ibrahim, *Holtz* 1151; N of Wami R., Momwere, *Holtz* 1142 & Kilwa Dist., Matumbi, Kibata, *Holtz* 3127 (all B† syntypes, BM).

   Pterocarpus tinctorius Welw. in Ann. Consel. Ultramar. 1858: 584 (1859) as "*megalocarpus*" race sensu Polhill in F.T.E.A., Legum., Pap.: 87 (1971).

Evergreen tree up to 20 m tall. Stem with buttresses at the base (*Pedro & Pedrógão* 5272); bark dark, flaking off in plates; crown spreading or round. Branchlets terete, bark brown, appressed-pubescent, interspersed with ± dense hispid reddish glandular hairs, 1–2 mm long. Leaves 18–22 cm long; petiole 2–3 cm long, rachis 9–11 cm long, hispid-glandular; petiolules 5–6 mm long, appressed-pubescent; leaflets 3–4 on each side of the rachis, alternate, 2–9 × 2.5–4 cm, ovate or elliptic to obovate, usually abruptly and shortly acuminate, often slightly emarginate at the apex, thin and stiff, deep green and glabrous on upper surface (when mature), paler and sparsely strigulose beneath; lateral nerves in 7–12 pairs, reticulation rather close, conspicuous on both surfaces; stipules caducous. Inflorescence (in East African material) of panicles in axils of upper leaves, 10–19 cm long, with a few lax rather spreading primary branches, many-flowered; bracts 1–3 mm long, lanceolate to ovate, caducous; bracteoles 3–6 mm long; pedicel small and narrow. Calyx 5.5–7 mm long, ± densely covered with short appressed brownish hairs outside and on the lobes inside. Corolla 11–13 mm long, cream or yellow, fragrant; standard broadly ovate; wings broadly expanded, enveloping the keel and nearly as long as the standard. Pod 1-seeded, stipitate, circular, 14.5 cm in diameter (Mozambique material); wing papyraceous, undulate and flaccid, sparsely pubescent, sometimes with a few scattered bristly hairs, asymmetric at the base, style-insertion bent down overlapping the stipe and producing a deep narrow sinus, thickened-umbonate over the seed, quite glabrous (when ripe).

Mozambique. N: Montepuez Dist., Macondes, Nairoto, fr. 18.ix.1948, *Pedro & Pedrógão* 5272 (LMA, PRE, LISC photo).

Also in SE Tanzania. Coastal and riverine forest; 0–600 m.

Conservation notes: Within the Flora Zambesiaca area known from only a single collection, and from only a few collections overall. Probably best assessed as Vulnerable B1.

See note on species 2–4 at the beginning of the genus.

A planted tree in Maputo (Parque José Cabral, fl.& fr. 27.x.1954, *Carvalho* 100 (LISC) and fr. 11.ix.1972, *Balsinhas* 2432 (K, LISC)) also appears to be this species.

3. **Pterocarpus stolzii** Harms in Bot. Jahrb. Syst. **53**: 475 (1915). —Baker f., Legum. Trop. Africa: 542 (1929). —Brenan, Check-list For. Trees Shrubs Tang. Terr.: 438 (1949). —White, Dowsett-Lemaire & Chapman, Evergr. For. Fl. Malawi: 332, fig.118A, B (2001). Syntypes: Tanzania, Rungwe Dist, Kyimbila, *Stolz* 529 & Mwakaleli, *Stolz* 1660 (K).

   Pterocarpus zimmermannii Harms in Bot. Jahrb. Syst. **53**: 474 (1915). —Baker f., Legum. Trop. Africa: 542 (1929). Types: Tanzania, E Usambara Mts., Amani, *Zimmermann* 894 (EA, K syntype) & *Engler* 3422 (EA syntype), and Bumbuli, *Meinhof* (B† syntype).

   Pterocarpus tinctorius Welw. in Ann. Consel. Ultramar. 1858: 584 (1859) as "*stolzii*" race sensu Polhill in F.T.E.A., Legum., Pap.: 84 (1971).

Evergreen tree 9–25 m tall; crown spreading; bark dark grey, becoming rough and fissured. Youngest parts appressed-puberulous, but branchlets early-glabrescent. Leaves 13–25 cm long; petiole 2–6 cm long, rachis 4–12 cm long, glabrous; petiolules 3–7 mm long; leaflets 2–4 on each side of the rachis, 3–10 × 2.5–7.5 cm, elliptic to broadly ovate-elliptic, rarely slightly obovate, shortly and abruptly acuminate at the apex, with the tip rounded to emarginate, thinly coriaceous, dark green, very closely reticulate, shiny on upper surface, duller and glabrous to sparsely puberulous beneath; lateral nerves in 10–14 pairs; stipules early-caducous. Inflorescence of panicles in the axils of upper leaves with a few lax rather spreading primary branches, up to 15 cm long in flower, puberulous; bract and bracteoles 1–2.5 mm long, caducous. Flowers 12–14 mm long, rather laxly inserted; pedicels 3–5 mm long, slender. Calyx 4–7 mm long, tomentellous with very short brown hairs; teeth short, lower one 1–1.5 mm long. Standard c.12 mm in diameter, subcircular, emarginate, base decurrent into a claw 3 mm long; wings a little shorter than the standard, auricle ± rounded, claw c.4 mm long; keel petals shorter than the wings, dorsally slightly connate. Stamens in a single bundle. Ovary 2-ovulate, shortly stipitate, densely brown-hairy; style glabrous. Fruit 1-seeded, (4)6–9.5 cm in diameter, subcircular, stipitate, thickened umbonate in the centre, puberulous to almost glabrous; style remnant bent down to the stipe.

Malawi. N: Ifumbo stream, "N Nyasa", fr. n.d., *Lewis* 49 (FHO).

Also in southern and eastern Tanzania. Riverine forest.

Conservation notes: Quite widely distributed in forests of E and S Tanzania as well as N Malawi. Lower Risk, Least Concern.

Other collections, all from N Malawi, include *Pawek* 9933 (fallen fruits only: K, MAL, MO, SRGH), *Chapman* 2281 (st., FHO) and probably *Topham* 824 (st., FHO). All resemble the plants treated as the '*stolzii*' race by Polhill (1971). The distribution of this taxon – mainly in S Tanzania – is consistent with its occurrence in N Malawi. More material is needed, and collectors should look out for *Pterocarpus* with glabrous or sparsely pubescent fruits 5.5–9.5 cm across, leaves with 10–14 primary nerves on either side of the midrib, and shortly brownish pubescent (not glandular) inflorescence axes. See note on species 2–4 at the beginning of the genus.

4. **Pterocarpus chrysothrix** Taub. in Engler, Pflanzenw. Ost-Afrikas **C**: 218 (1895). — White, F.F.N.R.: 162, fig.33 (1962). Type: Tanzania, Tabora Dist., Kakoma, *Boehm* 1a (B† holotype, US, Z).

   Pterocarpus tinctorius as used by many. —sensu Baker in F.T.A. **2**: 239 (1871). —sensu De Wild. in Ann. Mus. Congo, Bot. **1**: 48 (1903); in Ann. Mus. Congo, Bot. **2**: 152 (1907). —

sensu Baker f., Legum. Trop. Africa: 541 (1929). —sensu Hauman in F.C.B. **6**: 20 (1954). —sensu Sousa in C.F.A. **3**: 358 (1966). —sensu Gonçalves in Garcia de Orta, sér.Bot. **5**: 105 (1982), non Welw. in Ann. Consel. Ultramar. 1858: 584 (1859) sensu stricto.

Pterocarpus tinctorius var. *chrysothrix* (Taub.) Hauman in F.C.B. **6**: 21 (1954). —Barbosa & Torre in Garcia de Orta **5**: 126 (1957).

Evergreen tree up to 20 m tall; crown spreading; bark greyish-brown, shallow-fissured, flaking off in scales. Young parts fulvous- or rufous-subsericeous, often glandular-hispidulous; branchlets puberulous-glabrescent, bark dark- to reddish-brown. Leaves 12–25 cm long; petiole 3–6 cm long, rachis 12–15 cm long, glabrous or rarely greyish-tomentose; petiolules 3–7 mm long; leaflets 2–4 on each side of the rachis, 2–7 × 2.5–4.5 cm, broadly ovate to subelliptic, shortly and abruptly acuminate to acuminate-subacute at the apex, usually emarginate, coriaceous, dull green, very closely reticulate, shining on upper surface, paler and glabrous to appressed-puberulous beneath; lateral nerves in 6–8 pairs, scarcely visible on either surface; stipules 4–6 mm long, caducous. Inflorescence of axillary and terminal panicles, usually shorter than the subtending leaf, rufous-tomentellous interspersed with glandular-hispidulous hairs; bracts and bracteoles caducous. Flowers 14–17 mm long, ± congested; pedicels 1–3 mm long. Calyx 5–7 mm long, puberulous, often minutely glandular-hispidulous; upper teeth slightly emarginate, lower ones 0.5–1.5 mm long, obtuse or acute. Standard c.15 mm in diameter, circular, emarginate, base narrowly decurrent in a short claw; wings as long as the standard, auricle rounded, claw c.4 mm long; keel petals shorter than the wings, dorsally slightly connate. Stamens in a single bundle. Ovary 2-ovulate, shortly stipitate, densely pilose; style glabrous. Pod 1-seeded, (5.5)6–10 cm in diameter, circular or subcircular, shortly stipitate, thickened-umbonate in the centre, densely pilose with interspersed coarse yellow viscid hairs over the seed; wing broad, undulate, papyraceous, pubescent or subvelutinous; style remains bent down to near the stipe. Ripe seed not seen.

Zambia. B: Zambezi Dist., Chavuma, fr. 14.x.1952, *White* 3510 (BM, FHO, K). N: Kaputa Dist., N end of L. Mweru Wantipa (Mweru-wa-Ntipa), fl. 18.iv.1961, *Phipps & Vesey-FitzGerald* 3282 (K, LISC, SRGH). E: Lundazi Dist., Bulumuti's Village (Mbulumuti), fr. 21.viii.1938, *Greenway & Trapnell* 5621 (EA, FHO, K). **Malawi**. N: Chitipa Dist., Stevenson road, 70 km W of Karonga, fl. 26.iv.1977, *Pawek* 12693 (K, MAL, MO, SRGH, UC). **Mozambique**. T: Chiuta Dist., Kazula (Cazula), fr. 7.vii.1949, *Andrada* 1722 (COI, LISC).

Also in Angola, Congo and Tanzania. In miombo woodland and dry forest; 450–1750 m.

Conservation notes: A widespread taxon. Lower Risk, Least Concern.

There is some variation in leaf indumentum; the leaves of plants from Malawi and Mozambique are generally appressed-puberulous beneath while those from Zambia are glabrous or nearly so. See note on species 2–4 at the beginning of the genus.

5. **Pterocarpus brenanii** Barbosa & Torre in Garcia de Orta **5**: 124, t.1 (1957). — Gomes e Sousa, Dendrol. Moçamb. **5**: 194 (1960). —White, F.F.N.R.: 162 (1962). —K. Coates Palgrave, Trees Sthn. Africa, ed.2: 322 (1988).—M. Coates Palgrave, Trees Sthn. Africa: 390 (2002). Type: Mozambique, between Tete and Kazula (Casula), 12.x.1943 *Torre* 6015 (LISC holotype, BM, BR, COI, FHO, K, LMU, M, P, SRGH, WAG).

Pterocarpus sp. of Brenan in Mem. New York Bot. Gard. **8**: 424 (1954). —Wild in Rhod. Agric. J. **50**: 416 (1953).

Deciduous tree up to 10 m tall; bark grey-brown, longitudinally fissured, flaking off in strips. Young parts glabrous; branchlets smooth, bark reddish. Leaves 12–33 cm long; petiole 3–6 cm long, rachis 7–10 cm long; petiolules 3–7 mm long; leaflets (1)2–3 on each side of the rachis, usually opposite, 5–12 × 4.5–10 cm (to 19 × 16 cm on coppice shoots), rounded or sometimes broadly retuse and mucronate at the apex, rounded to subcordate at the base, chartaceous, light green on upper surface, paler beneath; lateral nerves in 6–10 pairs, slightly prominent, reticulation soon disappearing; stipules leaf-like, up to 2.5 cm long, 0.8–3 cm broad, ovate or subcordate, persistent. Inflorescence a terminal panicle 10–25 cm long; bracts ovate-acute to narrowly lanceolate, usually present at anthesis, bracteoles linear, caducous. Flowers fragrant, 16–18 mm long; pedicels 6–14 mm long. Calyx 5–8 mm long; upper teeth obtuse, others 1.5–2.5 mm long, acute. Standard 12–16 mm in diameter, claw 3–5 mm long; wings as long as the standard, auricle acute, claw 4–5 mm long; keel petals shorter than the wings, dorsally sometimes slightly connate. Stamens in a single bundle, or the upper one sometimes free. Ovary 2–3-ovulate, sparsely pilose on the midrib and suture; style c.4 mm long; stipe c.2 mm long, sparsely pilose. Pod 1(2)-seeded, 7–10.5 × 6.5–10 cm, broadly obovate to subcircular, moderately thickened, smooth, wing broad, coriaceous, covered in a bloom; style remnant near the apex, very small. Seed 10–12 × 5–6 × 2.5 mm; testa reddish-brown. Seedling (from *Fanshawe* 7600 (K)): epigeous, cotyledon-leaves 2–2.5 cm long, obovate; first leaves simple; stipules ovate-acute; petiole 4–8 mm long, slightly puberulous; lamina 2.5–3.5 × 2–4 cm, broadly ovate, obtuse to rounded at the apex, subcordate at the base, glabrous.

Zambia. C: Luangwa Dist., Katondwe Mission, fr. 5.ii.1963, *Grout* 283 (FHO). E: Petauke Dist., Luembe, fl. 14.xii.1958, *Robson* 943 (BM, K, LISC, SRGH). S: Gwembe Dist., L. Kariba, fr. 18.vi.1961, *Angus* 2924 (FHO, K). **Zimbabwe**. N: Uzumba-Maramba-Pfungwe Dist., Pfungwe C.L. (Fungwi Reserve), Nyadire (Nyaderi) R., fl. 19.x.1955, *Lovemore* 456 (K, LISC, SRGH). E: Chipinge, fl. 12.xi.1957, *Phelps* 196 (K, SRGH). S: Lower Save (Sabi), st. 2.xii.1959, *Wild* 2338 (SRGH). **Malawi**. S: Chikwawa, 5.x.1946 *Brass* 17989 (K). **Mozambique**. T: Chiuta Dist., between Tete and Kazula (Casula), fl.& fr. 12.x.1943, *Torre* 6015 (BM, BR, COI, FHO, K, LISC, LMU, M, P, SRGH, WAG). MS: Gorongosa Nat. Park, 15 km from Chitengo camp towards Park entrance, c.60 m, fl. 11.xi.1963, *Torre & Paiva* 9174 (BR, FHO, LISC, M, PRE, WAG).

Confined to the Flora area, in hot dry areas at low altitudes, in mopane, miombo and mixed deciduous woodlands, and in floodplain grassland and thickets; not recorded above 700 m.

Conservation notes: Endemic to the Flora Zambesiaca area but apparently frequent in the dry woodlands of the Zambezi Valley. Lower Risk, Least Concern.

6. **Pterocarpus rotundifolius** (Sond.) Druce in Bot. Soc. Exch. Club Brit. Isles **4**: 642 (1917).—Burtt Davy & Hoyle, Check-list For. Trees Shrubs, Nyasaland: 61 (1936). —Miller in J. S. Afr. Bot. **18**: 36 (1952). —Pardy in Rhod. Agric. J. **49**: 215 (1952). —Williamson, Useful Pl. Nyasaland: 102 (1955). —White, F.F.N.R.: 162 (1962). — Sousa in C.F.A. **3**: 360 (1966). —Polhill in F.T.E.A., Legum., Pap.: 84 (1971). — Palmer & Pitman, Trees Sthn. Africa: 941–945 (1972). Type: South Africa, Gauteng, Aapies R., *Burke & Zeyher* s.n. (K).

Dalbergia rotundifolia Sond. in Linnaea **23**: 35 (1850).

Deciduous tree up to 18 m tall; crown rounded; bark grey or brown, flaking off in scales. Young parts greyish-sericeous or fulvous-pubescent; branchlets pubescent-glabrescent or glabrous; bark grey or reddish, smooth. Leaves 12–35 cm long; petiole 3–7 cm long, rachis 6–15 cm long, pubescent or glabrous; petiolules 2–20 mm long, puberulous or glabrous; leaflets 3–19(21), alternate, opposite or subopposite, 2–10 × 2–7 cm, circular to elliptic or ovate, occasionally obovate or oblong-subelliptic, rounded or broadly retuse or shortly and bluntly acuminate to acuminate-acute at the apex, rounded or obtuse at the base, coriaceous or thin and stiff, glabrous on upper surface, paler and appressed-pubescent beneath; lateral nerves in 6–14 pairs; stipules 4–8 mm long, caducous. Inflorescence a terminal (rarely also axillary) panicle up to 30 cm long, appressed-pubescent or glabrous; bracts and bracteoles caducous. Flowers 14–17 mm long; pedicels 2–14(18) mm long, puberulous or glabrous. Calyx 5–7 mm long, pubescent inside the teeth, otherwise glabrous or puberulous; upper teeth shallow, rounded, the lateral ones obtuse, the lower one c.2.5 mm long, acute. Standard 13–15 mm in diameter, circular, claw c.4 mm long; wings as long as the standard, auricle obtuse, claw 4–5 mm long; keel petals shorter than the wings, arcuate, claw 4.5 mm long. Stamens 10, all fused. Ovary 3–4-ovulate, glabrous, or sparsely pilose on the margins, or densely pubescent; stipe c.4 mm long, appressed-pubescent. Pod 1–2-seeded, 3–6 cm long, obovate to broadly elliptic or oblong, somewhat incurved, rounded and raised-nerved in centre; style remains at the apex or nearly so. Seeds 6–7 × 3–4 × 2.5 mm; testa dark brown, smooth.

1. Ovary sericeous-pubescent on all surfaces; calyx sparsely pubescent; leaflets usually 9–13 · iii) subsp. *martinii*
– Ovary sericeous-pubescent only on the sutures; calyx glabrous · · · · · · · · · · · 2
2. Leaflets usually 3–7, suborbicular · · · · · · · · · · · · · · · · · i) subsp. *rotundifolius*
– Leaflets usually 13–19, ovate · · · · · · · · · · · · · · · · · · · ii) subsp. *polyanthus*

i) Subsp. **rotundifolius** —K. Coates Palgrave, Trees Sthn. Africa, ed.2: 324 (1988). — M. Coates Palgrave, Trees Sthn. Africa: 392 (2002).

> *Pterocarpus sericeus* Benth. in J. Proc. Linn. Soc., Bot. **4**, Suppl.: 75 (1860). —Harvey in F.C. **2**: 264 (1862). —Baker f. in J. Linn. Soc., Bot. **40**: 59 (1911); Legum. Trop. Africa: 540 (1929). —Harms in Engler, Pflanzenw. Afrikas **3**(1): 633 (1915). —Eyles in Trans. Roy. Soc. S. Africa **5**: 380 (1916). —Henkel in Proc. Rhod. Sci. Assoc. **30**: 16 (1931). Type from South Africa (Gauteng Province).

> *Pterocarpus mellifer* Baker in F.T.A. **2**: 239 (1871), as "*melliferus*". —Burkill in Append. II to Johnson, Brit. Cent. Africa: 244 (1897). —Eyles in Trans. Roy. Soc. S. Africa **5**: 380 (1916). Type from Angola (Cuanza Norte).

> *Pterocarpus buchananii* Schinz in Bull. Soc. Bot. Genève **6**: 66 (1899). —Burtt Davy & Hoyle, Check-list For. Trees Shrubs, Nyasaland: 61 (1936). Type: Malawi, Blantyre, *Buchanan* 38 bis (P holotype, K).

Leaflets 3–7(9), usually suborbicular, rounded to retuse at the apex and broadly cuneate to rounded or subcordate at the base; petiolules 7–18 mm long. Calyx glabrous outside. Ovary glabrous, or sericeous on the margins.

Zimbabwe. N: Bindura Dist., by dam 140 m N of boundary of farms Vale and Glamorgan, fl. 1.i.1973, *Liddle* in *GHS* 222179 (K, SRGH). W: Matobo Dist., Farm Besna Kobila, fl. xii.1953, *Miller* 2047 (K, SRGH). C: Harare–Chegutu road, fl. i.1949, *Davies* 350 (K, SRGH). E: Chipinge Dist., Chirinda, fl. i.1906, *Swynnerton* 28 (BM, K, SRGH). S: Masvingo (Fort Victoria), fl.& fr. 1905, *Monro* 750 (BM, SRGH). **Malawi**. C: Lilongwe Agric. Res. Station, fl. 16.xi.1951, *G. Jackson* 628 (BM). S: Blantyre, fl.& fr., *Buchanan* 38 (BM), 38 bis (K, P). **Mozambique**. Z: Morrumbala, Alto Chindio, st. 17.vii.1972, *Bond* W11 (LISC, SRGH). T: Macanga Dist., between Furancungo and Angónia, fr. 15.vii.1949, *Andrada* 1769 (COI, K, LISC, PRE). MS: Manica Dist., slopes of mountains of Manica (Macequece), fl. 15.xi.1943, *Torre* 6203 (BM, COI, K, LISC, LMA). M: Namaacha Dist., Montes de Goba, near spring (Fonte), fl. 22.xi.1944, *Mendonça* 3042 (BM, COI, FHO, K, LISC, LMA).

Also in Angola, Namibia, South Africa (Limpopo, Gauteng, Mpumalanga and KwaZulu-Natal Provinces) and Swaziland. In miombo and open mixed deciduous woodland, wooded grassland, in riverine fringes and along water courses and on stony hills; up to 1500 m.

Conservation notes: A widespread taxon. Lower Risk, Least Concern.

ii) subsp. **polyanthus** (Harms) Mendonça & E.P. Sousa in Bol. Soc. Brot., sér.2, **42**: 270 (1968). —Polhill in F.T.E.A., Legum., Pap.: 85, fig.16/3 (1971). —K. Coates Palgrave, Trees Sthn. Africa, ed.2: 324 (1988).—M. Coates Palgrave, Trees Sthn. Africa: 392 (2002). Types: Tanzania, Morogoro Dist., Mgeta-Mlali, *Holtz* 3164; Milengeza, *Holtz* 1252 and Kilwa Dist., Liwale, *Herb. Amani* 5725 & *Lommel* 715 (all B† syntypes).

Pterocarpus polyanthus Harms in Bot. Jahrb. Syst. **53**: 473 (1915). —Baker f., Legum. Trop. Africa: 541 (1929). —Burtt Davy & Hoyle, Check-list For. Trees Shrubs, Nyasaland: 61 (1936). —Brenan, Check-list For. Trees Shrubs Tang. Terr.: 438 (1949). —Hauman in F.C.B. **6**: 18 (1954). —Williamson, Useful Pl. Nyasaland: 102 (1955). —Torre & Barbosa in Garcia de Orta **5**: 125 (1957).

Leaflets (9)13–17(21), usually broadly ovate or broadly elliptic, usually acute at the apex and cuneate at the base; petiolules 2–8 mm long. Calyx glabrous outside. Ovary glabrous, or sericeous on the margins.

Zambia. C: Lusaka, 25.v.1934, *R.G. Miller* 37 (FHO). E: Chama Dist., Luangwa Valley, 10°34'S, 33°02'E, fl. 6.i.1938, *Trapnell* 1817 (K). S: Mazabuka, fl. i.1930, *Browne* 50 (FHO). **Zimbabwe**. N: Hurungwe Dist., Zambezi Escarpment, main Harare (Salisbury) to Lusaka road, fl. 31.i.1958, *Drummond* 5397 (K, SRGH). **Malawi**. C: Lilongwe–Nkhotakota road beyond Dowa turnoff, fl. 23.i.1982, *Chapman* 6104 (K, MAL). S: Mulanje Dist., Likhubula, fl. 19.i.1958, *Chapman* 561 (BM, FHO, K). **Mozambique**. N: Malema Dist., Mutuáli, fl. 14.ii.1958, *Gomes e Sousa* 4190 (COI, EA, K, LISC, LMA, SRGH). Z: Milange, fl. 24.ii.1943, *Torre* 4819 (BM, COI, K, LISC). T: Marávia Dist., Fíngoè, fr. 26.vi.1949, *Barbosa & Carvalho* 3299 (LISC, LMA, SRGH). MS: Nhamatanda Dist., montes de Xiluvo (Chiluvo), fr. 16.iv.1948, *Mendonça* 3984 (BM, COI, K, LISC, SRGH).

Also in Tanzania, SW Kenya and Congo (Katanga). In miombo woodland and fringing forests; up to 1400 m.

Conservation notes: A widespread taxon. Lower Risk, Least Concern.

iii) subsp. **martinii** (Dunkley) Lock in Kew Bull. **54**: 208 (1999). Type: Zambia, Bombwe, *Martin* 563 (K lectotype selected by Rojo 1972; BM, FHO).

    Pterocarpus martinii Dunkley in Bull. Misc. Inform., Kew **1935**: 260 (1935). —Miller in J. S. African Bot. **18**: 35 (1952).

    Pterocarpus rotundifolius subsp. *polyanthus* var. *martinii* (Dunkley) Mendonça & E.P. Sousa in Bol. Soc. Brot., sér.2, **42**: 272 (1968). —Drummond in Kirkia **8**: 225 (1972); in Kirkia **10**: 246 (1975). —Gonçalves in Garcia de Orta, sér.Bot. **5**: 105 (1982).

Leaflets 9–11(15). Calyx sparsely pubescent outside. Ovary sericeous-pubescent over its whole surface.

Caprivi Strip. Mpalela Is. (Mpilila), fl. 13.i.1959, *Killick & Leistner* 3376 (K, SRGH). **Botswana**. N: Namabale Pan, fl. iv.1937, *Miller* B146 (BM, FHO). **Zambia**. S: Kalomo Dist., Bombwe Forest, fl. 18.ii.1933, *Martin* 524/33 (BM, EA, K). **Zimbabwe**. N: Gokwe South Dist. (Sebungwe), Sengwa Gorge, fr. 7.viii.1951, *Whellan* 518 (K, SRGH). W: Umguza Dist., Nyamandhlovu Pasture Res. Station, fr. 16.i.1952, *Plowes* 1402 (K, LISC, SRGH). **Mozambique**. T: Mágoè Dist., 35 km from Chicoa to Mphende (Mágoè), fr. 19.ii.1970, *Torre & Correia* 18063 (LISC).

Confined to the Flora Zambesiaca area, mostly on the plateau areas and escarpments bordering the middle reaches of the Zambezi Valley, in miombo, mopane and mixed deciduous woodlands and wooded grassland, on sandy soils including Kalahari sands and rocky hill slopes; 250–1200 m.

Conservation notes: Endemic to the Flora Zambesiaca area. Apparently frequent in the dry woodlands of the Zambezi Valley. Lower Risk, Least Concern.

7. **Pterocarpus lucens** Guill. & Perr. in Guillemin, Perrottet & Richard, Fl. Seneg. Tent.: 228 (1832). Type: Senegal, Galam, near Bakel, *Leprieur* s.n. (P holotype).

Subsp. **antunesii** (Taub.) Rojo, Phan. Monogr. **5**: 50 (1972). —K. Coates Palgrave,

Trees Sthn. Africa, ed.2: 323 (1988). —M. Coates Palgrave, Trees Sthn. Africa: 391 (2002). Type from Angola (Huíla).

Calpurnia antunesii Taub. in Bot. Jahrb. Syst. **23**: 173 (1896).

Pterocarpus antunesii (Taub.) Harms in Bot. Jahrb. Syst. **30**: 89 (1901); in Engler, Pflanzenw. Afrikas **3**(1): 633 (1915). —Baker f., Legum. Trop. Africa: 540 (1929). —Brenan in Mem. New York Bot. Gard. **8**: 424 (1954). —Barbosa & Torre in Garcia de Orta **5**: 122 (1957). —White, F.F.N.R.: 162 (1962). —Sousa in C.F.A. **3**: 361 (1966). —Schreiber in Merxmüller, Prodr. Fl. SW Afrika, fam. 60: 97 (1970). —Drummond in Kirkia **8**: 225 (1972); in Kirkia **10**: 246 (1975). —Palmer & Pitman, Trees Sthn. Africa: 940–944 (1972). —Lock, Leg. Afr. Check-list: 241 (1989).

Pterocarpus stevensonii Burtt Davy in Bull. Misc. Inform., Kew **1932**: 262 (1932). —Burtt Davy & Hoyle, Check-list For. Trees Shrubs, Nyasaland: 61 (1936). —Miller in J. S. African Bot. **18**: 36 (1952). —Mendonça & Torre in Mem. Junta Invest. Ultramar, sér.Bot. **1**: 30 (1950). Type: Zimbabwe (?Victoria Falls), *Allen* 85 (K holotype).

Pterocarpus lucens sensu Baker in F.T.A. **2**: 238 (1871) as regards specimen *Kirk* s.n. (K).

Deciduous tree, often branched from the base, up to 12 m tall; crown spreading or irregular; bark greyish, flaking off in plates. Young parts appressed-pubescent, soon glabrous; branchlets glabrous, bark reddish-brown. Leaves up to 12 cm long; petiole 1–3 cm long, rachis 3–6 cm long, sparsely puberulous or glabrous; petiolules 1.5–3 mm long; leaflets 1–9(11), subopposite to alternate, 2.5–5 × 1.5–3.5 cm, ovate to elliptic or oblong-ovate, rounded or obtuse or acute and sometimes emarginate at the apex, rounded or obtuse or cuneate at the base, glabrous on upper surface, minutely appressed-puberulous or glabrous beneath; lateral nerves in 4–6 pairs, these and the reticulation scarcely visible on either surface; stipules 5–8 mm long, linear, caducous. Inflorescence of axillary racemes 4–16 cm long, hairy or glabrous; bracts and bracteoles caducous. Flowers 15–18 mm long; pedicels rather slender, 10–18 mm long. Calyx 3.5–5 mm long, thinly appressed-pubescent outside, ciliolate on the teeth-margin; teeth obtuse to acute. Standard circular, 12–15 mm in diameter, claw c.3.5 mm long; wings nearly as long as the standard, auricle rounded, claw c.4 mm long; keel petals shorter than the wings, free, dorsally slightly overlapping. Stamens 9 fused, the upper one free for much of its length. Ovary 2–3-ovulate, appressed-pilose on the sutures; style 3–4 mm long, glabrous; stipe short, puberulous. Pod 1(2)-seeded, 3–4 × 2.5–5 cm, obovate or oblong-obovate, often somewhat incurved, narrowly winged, apex rounded, base obtuse or cuneate, veined or smooth over the seed; style-base at the apex or nearly so.

Caprivi Strip. Singalamwe, fr. 1.i.1959, *Killick & Leistner* 3245 (PRE, SRGH). **Botswana**. N: Chobe R., fr. i.1945, *Miller* B322 (PRE). **Zambia**. B: Sesheke Dist., Machili, fl. 18.xii.1952, *Angus* 942 (BM, K). W: Gwembe, fl. 21.xii.1955, *Bainbridge* 202/55 (FHO, K). C: Chisamba Dist., Chisamba Forest Reserve, fl. 5.xii.1957, *Fanshawe* 4121 (FHO, K, SRGH). S: Livingstone, fr. 29.xii.1930, *Pardy* Herb. 4811 (K, SRGH). **Zimbabwe**. N: Hurungwe (Urungwe), fl. 26.xi.1952, *Lovemore* 314 (K, SRGH). W: Hwange (Wankie), fl. 30.xii.1934, *Eyles* 8293 (K, SRGH). C: Kadoma Dist., Lodestar Ranch, fl. 29.xi.1976, *Tiffin* in *GHS* 250502. E: Chipinge Dist., E escarpment of Save (Sabi) R., near Mwangazi Gap, fr. 29.i.1975, *Pope, Biegel & Russell* 1428 (K, SRGH). S: Save-Runde Junction, Chitsa's Kraal, fr. 8.vi.1950, *Wild* 3394 (K, SRGH). **Malawi**. S: Chikwawa, fr. 2.x.1946, *Brass* 17894 (K, SRGH). **Mozambique**. Z: Morrumbala Dist., between Águas Quentas and Morrumbala, 9.4 km from Águas Quentas, st. 28.vi.1949, *Barbosa & Carvalho* 3772 (K, LISC, LMA). T: Mutarara Dist., Dona Ana, fr. 4.x.1944, *Mendonça* 2345 (BM, COI, EA, LISC, LMA, PRE, SRGH, WAG). MS: Gorongosa Dist., Gorongosa Nat. Park, fr. 6.xi.1963, *Torre & Paiva* 9071 (LISC). GI: Guijá Dist., between Caniçado and Nalazi (Nalasi), fl. 3.xii.1944, *Mendonça* 3237 (BM, COI, K, LISC, LMU, SRGH).

Also in S Angola and Namibia. In open woodlands and *Baikiaea* forest, often on Kalahari sands and in rocky places; 40–1200 m.

Conservation Status: A widespread taxon. Lower Risk, Least Concern.

Subsp. *lucens* occurs in Sahelian and Sudanian woodland from Senegal to Ethiopia and Uganda. The two subspecies, although widely disjunct, are extremely similar, differing mainly in the calyx which is pubescent in subsp. *antunesii* and glabrous in the typical subspecies, which also tends to have larger and more discolorous leaflets.

Dawe 462 (K) from Mozambique GI: is abnormal and is discussed by Rojo; the partially paniculate inflorescence is probably fasciated, and the flowers also appear abnormal.

9. DALBERGIA L.f.

Dalbergia L.f., Suppl. Pl.: 52 (1781) nom. conserv. —Bentham in Bentham & Hooker, Gen. Pl. **1**: 544 (1865).

Trees, shrubs or lianes, occasionally spiny. Leaves imparipinnate, rarely unifoliolate (not in Flora area); leaflets alternate to subopposite; stipules caducous or sometimes subpersistent. Flowers white or violet-purple, fragrant, in terminal or axillary and terminal panicles, rarely in cymose racemes; bracts and bracteoles caducous or subpersistent. Calyx 5-dentate, with upper teeth broader than the lower ones. Petals clawed; standard circular to obovate; wings oblong-obovate; keel petals usually dorsally connate at the apex, sometimes free. Stamens connate into a dorsally split sheath, usually the upper one free or absent or stamens in 2 bundles of 5; anthers small, erect, didymous, the thecae dehiscing by an apical slit. Ovary stipitate, few-ovulate; style incurved; stigma small, terminal. Pod indehiscent, flat, oblong or linear, ± thickened over the seeds. Seed reniform, compressed; radicle inflexed.

A genus of probably more than 100 species in the tropics and subtropics of the Old and New Worlds, much in need of revision. About 30 species in Africa, of which 13 are known from the Flora area.

1. Small tree; leaflets 1–2 on each side of the rachis, suborbicular with an abrupt acumen; wing and keel petals without a basal auricle · · · · · · · · · · · · · **13.** *sissoo*
   – Trees, shrubs or woody climbers; leaflets usually 3 or more on each side of the rachis, but if less then not suborbicular; keel petals with a basal auricle · · · · · 2
2. Leaflets 8–16 on each side of the rachis, not exceeding 10 mm long · · **2.** *armata*
   – Leaflets less numerous, usually more than 10 mm long · · · · · · · · · · · · · · · · 3
3. Indumentum including at least some large glandular hairs with swollen bases · 4
   – Indumentum without such large glandular hairs · 6
4. Leaflets more than 5 on each side of the rachis · · · · · · · · · · · · · · · · · **5.** *fischeri*
   – Leaflets (1)2–4 on each side of the rachis · 5
5. Flowers 7–8 mm long · **6.** *martinii*
   – Flowers c.12 mm long · **7.** *acutifoliolata*
6. Inflorescences borne on short shoots which are often spine-tipped · **1.** *melanoxylon*
   – Inflorescences axillary or terminal on normal shoots · · · · · · · · · · · · · · · · 7
7. Standard suborbicular; claw c.1 mm long; flowers blue, mauve or purple · · · · 8
   – Standard oblong, spathulate or obovate; claw 2–3 mm long; flowers white or cream · 9
8. Leaflets 3–6 on each side of the rachis · · · · · · · · · · · · · · · · · · · **11.** *lactea*
   – Leaflets 7–10 on each side of the rachis · · · · · · · · · · · · · · · · **12.** *malangensis*
9. Calyx 3-lobed, the lateral lobes scarcely developed · · · · · · · · · · · **9.** *bracteolata*
   – Calyx 5-lobed · 10
10. Calyx glabrous or appressed-puberulous · 11
    – Calyx pubescent to tomentose · 12
11. Pedicels 2–6 mm long; fruits scarcely thickened over the seed; leaflets (3)4–6(7) on each side of the rachis · **3.** *boehmii*

- Pedicels 1 mm long or less; fruits thickened and raised reticulate over the seed; leaflets 2–3(4) on each side of the rachis ···················· **10.** *obovata**
12. Calyx coarsely pubescent ·································· **8.** *arbutifolia*
- Calyx shortly tomentose ··································· **4.** *nitidula*

1. **Dalbergia melanoxylon** Guill. & Perr. in Guillemin, Perrottet & Richard, Fl. Seneg. Tent.: 227, fig.53 (1832). —Klotzsch in Peters, Naturw. Reise Mossamb. **6**(1): 27 (1861). —Baker in F.T.A. **2**: 233 (1871). —Taubert in Engler, Pflanzenw. Ost-Afrikas **C**: 217 (1895). —Harms in Bot. Jahrb. Syst. **28**: 407 (1900); in Engler, Pflanzenw. Afrikas **3**(1): 629 (1915). —Sim, For. Fl. Port. E. Africa: 41 (1909). — Baker f. in J. Linn. Soc., Bot. **40**: 61 (1911); Legum. Trop. Africa: 520 (1929). — Eyles in Trans. Roy. Soc. S. Africa **5**: 379 (1916). —Burtt Davy & Hoyle, Check-list For. Trees Shrubs Brit. Emp., Nyasaland: 59 (1936). —Miller in J. S. Afr. Bot. **18**: 30 (1952). —Gomes e Sousa, Dendrol. Moçamb. Estudo Geral **1**: 280 (1966). —Mendonça & Torre in Mem. Junta Invest. Ultramar, Sér. Bot. **1**: 28 (1950). — Pardy in Rhod. Agric. J. **52**: 514 (1955). —Cronquist in F.C.B. **6**: 54 (1954). — White, F.F.N.R.: 148 (1962). —Sousa in C.F.A. **3**: 347 (1966). —Polhill in F.T.E.A., Legum., Pap.: 100, fig.21 (1971). —Palmer & Pitman, Trees Sthn. Africa: 934–936 (1972). —K. Coates Palgrave, Trees Sthn. Africa, ed.2: 319 (1988). —M. Coates Palgrave, Trees Sthn. Africa: 386, fig.105 (2002). Types: Senegal, *Leprieur* (P, BM syntypes). FIGURE 3.3.**12**/A.

Shrub or straggling tree up to c.10 m high, much-branched, with spines derived from modified shoots; crown irregular; bark rough, greyish, flaking off in small scales. Young parts minutely crisped-puberulous, soon glabrous; bark smooth and pale grey. Leaves 8–20 cm long; petiole 1.5–5 cm long, rachis 4–13 cm long; petiolules 1.5–2 mm long; leaflets 6–12 or more, 1–3.5 × 1–2.5 cm, circular to broadly elliptic or obovate or subspathulate, rounded to truncate and often emarginate or retuse at the apex, rounded or slightly cordate or cuneate at the base; lateral nerves scarcely visible; stipules 2.5–5 mm long, caducous. Inflorescence of axillary and terminal panicles 2.5–8 cm long, usually borne on reduced short, spine-tipped shoots; bracts and bracteoles c.1 mm long, usually present at anthesis. Flowers white, sweet-scented, 3.5–5 mm long, secund; pedicels 1–3 mm long. Calyx 2.5–3 mm long, slightly puberulous; upper teeth shallow and rounded, the lateral ones rounded, the lower one triangular-acute and longer than the rest. Standard 3.5–4.5 mm long, broadly obovate, emarginate or retuse, claw c.1 mm long; wings oblong-obovate, claw c.1.5 mm long; keel petals shorter and broader than the wings, dorsally slightly connate or free. Stamens 9 in a single bundle, the upper one free or absent, or fused in 2 bundles of 5. Ovary 2–4-ovulate, puberulous; style short, oblique, stipe c.1.5 mm long. Pod 1–3-seeded, 2.5–7 × 1–1.8 cm, thin, stiff, conspicuously reticulate over the seeds. Seed 8–10 × 7–8 × 1.5 mm; testa reddish-brown, glossy.

Botswana. N: Chobe Dist., Gubatsaa Hills, Chobe Nat. Park, fl. 23.x.1972, *Pope, Biegel & Russell* 839 (K, SRGH). **Zambia**. C: Lusaka, fl. 17.ii.1956, *Simwanda* 71 (FHO, K, SRGH). E: Chipata Dist., Luangwa Valley, fl. 11.x.1958, *Robson & Angus* 50 (BM, K, LISC, SRGH). S: Choma, fl. 9.xi.1955, *Bainbridge* 177 (FHO, K, SRGH). **Zimbabwe**. N: Mount Darwin, immat. fr. 23.i.1960, *Phipps* 2427 (K, SRGH). W: Bulawayo, fl. x.1931, *Eyles* 7025 (K, SRGH). C: Chegutu (Hartley), fl. 1924, *Jack* 5313 (BM, SRGH). E: Mutare (Umtali), fl. xi.1945, *Chase* 183 (BM, K, SRGH). S: Chipinge Dist., lower Save, fr. 28.i.1948, *Wild* 2324 (K, LISC, SRGH). **Malawi**. S: Chikwawa, fl. 5.x.1946, *Brass* 17988 (K, SRGH). **Mozambique**. N: Nampula, near Estação Zootécnico, fl. 2.xi.1948, *Andrada* 1456 (BR, COI, K, LISC, LMA, LMU, SRGH). Z: Mocuba, fr. 2.vi.1949, *Barbosa & Carvalho* 2969 (LISC, LMA). T: Moatize, fr. 8.v.1948, *Mendonça* 4136 (COI,

---

* If pedicels 1.5–4 mm long, fruit thickened over the seed and lateral leaflets 9–10 per side; see also occasional forms of *D. fischeri* without glandular hairs.

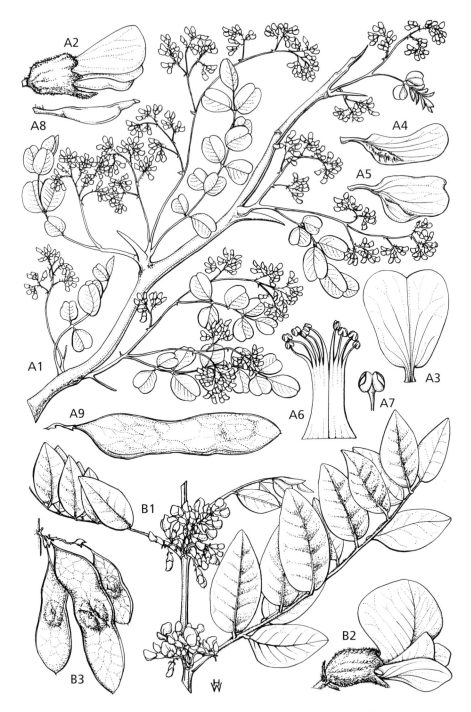

Fig. 3.3.**12**. A. —DALBERGIA MELANOXYLON. A1, flowering branch (× ²/₃); A2, flower (× 7); A3, standard (× 7); A4, wing (× 7); A5, keel (× 7); A6, stamens, spread out (× 7); A7, anther (× 20); A8, gynoecium (× 7), A1–A8 from *Milne-Redhead & Taylor* 7548; A9, pod (× 7), from *Greenway & Polhill* 11467. B. —DALBERGIA NITIDULA. B1, portion of flowering branch (× ²/₃); B2, flower (× 3), B1 & B2 from *Carmichael* 878; B3, infructescence (× 1), from *Tanner* 5057. Drawn by Heather Wood. From F.T.E.A.

LISC, LMA, P). MS: Gorongosa, fl. 19.xi.1956, *Gomes e Sousa* 4336 (COI, FHO, K, LMA, PRE). GI: Govuro, between Mambone and Vilanculos, fr. 27.v.1941, *Torre* 2760 (LISC). M: Magude Dist., between Panjane and Macaene, st. 28.i.1948, *Torre* 7233 (LISC).

Widespread from Senegal and Guinea to Eritrea and Ethiopia, southwards to Mozambique and South Africa (Limpopo and Mpumalanga Provinces) and S Angola. At medium to low altitudes, in seasonally dry mixed woodland, bushland and thicket; up to 1450 m.

Conservation notes: A very widespread taxon, globally Lower Risk, Least Concern. However, the heavy, fine-grained black heartwood (African Blackwood) is much sought-after for carvings and for high quality woodwind instruments, and larger specimens are now locally scarce enough for the species to have been proposed for CITES listing as Endangered in some coastal eastern African countries.

Flowering is often precocious, so that most flowering collections have only immature leaves. Plants are vegetatively very variable, probably in response to water availability.

2. **Dalbergia armata** E. Mey., Comment. Pl. Afr. Austr. **1**: 152 (1836). —Harms in Engler, Pflanzenw. Afrikas **3**(1): 629 (1915). —Gomes e Sousa, Dendrol. Moçamb. **2**: 23 (1949). —Polhill in F.T.E.A., Legum., Pap.: 105, fig.21 (1971). — Palmer & Pitman, Trees Sthn. Africa: 932 (1972). —K. Coates Palgrave, Trees Sthn. Africa, ed.2: 317 (1988). —M. Coates Palgrave, Trees Sthn. Africa: 385 (2002). Types: South Africa, KwaZulu-Natal, Durban (Port Natal), *Krauss* s.n., and E Cape Prov., R. Omtento to R. Omsamculo, *Drège* s.n. (all LUB† syntypes, K).

Dalbergia myriantha Meisn. in Hooker, London J. Bot. **2**: 100 (1843). Type from South Africa (KwaZulu-Natal).

Robust climber, scrambling shrub or liane to 8 m high, with woody spines (modified shoots) on the stem, briefly deciduous; side branches often becoming hooked or coiling and woody tendril-like toward the apex (but these, and the spines, often absent in herbarium material); spines robust up to c.10 cm long, usually in clusters ± encircling the stem, or single and 2–4 cm long on branches, or rarely spines absent. Young parts minutely fulvous-puberulous, soon glabrous. Leaves up to 6(8) cm long; petiole 3–6 mm long, rachis 3–6 cm long; petiolules 0.5–0.8 mm long; leaflets 8–16(20) on each side of the rachis, 4–10 × 2–4 mm, elliptic to oblong-elliptic, rounded and sometimes emarginate at the apex, rounded at the base, dull blue-green on upper surface, paler with darker reticulation beneath; stipules caducous. Inflorescence of corymbose axillary and terminal congested panicles 1.5–3(10) cm long, minutely fulvous-puberulous; bracts and bracteoles 0.5–0.8 mm long, ovate, present at anthesis. Flowers white, 3.5–4.5 mm long; pedicels 0.5–1 mm long. Calyx 2–2.5 mm long, sparsely puberulous; upper teeth shallow and rounded, the lateral ones obtuse, the lower one c.1 mm long, triangular-acute. Standard 3.5–4 mm long, broadly obovate, emarginate or retuse, claw c.0.5 mm long; wings slightly shorter than the standard, oblong-obovate, auricle rounded, claw c.0.8 mm long; keel petals broader than the wings, dorsally slightly connate or free, auricle rounded, claw c.1 mm long. Stamens 9, in 1–2 bundles, the upper one free or absent. Ovary 2–4-ovulate, puberulous on the midrib and suture; style glabrous; stipe c.1.5 mm long, puberulous. Pod 1–3-seeded, flat, 3–5.5 × 1–1.5 cm, thin and stiff, glabrous, slightly thickened, blackish-brown and reticulate over the seeds; stipe 6–9 mm long. Seed 7–8 × 5–6 mm, rather thin; testa blackish-brown.

Mozambique. M: Matutuíne Dist., left bank of R. Maputo, near Salamanga, 50 km S of Maputo (Lourenço Marques), fl. 10.x.1947, *Gomes e Sousa* 3627 (COI, K, LISC); Namaacha Dist., Goba Fronteira, 1 km from Posto Fiscal, fr. 22.xii.1944, *Mendonça* 3437 (BM, COI, K, LISC).

Also in S Tanzania, South Africa (Mpumalanga and KwaZulu-Natal) and Swaziland. At low altitude, in forest and riverine vegetation; up to c.300 m.

Conservation notes: Virtually confined to White's Tongaland-Pondoland Centre of Endemism, but appears to be widespread within this region. Lower Risk, Least Concern.

The Tanzanian material has laxer panicles, a shorter lower lobe to the calyx, and a longer claw to the standard.

3. **Dalbergia boehmii** Taub. in Engler, Pflanzenw. Ost-Afrikas **C**: 218 (1895). —Baker f., Legum. Trop. Africa: 523 (1929). —Cronquist in F.C.B. **6**: 56 (1954). —Hepper in F.W.T.A., ed.2, **1**: 516 (1958). —White, F.F.N.R.: 148 (1962). —Sousa in C.F.A. **3**: 348 (1966). —Polhill in F.T.E.A., Legum., Pap.: 105 (1971). —K. Coates Palgrave, Trees Sthn. Africa, ed.2: 318 (1988). —M. Coates Palgrave, Trees Sthn. Africa: 385 (2002). Type: Tanzania, Tabora Dist., near Igonda, *Boehm* 151a (B† holotype).

Dalbergia bracteolata sensu Baker in F.T.A. **2**: 254 (1871) in part as regards this area.

Dalbergia elata Harms in Bot. Jahrb. Syst. **26**: 296 (1899); in Engler, Pflanzenw. Afrikas **3**(1): 630 (1915). —Baker f., Legum. Trop. Africa: 522 (1929). —Burtt Davy & Hoyle, Check-list For. Trees Shrubs, Nyasaland: 69 (1936). —Pedro & Barbosa, Esb. Rec. Ecol.-Agric. Moçamb. **2**: 180 (1955). Types: Tanzania, Morogoro Dist., Uluguru Mts, Kibambira, *Stuhlmann* 9190 and Tununguo, *Stuhlmann* 8955 (both B† syntypes).

Unarmed tree or erect shrub 4–15 m high; crown spreading or irregular; bark pale brown, longitudinally shallow-fissured. Young parts glabrous, rarely hairy. Leaves 14–26 cm long; petiole 2–5.5 cm long, rachis 8–18 cm long; petiolules 3–6 mm long; leaflets (3)4–6(7) on each side of the rachis, occasionally subopposite, 2–6(8) × 1.5–3.5(4) cm, ovate to oblong-ovate or elliptic, obtuse or acute at the apex, broadly obtuse or cuneate at the base; lateral nerves in 4–7 pairs, reticulation on both surfaces scarcely visible; stipules 8–12 mm long, elliptic, acute, falling very early. Inflorescences of axillary and terminal spreading panicles, 4–10 cm long; bracts and bracteoles 1.5–2 mm long, falling before anthesis. Flowers white or cream, 5–6 mm long, sweet-scented; pedicels 1.5–4 mm long. Calyx 3–4 mm long, glabrous; upper teeth shallow-lobed, the lateral ones c.1 mm long, the lower one 1.5–2 mm long, triangular-acute. Standard subcircular to obovate, claw 2 mm long; wings as long as the standard, obovate, with auricle rounded and claw 2 mm long; keel petals broader than the wings, dorsally slightly connate at the apex, with auricle subacute and claw c.3 mm long. Stamens 9–10, connate in a single phalange. Ovary 3–5-ovulate, glabrous; style 1.5 mm long; stipe 2.5 mm long. Pod 1–3-seeded, flat, thin, 3.5–8(10) × 1–1.5 cm, coriaceous, slightly impressed-reticulate or smooth, slightly thickened and blackish over the seeds, rounded or obtuse to acuminate-acute at the apex, narrowly cuneate or narrowed at the base; stipe 6–10 mm long. Seed 5 × 4 × 1.5 mm, reniform; testa blackish-brown. Seedling (*Fanshawe* 4178): germination epigeal; cotyledons 10 × 4 mm, oblong; first 2–3 true leaves 3-foliolate, the terminal leaflet larger than the laterals; next leaves 5-foliolate.

Zambia. N: Nchelenge Dist., L. Mweru, fr. 12.xi.1957, *Fanshawe* 3951 (EA, K, SRGH). W: Mwinilunga, fl. 20.ix.1952, *Holmes* 878 (FHO, K). C: 16 km S of Lusaka, fr. 15.xi.1957, *Angus* 1785 (FHO, K, SRGH). E: Chipata Dist., Jumbe, fl. 19.x.1938, *Trapnell* 1946 (EA). S: Mazabuka Dist., c.17.5 km from Mazabuka on Kafue road, fl. 13.x.1957, *Angus* 1766 (FHO, K, SRGH). **Zimbabwe**. E: Mutare (Umtali), Commonage, fl. 6.xi.1945, *Chase* 165 (BM, K, SRGH). **Malawi**. C: Dedza Dist., Mua, N of Sosola Resthouse, fr. 6.i.1965, *Banda* 611 (K, SRGH). S: Mulanje Mt., Thuchila (Tuchila) R., fr. 15.viii.1957, *Chapman* 410 (BM, FHO, K). **Mozambique**. N: Maúa (Mahua), fl.& fr. 15.x.1942, *Mendonça* 859 (BM, COI, K, LISC, SRGH). Z: Gurué Dist., Lioma, fl. 29.ix.1944, *Mendonça* 2306 (COI, K, LISC, SRGH). T: Cahora Bassa Dist., between Chitima (Estima) and Marueira (Maroeira), fl. 20.x.1973, *Macêdo* 5303 (LISC). MS: Dondo, Nhamatanda (Vila Machado), fl. 22.ix.1943, *Torre* 5920 (BR, LISC, M, P).

From Guinea Bissau eastwards to Kenya and Tanzania, and in Angola (Lunda). In miombo and mixed deciduous woodlands, often on rocky hillsides or on termite mounds, and in riverine forests; 50–1400 m.

Conservation notes: A very widespread species. Lower Risk, Least Concern.

Polhill (1971) distinguished two subspecies. Only the nominate form occurs in the

Flora Zambesiaca area; subsp. *stuhlmannii* (Taub.) Polhill is known only from C
Tanzania. Most specimens from the Flora area are glabrous or very nearly so, and the
stipules fall very early.

4. **Dalbergia nitidula** Baker in F.T.A. **2**: 235 (1871). —Harms in Engler, Pflanzenw.
   Afrikas **3**(1): 631 (1915). —Baker f., Legum. Trop. Africa: 532 (1929). —Burtt
   Davy & Hoyle, Check-list For. Trees Shrubs, Nyasaland: 59 (1936). —Cronquist
   in F.C.B. **6**: 65 (1954). —Gomes e Sousa, Dendrol. Moçamb. Estudo Geral **1**: 281
   (1966). —White, F.F.N.R.: 149 (1962). —Sousa in C.F.A. **3**: 352 (1966). —Polhill
   in F.T.E.A., Legum., Pap.: 109 (1971). —Palmer & Pitman, Trees Sthn. Africa:
   936 (1972). —K. Coates Palgrave, Trees Sthn. Africa, ed.2: 320 (1988). —M.
   Coates Palgrave, Trees Sthn. Africa: 387, fig. 106 (2002). Type: Angola, Cuanza
   Norte, Ambaca, *Welwitsch* 1885 (LISU holotype, BM). FIGURE 3.3.**12**/B.

   Dalbergia mossambicensis Harms in Bot. Jahrb. Syst. **26**: 295 (1899); in Engler, Pflanzenw.
   Afrikas **3**(1): 631 (1915). —Burtt Davy & Hoyle, Check-list For. Trees Shrubs Brit. Emp.,
   Nyasaland: 59 (1936). —Baker f., Legum. Trop. Africa: 532 (1929). Type: Mozambique, Mt
   Gorongosa, *M.R.P. Carvalho* s.n. (B† holotype, ?COI).

   Dalbergia dekindtiana Harms in Bot. Jahrb. Syst. **26**: 298 (1899). —Monro in Proc. Rhod.
   Sci. Assoc. **8**: 60 (1908). —Eyles in Trans. Roy. Soc. S. Africa **5**: 379 (1916). —Baker,
   Legum. Trop. Africa: 531 (1929). Type from Angola (Huíla).

   Dalbergia swynnertonii Baker f. in J. Linn. Soc., Bot. **40**: 60 (1911); Legum. Trop. Africa:
   529 (1929). —Eyles in Trans. Roy. Soc. S. Africa **5**: 379 (1916). Types: Zimbabwe,
   Chimanimani (Melsetter), Inyamadzi Hills, *Swynnerton* 1316 & 1317 (BM syntypes).

Unarmed slender shrub or small straggling tree up to 10 m tall; crown spreading; bark
rough, grey or dark brown, deeply fissured. Young parts fulvous-subsericeous or shortly
tomentose; branchlets with reddish cortex. Leaves up to 16 cm long; petiole 1.5–2 cm long,
fulvous-puberulous, glabrescent; petiolules 1.5–3.5 mm long; leaflets (3)4–6(7) on each side
of the rachis, 1.5–5(6) × 0.8–3 cm, ovate to elliptic or oblong-ovate, rounded or obtuse or
acute at the apex, rounded at the base, thinly appressed-puberulous or shortly tomentose to
subsericeous on both surfaces; lateral nerves 4–6 pairs and reticulation scarcely visible;
stipules caducous. Inflorescence usually an aggregate of small showy panicles or cymose
racemes 3.5–6 cm long, borne precociously in the axils of fallen leaves, rarely also in axils of
leaves of new shoots, fulvous-tomentellous or puberulous; bracts 1 mm long, bracteoles
linear, 1.5 mm long, often present at anthesis. Flowers white, 8–10 mm long, sweet-scented;
pedicels 3–6 mm long. Calyx 3–4.5 mm long, pubescent-glabrescent; upper teeth rounded,
the others acute to subacute. Standard 7–8 mm in diameter, subcircular, apex emarginate or
retuse, base narrowly decurrent into a short claw; wings obovate, as long as the standard,
auricle rounded, claw c.3 mm long; keel petals shorter and broader than the wings,
incurved, usually free, slightly overlapping at the apex, claw c.3 mm long. Stamens in a
single bundle or in 2 bundles of 5. Ovary 2–3-ovulate, tomentose or glabrous; style glabrous;
stipe 2–3 mm long. Pod flat, 1–3-seeded, 3–5.5 × 1.7 cm, fulvous-velutinous or glabrous. Ripe
seeds not seen.

Zambia. B: Senanga, fl. 2.viii.1952, *Codd* 7337 (BM, COI, EA, K, SRGH). N: Samfya
Dist., c.7 km S of Samfya, L. Bangweulu fl. 18.viii.1952, *Angus* 226 (BM, COI, EA,
FHO, K). W: Solwezi R. gorge, fr. 12.ix.1952, *Angus* 427 (BM, FHO, K). C: Lusaka,
fr. 1.x.1956, *Angus* 1407 (BM, FHO, K). E: Katete, St. Francis' Hospital, fl.
12.ix.1955, *Wright* 21 (K). S: Kalomo Dist., Katombora, fl. 21.viii.1955, *Gilges* 419
(EA, K, SRGH). **Zimbabwe**. N: Hurungwe Dist., Sanyati R., fl. 12.viii.1953, *Lovemore*
369 (K, SRGH). W: Binga Dist., Sebungwe, fl. ix.1955, *Davies* 1529 (SRGH). C:
Goromonzi Dist., Domboshawa, fl. 5.x.1945, *Wild* 186 (K, SRGH). E: Chimanimani
Dist., Inyamadzi Hills, fl. 27.ix.1906, *Swynnerton* 1793 (BM). S: Masvingo (Fort
Victoria), fl. viii.1949, *Hodgson* 45/49 (FHO, SRGH). **Malawi**. N: Mzimba, fl.
2.xi.1941, *Greenway* 6392 (EA, K). C: Nkhotakota Game Res. boundary, Mbobo

Village, fl. 17.ix.1978, *Patel* 313 (MAL, SRGH). S: Blantyre Dist., Mpemba Hill, fl. 29.viii.1978, *Patel* 246 (MAL, SRGH). **Mozambique**. N: Lichinga Dist., between Lichinga (Vila Cabral) and Litunde, fl. 9.x.1942, *Mendonça* 690 (K, LISC, P, Z). Z: Milange, fl. 2.x.1944, *Mendonça* 2327 (BM, COI, K, LISC, LISU, LMU, WAG). T: Moatize Dist., Zóbuè, fr. 26.ix.1942, *Torre* 4566 (BM, LISC, LISU). MS: Gorongosa Dist., between Rio Urema and Derunde (Durundi), fl. 10.ix.1942, *Mendonça* 180 (BR, COI, EA, FHO, LISC, LMU, M, WAG). GI: Panda, fl. ix.1936, *Gomes e Sousa* 1821 (COI, K, LISC). M: Matutuíne Dist., Bela Vista, fl. 25.viii.1948, *Gomes e Sousa* 3806 (COI, K, LISC, SRGH).

Also in Uganda, Kenya, Tanzania, Congo, Angola and South Africa (Limpopo, Mpumalanga and KwaZulu-Natal Provinces). Locally frequent in woodland, and on rocky outcrops; 30–1350 m.

Conservation notes: A widespread taxon with no evidence of threat; Lower Risk, Least Concern.

This species is very variable in indumentum. The ovary may be glabrous or pubescent, as may be the fruit. As in East Africa, there does not seem to be any significant correlation between pubescence and habitat or geographical distribution.

The axillary buds (probably usually the inflorescence buds) are frequently transformed into showy red long-setose galls. Less frequently a similar but almost smooth subspherical gall is formed in the same place.

Greenway 5751 (Zambia N: W shores of L. Ishiba Ngandu (L. Young, Shiwa Ngandu), fl. 20.ix.1938 (K)) keys out as this species or *D. obovata*, but is clearly a climber with the leaves sparsely strigose beneath and flowers described as "small, yellow". The flowers are still in bud and there are no fruits. It may be an extreme form of *D. nitidula* or a new taxon.

5. **Dalbergia fischeri** Taub. in Engler, Pflanzenw. Ost-Afrikas **C**: 218 (1895). —Harms in Bot. Jahrb. Syst. **28**: 407 (1900). —Baker f., Legum. Trop. Africa: 524 (1929). —Burtt Davy & Hoyle, Check-list For. Trees Shrubs, Nyasaland: 59 (1936).— White, F.F.N.R.: 433 (1962). —Polhill in F.T.E.A., Legum., Pap.: 110 (1971). — White, Dowsett-Lemaire & Chapman, Evergr. For. Fl. Malawi: 325, fig.115D–F (2001). Type: Tanzania, Singida Dist., Usure, *Fischer* 191 (B† holotype).
   Dalbergia sp. 2 of White, F.F.N.R.: 149 (1962).

Robust spiny climber or liane to 12 m high reaching the canopy, or scrambling shrub with long arching branches, briefly deciduous. Lower part of stems with clusters of sharp woody spines, the middle spines longest, up to 7.5 cm long; side branches often becoming hooked or coiling and woody, tendril-like toward the apex. Branchlets, often fulvous-pubescent when young, glabrescent; cortex reddish, lenticellate; lenticels greyish, oblong. Leaves 10–15 cm long; petiolules c.2 mm long; leaflets 6–9(10) on each side of the rachis, 1–3.5 × 0.6–1.8 cm, elliptic or oblong-elliptic or subspathulate, rounded or subtruncate and often emarginate at the apex, obtuse at the base, sparsely appressed-pubescent on lower surface; lateral nerves in 4–6 pairs, scarcely visible on both surfaces; stipules 5–8 mm long, oblong, caducous. Inflorescence a terminal and occasionally also axillary panicle 4–8 cm long, fulvous-pubescent or shortly tomentose, usually sparsely interspersed with glandular-verrucose or fusiform plumose hairs; bracts and bracteoles 1.5–3 mm long, broadly ovate, puberulous, often present at anthesis. Flowers white or yellowish, 7–8 mm long; pedicels 1.5–4 mm long. Calyx 3.5–4 mm long, pubescent, occasionally with some glandular-fusiform hairs; upper teeth shallow and rounded, the lateral ones c.1 mm long, obtuse, the lower one 2 mm long, subulate. Standard 5.5 × 4 mm, oblong, retuse at the apex, sparsely pubescent on the back at the basal angle, with claw 1.5–2 mm long; wings as long as the standard, with auricle obtuse and claw c.3 mm long; keel petals shorter than the wings, free, overlapping near the apex, with claw c.3.5 mm long. Stamens usually in 2 bundles of 5, rarely in a single bundle. Ovary 3–5-ovulate, pilose on the midrib and suture; style very short; stipe 3 mm long. Pod flat, 1–4-seeded, 4–12 × 1.8–2.5 cm, thin and stiff, thickened-umbonate, reticulate and blackish over the seeds, rounded or obtuse or acute at the apex, cuneate to narrowed at the base; stipe 6–9 mm long. Seeds 8–10 × 5–6 mm; testa reddish, smooth.

Zambia. N: Mbala Dist., Sunzu Hill, imm.fr. 18.xi.1952, *Angus* 785 (BM, COI, EA, FHO, K, PRE). C: Lusaka, fl. 6.xi.1962, *Fanshawe* 7147 (FHO, K). E: Petauke Dist., Great East Road, between Hofmeyer turn-off and Kachalola, fl. 12.xii.1958, *Robson* 906 (BM, K, LISC, PRE, SRGH). S: Mazabuka Dist., Kafue Gorge, imm.fr. 9.ii.1963, *van Rensburg* 1361 (K, SRGH). **Zimbabwe**. E: Mutare (Umtali), fl. 21.xii.1959, *Chase* 7206 (BM, K, LISC, PRE, SRGH). **Malawi**. N: Nkhata Bay Dist., Roseveare's, 8 km E of Mzuzu, fr. 18.vi.1976, *Pawek* 11375 (K, MAL, MO, SRGH). S: Mulanje Dist., junction of Namajani stream and Phalombe (Palombe) R., at bottom of Fort Lister Gap, fr. 19.v.1958, *Chapman* 574 (EA, FHO, K, PRE, SRGH). **Mozambique**. N: suburbs of Nampula, fl. 20.xi.1948, *Andrada* 1466 (BR, COI, K, LISC, LMA, LMU). Z: Ile Dist., margin of R. Luá, edge of Gurué–Mocuba road, fl.& fr. 9.xi.1942, *Mendonça* 1342 (COI, K, LISC, LMU, SRGH). T: Cahora Bassa Dist., Songo, Posto de Repetiçâo, fl. 5.ii.1973, *Torre, Carvalho & Ladeira* 19008 (LISC). MS: Guro Dist., between Mungári and Luenha (Changara), fl. 26.x.1942, *Torre* 6087 (BM, COI, EA, FHO, K, LISC, LMA, LMU).

Also in Kenya and Tanzania. In riverine forest, montane rain forest, miombo woodland and dry semi-deciduous thickets; 400–1800 m.

Conservation notes: A widespread taxon, with no evidence of threat; Lower Risk, Least Concern.

Some specimens from the Viphya Plateau in Malawi, including *Chapman* 2263, *Pawek* 1623 & 3002 and *Phillips* 3827, have very few, or no, large prickle-like hairs on the inflorescence axis but otherwise resemble this species. They might be confused with *D. arbutifolia* but the pods of that species are larger and acute, not rounded, at the apex and base. *D. arbutifolia* usually flowers precociously, its leaflets are thicker in texture and the leaflet margins are usually recurved.

6. **Dalbergia martinii** White, F.F.N.R.: 148, 455, fig.27E (1962). —K. Coates Palgrave, Trees Sthn. Africa, ed.2: 318 (1988).—M. Coates Palgrave, Trees Sthn. Africa: 386 (2002). Type: Zambia, Bombwe, 21.xii.1931, *Martin* 164 (K holotype, FHO).

> *Dalbergia glandulosa* Dunkley in Bull. Misc. Inform., Kew **1935**: 260 (1935) non *D. glandulosa* Benth. (1860). Type as above.

Unarmed many-stemmed climbing shrub 3–10 m high or occasionally a small tree to 4 m high, with some side branches coiling and woody, tendril-like. Stems and branches long, arching. Branchlets and inflorescences fulvous-tomentellous ± densely interspersed with glandular-verrucose prickle-shaped hairs. Leaves up to 15 cm long; petiole 1–2 cm long, rachis 4–9 cm long, pubescent like the branchlets; leaflets (1)2–3 on each side of the rachis, 1.5–6 × 1–2.5 cm, ovate to elliptic or obovate, rounded or obtuse and mucronate or sometimes emarginate at the apex, rounded or obtuse at the base, thin and stiff, appressed-pubescent and glabrescent upper surface, hairy beneath; stipules 6–10 mm long, falcate, with 4–5 longitudinal prominent nerves, puberulous. Inflorescence of axillary and terminal panicles 3–8 cm long, fulvous-tomentellous with sparse glandular hairs; bracts and bracteoles small, sometimes present at anthesis. Flowers white, 7–8 mm long; pedicels 1–2 mm long. Calyx 4–4.5 mm long, fulvous-pubescent; upper teeth shallow rounded, the lateral ones c.1 mm long, obtuse to acute, the lower one 2–2.5 mm long, triangular-acute. Standard c.7 mm in diameter, oblong to subcircular, emarginate, claw c.2 mm long; wings obovate, auricle obsolete, claw 2.5 mm long; keel petals shorter and broader than the wings, connate-cucullate at the apex, claw 3 mm long. Stamens in a single bundle. Ovary 2–3-ovulate, pilose; style glabrous; stipe short. Pod flat, 1–2-seeded, 5–10 × 1.5–2 cm, obtuse or acute at the apex, cuneate to narrowed at the base, coriaceous, pubescent with a mixture of simple hairs and large glandular hairs (or their scars) glabrescent, reticulate, particularly over the seeds. Seed 12–14 × 6–7 × 1.5 mm; testa light brown, glossy.

Caprivi Strip. From Singalamwe to Katima Mulilo, fr. 3.i.1959, *Killick & Leistner* 3283 (K, PRE, SRGH). **Zambia**. B: Sesheke, fr. 9.iv.1955, *Exell, Mendonça & Wild* 1447 (BM, LISC, SRGH). C: Lusaka Dist., Kafue R. bridge, fr. 20.iii.1952, *White* 2310 (FHO, K). S: Kalomo Dist., Bombwe, fl. 21.xii.1931, *Martin* 164 (FHO, K). **Zimbabwe**. N: Hurungwe Dist., Zambezi Valley, Chirundu road, fr. 23.ii.1952, *Wild* 4037 (K, SRGH). W: Hwange (Wankie), fl. 26.xi.1951, *Lovemore* 206 (K, SRGH).

Known only from the Flora area. In *Baikiaea* mutemwa (thicket) and dry forest on Kalahari sands, in deciduous thickets in Zambezi, Kafue and Luangwa Valleys and in mopane/*Combretum* woodlands on stony ground and rocky hill sides; 500–1050 m.

Conservation notes: Endemic to the Flora Zambesiaca area. Not many collections, and occupies vegetation types that are vulnerable to clearance; probably best assessed as Lower Risk, Near Threatened.

7. **Dalbergia acutifoliolata** Mendonça & E.P. Sousa in Bol. Soc. Brot., sér.2, **42**: 263, t.1 (1968). Type: Zambia, Kawambwa, Kolwe stream, fl. 15.vii.1962, *Lawton* 963 (FHO holotype, K)

Climbing shrub. Young parts minutely fulvous-pubescent, glabrescent. Branchlets with rather sparse glandular-verrucose hairs, glabrescent, and ± densely lenticellate, lenticels greyish, oblong. Leaves 3.5–6 cm long; rachis including the petiole 1.5–3 cm long, minutely puberulous, soon glabrous; petiolules 1–2.5 mm long; leaflets (1)2–4 on each side of the rachis, alternate or subopposite, 1–5 × 0.5–1.5 cm, lanceolate to subelliptic, acute to obtuse at the apex, obtuse to cuneate at the base, thin and stiff, dull green and glabrous on upper surface, pale, minutely appressed-puberulous and glabrescent beneath, midrib somewhat impressed above, prominent beneath, lateral nerves obsolete; stipules caducous. Inflorescence an axillary panicle up to 3.5 cm long, minutely fulvous-pubescent; bracts caducous; bracteoles c.1 mm long, present at anthesis. Flowers white, c.11 mm long; pedicels 1.5–2.5 mm long. Calyx 4 mm long, thinly subsericeous; upper teeth shallow-lobed, obtuse, the others 1 mm long, acute. Standard c.10 mm long, reflexed-patent, obovate, emarginate, claw c.2.5 mm long; wings as long as the standard, arcuate, obtuse, auricle acute, claw 3 mm long; keel petals shorter than the wings, dorsally slightly connate, auricle acute, claw 3–3.5 mm long. Stamens in 2 bundles of 5. Ovary 2-ovulate, glabrous; style 2 mm long; stipe c.3 mm long. Pod unknown.

Zambia. N: c.10 km S of Kawambwa, 19.ix.1970, *Lawton* 1589 (K).

Only two collections known. In swamp forest. Further material may show that this is no more than an extreme form of *D. nitidula*.

Conservation notes: Endemic to one Province in the Flora area. The small number of collections and the restricted habitat suggest a rating of Endangered B1.

8. **Dalbergia arbutifolia** Baker in F.T.A. **2**: 232 (1871). —Harms in Engler, Pflanzenw. Afrikas **3**(1): 631 (1915). —Eyles in Trans. Roy. Soc. S. Africa **5**: 379 (1916). — Baker, Legum. Trop. Africa: 522 (1929). —Burtt Davy & Hoyle, Check-list For. Trees Shrubs, Nyasaland: 59 (1936). —Cronquist in F.C.B. **6**: 58 (1954). — Gomes e Sousa, Dendrol. Moçamb. Estudo Geral **1**: 282, t.76 (1966). —White, F.F.N.R.: 149 (1962). —Polhill in F.T.E.A., Legum., Pap.: 108 (1971). Type: Mozambique, Manyerere, near Chicoa (Chikoa), R. Zambezi, xi.1860, *Kirk* (K holotype). FIGURE 3.3.**13**.

Dalbergia multijuga sensu Baker f., Legum. Trop. Africa: 522 (1929) in part as regards Zambesian region excl. Natal.

Dalbergia sambesiaca Schinz in Bull. Herb. Boissier, sér.2, **2**: 998 (1902). —Baker f., Legum. Trop. Africa: 526 (1929). —Gonçalves in Garcia de Orta, Sér. Bot. **5**: 76 (1982). Type: Mozambique, Tete, Boroma, fl.& fr. 1.viii.1891, *Menyharth* 373 (Z holotype).

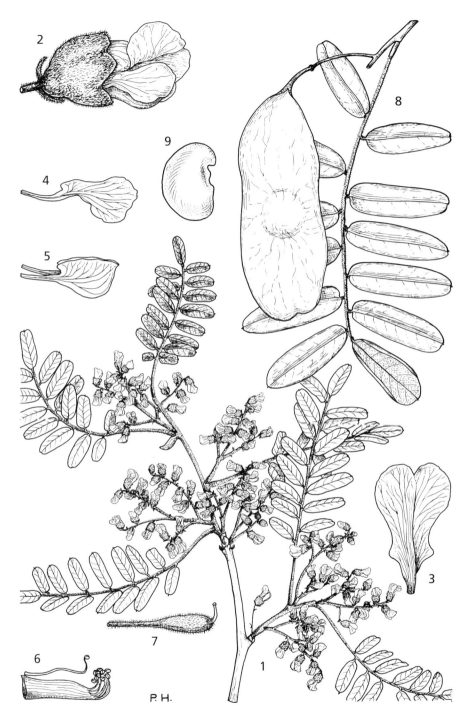

Fig. 3.3.**13**. DALBERGIA ARBUTIFOLIA. 1, flowering branch (× ²/₃); 2, flower (× 4); 3, standard (× 4); 4, wing (× 4); 5, keel (× 4); 6, stamens (× 4); 7, gynoecium (× 4), 1–7 from *Mendonça* 754; 8, fruiting branchlet (× ²/₃), from *Barbosa* 1811; 9, seed (× 2), from *Wild* 3342. Drawn by Pat Halliday.

Climbing or scandent shrub or liane up to 12 m tall, often with side branches on older wood becoming hooked or coiled and tendril-like (stem-tendrils), deciduous. Young parts fulvous- or rufous-tomentellous or pubescent-glabrescent. Leaves up to 16 cm long; petiole 1–3 cm long, rachis 6–12(14) cm long, fulvous-puberulous, glabrescent; petiolules 1–2.5 mm long; leaflets (5)6–9(10) on each side of the rachis, (1)1.5–4.5 × 0.7–1.5(2) cm, elliptic to oblong-elliptic or obovate-oblong, rounded to broadly obtuse or truncate at the apex, margin recurved, base often slightly asymmetric, rounded or subcordate, thin and stiff, dull green, glabrous on upper surface, paler, puberulous-glabrescent beneath, midrib impressed above, prominent beneath, lateral nerves in 5–8 pairs, scarcely conspicuous; stipules c.6 mm long, caducous. Inflorescence of axillary and terminal panicles, often also borne precociously in axils of fallen leaves, 2–5 cm long, rufous- or fulvous-tomentellous; bracts and bracteoles small, usually present at anthesis. Flowers white, 6–8 mm long; pedicels 1.5–5.5 mm long. Calyx 3.5–4 mm long, puberulous; upper teeth shallow-lobed, the lateral ones rounded, the lower one c.1.5 mm long, triangular-acute. Standard 6–7 mm long, obovate, with somewhat wavy margin, emarginate, claw c.1.5 mm long; wings obliquely obovate, auricle obtuse, claw c.2 mm long; keel petals shorter than the wings, dorsally slightly connate or free, claw 2.5 mm long. Stamens in a single bundle, occasionally the upper one free. Ovary 2–3-ovulate, pilose on midrib and suture; style 1.5 mm long; stipe c.3 mm long, puberulous. Pod flat, 1–2(3)-seeded, 6–12 × 2–3.5 cm, coriaceous, at first densely pubescent, glabrescent, apex rounded or obtuse, base cuneate to narrowed, rather thickened and eventually cracking over the seeds. Seed 10–15 × 6–7 × 4 mm; testa brown.

Zambia. N: Chinsali Dist., Mbesuma Ranch, Chambeshi R., fl. 5.ix.1961, *Astle* 885 (K, SRGH). C: South Luangwa Nat. Park, Mfuwe, fr. 23.iv.1965, *B.L. Mitchell* 2645 (K). E: Petauke Dist., W of Sasare (old Boma road), fr. 10.ii.1958, *Robson* 897 (BM, K, LISC, PRE, SRGH). **Zimbabwe**. N: Mutoko (Mtoko), Nyadire R., fr. 18.x.1955, *Lovemore* 446 (K, LISC, SRGH). E: Mutare (Umtali), Commonage, fl. 25.ix.1948, *Chase* 1604 (BM, COI, K, SRGH). S: Chiredzi Dist., Save-Runde (Sabi-Lundi) junction, Chitsa's Kraal, fr. 4.vi.1950, *Wild* 3342 (BM, K, LISC, SRGH). **Malawi**. N: Karonga Dist., c.21 km W of Karonga, North Rukuru Valley, fr. 31.xii.1976, *Pawek* 12124 (K, MAL, MO, SRGH). C: Salima Dist., Dowa, Chitala, fr. 29.x.1941, *Greenway* 6387 (EA, K, PRE). S: Chikwawa Dist., lower Mwanza R., fl.& fr. 6.x.1946, *Brass* 18009 (K, PRE, SRGH). **Mozambique**. N: Lago Dist., Metangula, margin of Lake Niassa (L. Malawi), fl.& imm.fr. 11.x.1942, *Mendonça* 754 (BR, COI, FHO, K, LISC, LMU, SRGH, WAG). Z: Lugela Dist., Mocuba, Namagoa Estate, fl. x.1946, *Faulkner* PRE 312 (COI, K, LISC, PRE, SRGH). T: Changara Dist., between Tete and Boruma, fr. 5.v.1948, *Mendonça* 4086 (COI, K, LISC). MS: Guro Dist., Mungári, fl.& fr. 1.ix.1943, *Torre* 5815 (B, EBV, FI, G, LISC). GI: Mabote Dist., between Zinave (Covane) and Massangena, st. 7.x.1947, *Pedro* 3478 (LMA).

Also in Tanzania and Congo (Katanga). Riverine thicket vegetation, often beside seasonal rivers in hot low altitudes, and open woodland; 30–1200 m.

Conservation notes: A widespread taxon. Lower Risk, Least Concern.

Subsp. *aberrans* Polhill, with only 4–6 pairs of leaflets which are more elliptic-oblong with less incurved margins and more steeply ascending lateral nerves, occurs on the central plateau of Tanzania.

9. **Dalbergia bracteolata** Baker in F.T.A. **2**: 234 (1871) in part, excl. Zambesian plants and pod description. —Baker f., Legum. Trop. Africa: 528 (1929). —Polhill in F.T.E.A., Legum., Pap.: 110, fig.20/2, 6 (1971). Type: Zanzibar Is, or mainland opposite, *Kirk* s.n. (K).

> *Dalbergia goetzei* Harms in Bot. Jahrb. Syst. **28**: 407 (1900). —Baker f., Legum. Trop. Africa: 522 (1929). Type: Tanzania, Uzaramo Dist., Kisangire, *Goetze* 31 (K).

Climbing or scandent shrub or liane up to 6 m tall, occasionally with side branches on older wood becoming hooked or coiled and tendril-like, deciduous; branchlets reddish-brown, smooth or with sparse rather small lenticels. Young parts ± sparsely pilose, very soon becoming glabrous. Leaves up to 12 cm long; petiole 1–2 cm long, rachis 4–8 cm long; petiolules 1.5–3 mm long; leaflets (2)3–5(6) on each side of the rachis, 1.8–5 × 1.5–3 cm, ovate to elliptic, acuminate-acute or blunt, sometimes emarginate at the apex, rounded at the base, sparsely appressed-pubescent beneath; lateral nerves in 3–5 pairs, these and the reticulation scarcely visible on either surface; stipules 3–8 mm long, subfalcate, caducous. Inflorescence of axillary and terminal panicles 3–8 cm long, very sparsely pilose, soon glabrous; bracts 1–1.5 mm long, bracteoles 1.5–3 mm long, caducous at anthesis. Flowers white, scented, 5–6 mm long. Calyx 4 mm long, deeply bilabiate, glabrous; upper teeth broad, shallow-lobed, the others very narrow, the lateral ones rather small ± hidden under the lower one which is c.2.5 mm long, acute. Standard c.5 mm long, broadly obovate to subcircular, emarginate, subcordate at the base, claw c.2 mm long; wings as long as the standard, claw 2.5 mm long; keel petals somewhat shorter than the wings, free, slightly overlapping at the apex, claw 3 mm long. Stamens connate in a single bundle. Ovary 3–4-ovulate, pilose on the midrib and suture; style 0.8 mm long; stipe c.3 mm long, puberulous. Pod flat, 1–2-seeded, 3.5–5.5 × 1.5 cm, thin and stiff to coriaceous, rounded or obtuse at the apex, cuneate at the base, smooth, yellowish-red over the seeds; stipe 3–4 mm long. Seed 6–8 × 4.5 × 2 mm; testa blackish-brown, glossy.

Mozambique. N: Eráti Dist., 10 km from Namapa towards Nacaroa, c.300 m, fl. 11.xii.1963, *Torre & Paiva* 9505 (COI, K, LISC, SRGH); Mocímboa da Praia Dist., outskirts of Mocímboa da Praia, c.10 m, fr. 14.iv.1964, *Torre & Paiva* 11915 (COI, K, LISC, LMU, SRGH).

Also in Tanzania, Kenya and Madagascar. Coastal thickets and bush grassland; 10–300 m.

Conservation notes: A widespread taxon, in a specific but extensive habitat. Lower Risk, Least Concern.

10. **Dalbergia obovata** E. Mey., Comment. Pl. Afr. Austr.: 152 (1836). —Harvey in F.C. **2**: 265 (1862). —Sim, For. Fl. Port. E. Africa: 41 (1909). —Harms in Engler, Pflanzenw. Afrikas **3**(1): 629 (1915). —Gomes e Sousa, Dendrol. Moçamb. **2**: 25 (1949). —Mogg in Macnae & Kalk, Nat. Hist. Inhaca Is.: 146 (1969). — Polhill in F.T.E.A., Legum., Pap.: 107 (1971). —Palmer & Pitman, Trees Sthn. Africa: 933–935 (1972). —Ross in Fl. Pl. Afr. **46**: t.1818 (1980). —K. Coates Palgrave, Trees Sthn. Africa, ed.2: 320 (1988).—M. Coates Palgrave, Trees Sthn. Africa: 388 (2002). Types: South Africa, Cape Prov., Bashee R., *Drège* s.n., and KwaZulu-Natal, near Durban, *Krauss* 193 (both LUB† syntypes, K).

Small tree or scrambling shrub 3–5 m high or an unarmed climber or liane to 30 m high reaching the canopy, often with side branches on older wood becoming hooked or coiled and tendril-like. Young parts minutely tomentose or puberulous. Leaves up to 12 cm long; petiole 0.6–1.5 cm long, rachis 3–7 cm long, puberulous-glabrescent; petiolules 1.5–3.5 cm long; leaflets (2)3–4 on each side of the rachis, 1.5–4.5 × 0.8–2.8 cm, obovate to elliptic, rounded or obtuse or acute at the apex, occasionally emarginate, rounded to obtuse at the base, undulate on the margin, glabrous on upper surface, sparsely puberulous beneath; lateral nerves in 6–10 pairs, visible on both surfaces; stipules 4–6 mm long, subfalcate-acute, early-caducous. Inflorescence of axillary and terminal panicles 3–10 cm long, often somewhat scorpioid, minutely fulvous-pubescent; bracts and bracteoles small, present at anthesis. Flowers white, scented, 6–7 mm long, congested; pedicels up to 1 mm long or flowers sessile. Calyx 3.5–4.5 mm long, appressed-puberulous or subtomentellous; upper teeth shallow-rounded, the lateral ones 1–1.5 mm long, obtuse, the lower one 2–2.5 mm long, oblong-ovate or acute. Standard 5–6 mm long including the claw c.1.5 mm long, subpandurate; wings as long as the standard, auricle obsolete, claw 2 mm long; keel petals slightly shorter than the wings, dorsally connate-cucullate at the apex, claw 2.5 mm long. Stamens all connate in a single bundle or in 2 bundles of 5. Ovary 2–3-ovulate, pilose on the midrib and suture; style 1 mm long, glabrous; stipe c.2 mm long, puberulous. Pod flat, 1–2-seeded, 3.5–7 × 1.2–1.8 cm, thin, stiff, fulvous, subsericeous, attenuate subacute at the apex, rounded to subacute or cuneate at the base, moderately thickened and raised-reticulate over the seeds. Seed 9 × 5 × 1.5 mm; testa brown, glossy.

Mozambique. GI: Zavala Dist., Quissico (Zavala), Zandamela, fl. 5.xii.1944, *Mendonça* 3266 (LISC, M, P, Z). M: Maputo (Lourenço Marques), Polana, fr. 20.i.1960, *Lemos & Balsinhas* 10 (COI, K, LISC, LMU).

Also in E Tanzania and E South Africa (KwaZulu-Natal). In evergreen littoral thickets, riverine vegetation and coastal forest.

Conservation notes: Confined to White's Tongaland–Pondoland Regional Mosaic, in a specific but extensive habitat. Lower Risk, Least Concern.

See note under *D. nitidula*.

11. **Dalbergia lactea** Vatke in Oesterr. Bot. Z. **29**: 251 (1879). —Taubert in Engler, Pflanzenw. Ost-Afrikas **C**: 217 (1895). —Harms in Engler, Pflanzenw. Afrikas **3**(1): 630 (1915). —Eyles in Trans. Roy. Soc. S. Africa **5**: 379 (1916). —Baker f. in J. Linn. Soc., Bot. **40**: 61 (1911); Legum. Trop. Africa: 525 (1929). —Burtt Davy & Hoyle, Check-list For. Trees Shrubs Nyasaland: 59 (1936). —Cronquist in F.C.B. **6**: 64 (1954). —Hepper in F.W.T.A., ed.2, **1**: 516 (1958). —White, F.F.N.R.: 148 (1962). —Polhill in F.T.E.A., Legum., Pap.: 111, fig.20/3, 7 (1971). —White, Dowsett-Lemaire & Chapman, Evergr. For. Fl. Malawi: 325, fig.115A–E (2001). Type: Kenya, Taita Hills, Ndara, *Hildebrandt* 2439 (W holotype).

Evergreen scrambling shrub or an unarmed climber or liane to 20 m high reaching the canopy, often with side branches on older wood becoming hooked or coiled and tendril-like. Branchlets and leaves fulvous-puberulous when young, soon glabrous; bark reddish-brown, smooth. Leaves up to 22 cm long; petiole 1.5–4.5 cm long, rachis 6–15 cm long; petiolules 1.5–5 mm long; leaflets (3)4–6 on each side of the rachis, 2–8 × 1.5–4.5 cm, ovate to elliptic, obovate or subspathulate, rounded or emarginate or truncate or retuse at the apex, rounded or obtuse or cuneate at the base, thin and stiff, light green, concolorous (when mature); lateral nerves in 8–12 pairs, these and the reticulation slightly prominent on both surfaces; stipules 4–6 mm long, oblong-acute, early-caducous. Inflorescence a robust panicle (6)10–60 (or more) cm long, axillary and terminal, occasionally borne on the top of previous season's stem, densely rufous- or fulvous-subvelutinous, glabrescent; bracts 1.5–2.5 mm long, bracteoles 1–1.5 mm long, caducous. Flowers 4–5.5 mm long, violet or purplish, congested on short cymose, often scorpioid, branchlets; pedicels 0.5–2 mm long. Calyx 2.5–3.5 mm long, minutely fulvous-puberulous; upper teeth shallow-rounded, the lateral ones c.1 mm long, obtuse, the lower one 1.5–2 mm long, triangular-acute. Standard 5–7 mm in diameter, subcircular, reflexed-patent, emarginate, claw 1–1.5 mm long; wings as long as the standard, auricle obtuse, claw c.1.5 mm long; keel petals shorter and broader than the wings, connate-cucullate at the apex, auricle rounded, claw 1.5 mm long. Stamens connate in 2 bundles of 5 or in a single bundle. Ovary 1–3-ovulate, slightly puberulous on the midrib and suture or glabrous; style slender, 1 mm long; stipe 2 mm long. Pod flat, 1-seeded, 8–13 × 2.5–4 cm, oblong-elliptic, membranaceous, reticulate, apex rounded or obtuse, base obtuse or cuneate, stipe 5–8 mm long. Ripe seed not seen.

Zambia. N: Isoka Dist., near Chisenga, in river gorge of Mafinga Mts., fl. 21.xi.1952, *Angus* 810 (FHO, K). W: Kitwe, fl. 24.vii.1958, *Fanshawe* 4629 (FHO, K, SRGH). C: Serenje Dist., Kanona Resthouse, fr. 30.xi.1952, *White* 3810 (FHO, K). E: Isoka Dist., Luangwa R., 1220 m, fr. ix.1902, *McClounie* 176 (K). **Zimbabwe**. E: Chimanimani (Melsetter), 1070 m, fl.& fr. ix.1961, *Goldsmith* 103/61 (K, LISC, SRGH). **Malawi**. N: Chitipa Dist., Misuku Hills, Wilindi Forest, 2000 m, fl. 12.xi.1958, *Robson & Fanshawe* 587 (BM, K, LISC, SRGH). S: Mulanje, 750 m, fl. 20.xi.1946, *Brass* 17880 (BM, K, PRE, SRGH). **Mozambique**. Z: Lugela Dist., Namagoa, fl. ix.1946, *Faulkner* PRE 123 (BM, BOL, COI, K, LISC, PRE, SRGH). T: Moatize Dist., Moatize (Muatiza), Mt. Zóbuè, fl. 2.x.1942, *Mendonça* 557 (BR, COI, LISC, LMU). MS: Gorongosa Dist., between R. Vandúzi and Gorongosa town (Vila Paiva), fl. 21.vii.1941, *Torre* 3137 (COI, LISC, LMA, PRE, SRGH, WAG).

Also in Congo, Uganda, Tanzania, Kenya, Ethiopia and westwards to Senegal. In Afromontane evergreen rain forest, riverine forests and high rainfall woodland and on termite mounds; up to 2000 m.

Conservation notes: A very widespread taxon. Lower Risk: Least Concern.

12. **Dalbergia malangensis** E.P. Sousa in Mem. Junta Invest. Ultramar, sér.2, **38**: 55 (1962); in C.F.A. **3**: 349 (1966). —Polhill in F.T.E.A., Legum., Pap.: 112 (1971). Type: Angola, Moxico, near Lucinda and Luao rivers, *Gossweiler* 12234 (LISC holotype, BM).

   Dalbergia sp. 1 of White, F.F.N.R.: 149 (1962). —Lock, Leg. Afr. Check-list: 240 (1989).

Evergreen, unarmed giant liane reaching the canopy at c.20 m high then descending to ground again, often with side branches on older wood becoming hooked or coiled and tendril-like, occasionally a much-branched tree. Stem exuding red gum when cut (see White 1962). Young parts fulvous-tomentellous or subvelutinous. Leaves up to 20 cm long; petiole 1.5–3 cm long, rachis 10–15 cm long, fulvous-tomentellous; leaflets 7–10 on each side of the rachis, 2.5–5.5 × 1–1.5(2) cm, elliptic to oblong-elliptic, rounded and sometimes emarginate or retuse at the apex, rounded or broadly obtuse or cuneate at the base, papyraceous, ± densely fulvous-subvelutinous; lateral nerves in 8–12 pairs, scarcely visible on either surface but more so beneath; stipules c.10 mm long, acute, caducous. Inflorescence a terminal panicle 10–12(14) cm long, densely fulvous-pubescent; bracts and bracteoles 2–3.5 mm long, caducous. Flowers violet-purple, 7–8 mm long; pedicels 2–3 mm long. Calyx c.4 mm long, minutely fulvous-puberulous; upper teeth shallow-rounded, the others c.1.5 mm long, acute. Standard c.7 mm in diameter, subcircular, emarginate, glandular-callose at the base, somewhat reflexed, claw c.1.5 mm long; wings as long as the standard, obliquely obovate, auricle obtuse and claw short; keel petals shorter than the wings, incurved, free, slightly overlapping at the apex, auricle rounded, claw 2 mm long. Stamens in 2 bundles of 5, persisting at the fruiting stage. Ovary 4-ovulate, slightly pilose on the midrib and suture; style 1 mm long, glabrous; stipe 1.5–2 mm long, puberulous. Pod flat, 1-seeded, 6–11 × 1.5–3.5(4) cm, membranaceous, reticulate, obtuse at apex, rounded at base. Ripe seed not seen.

Zambia. W: Mwinilunga Dist., Zambezi R., 6.4 km N of Kalene Mission, fr. 26.ix.1952, *White* 3382 (BM, COI, FHO, K, PRE).

Also in Angola and Tanzania. In evergreen riverine forest; c.1200 m.

Conservation notes: Within the Flora Zambesiaca area only known from a single site. Probably more widespread but poorly understood. Best assessed at a global scale as Data Deficient.

13. **Dalbergia sissoo** DC., Prodr. **2**: 416 (1825). —White, F.F.N.R.: 148 (1962). — Polhill in F.T.E.A., Legum., Pap.: 96 (1971). Type from India.

Tree to 10 m or more high; crown narrow. Young parts pubescent, glabrescent. Leaves 8–15 cm long, thinly pubescent, glabrescent; petiole 1.5–2.5 cm long, rachis 2.5–4 cm long; leaflets 1–2 on each side of the rachis, 3.8–8 × 3.3–6 cm, subcircular, rhombic-quadrate or very broadly obovate, abruptly acuminate at the apex, rounded or very broadly cuneate at the base, almost glabrous at maturity; stipules c.4 × 1 mm at the base, narrowly triangular, early-caducous. Inflorescences axillary, paniculate, 3.5–7 cm long, the ultimate branches often scorpioid, thinly pubescent; bracts and bracteoles 1–2 mm long, caducous at anthesis. Flowers creamy-white, 6–8 mm long; pedicels 0–2 mm long. Calyx 5–6 mm long, sparsely pubescent; upper teeth 1.5 mm long, rounded, lateral teeth 1–2 mm long, rounded, lower one 2.5 mm long, rounded. Standard 6–7 × 4.5 mm, including the c.1.5 mm long claw, spathulate, emarginate; wings c.5.5 × 1 mm, narrowly obovate, auricle absent, claw c.1.5 mm long; keel petals c.5 × 1 mm, free, auricle absent, claw 1 mm long. Stamens 10 in a single bundle, or in 2 bundles of 5. Ovary 1–2-ovulate, slightly pilose on the midrib and suture; style 0.5 mm long, glabrous; stipe 2 mm long, puberulous. Pods clustered, flat, 1–2-seeded, 4–5.5 × 1-1.2 cm, narrowly elliptic, reticulate, apex and base acute to rounded. Ripe seed flattened, kidney-shaped, c.5–7 x 4 mm, brown.

Zambia. W: Ndola Dist., Dola Hill, fl. 1.x.1958, *B.M. Savory* 835 (FHO). C: Lusaka Dist., Chilanga, Game & Tsetse Dept. HQ, cult., fr. 23.iii.1952, *White* 2328 (BM, FHO, K). **Zimbabwe**. N: Makonde Dist., Manyame (Hunyani) Range, roadside just N of Chinhoyi (Sinoia), fl. 28.ix.1965, *Corby* 1388 (K, SRGH). W: Matobo Dist., Matopos, fr. x.1930, *Eyles* 6621 (K, SRGH). C: Harare (Salisbury), fl. x.1910, *H.G. Mundy* in *GHS* 1025 (K, SRGH). E: Chimanimani Dist., Save (Sabi) Valley, Nyanyadzi Irrigation Scheme, cult., fl.& fr. 9.xii.1949, *Chase* 1857 (BM, COI, K, LISC, SRGH). S: Masvingo (Fort Victoria), fl.& fr. 5.ix.1955, *Chase* 5762 (BM, SRGH). **Malawi**. C: Lilongwe Dist., Lilongwe Nature Sanctuary Forest, Zone A, fl. 17.ix.1985, *Patel & Kwatha* 2657 (K, MAL). **Mozambique**. Z: Mocuba, Posto Agrícola, cult., fr. 6.vi.1949, *Barbosa & Carvalho* 2980 (K).

Native of India. Widely planted for firewood and shade; according to collectors' notes naturalized in Zimbabwe and even becoming invasive (*Fletcher* in *GHS* 234630; *Biegel* 3610 (K)). Probably more widespread than the collections suggest. The status of some of the cited collections is unclear but most are probably from planted trees.

Tribe 4. **MILLETTIEAE**

by R.K. Brummitt, J.M. Lock & B. Verdcourt*

Millettieae Miq., Fl. Indiae Batavorum **1**: 137 (1855).
  Lonchocarpeae Hutch., Gen. Fl. Pl. **1**: 380 (1964).
  Tephrosieae (Benth.) Hutch., Gen. Fl. Pl. **1**: 394 (1964).

Trees, shrubs, lianes and herbs. Leaves usually imparipinnate, rarely 1-foliolate or digitately 3–7-foliolate, often stipellate; leaflets usually opposite (alternate in *Craibia*). Flowers generally in clusters or on short lateral axes of extended inflorescences, forming pseudoracemes or pseudopanicles, or 1–several in the axils, less often well spaced in normal racemes and panicles; bracteoles usually present (absent in *Wisteria* and some species of *Tephrosia* and related genera). Calyx generally without a hypanthium beneath, with upper lobes often joined higher, sometimes subtruncate. Standard sometimes with calluses at the base, sometimes with well-developed auricles; keel petals valvately connate along the lower edge, commonly interlocked with the wings. Filaments connate, the upper filament free or joined with the others, often joined medially but free, arched and thickened to form opening near the base; anthers dorsifixed. Ovary generally with a nectarial disk around the stipe at the base, with c.2–12 ovules; styles curved, tapering, glabrous above, with a small stigma. Fruit generally a dehiscent, often rather woody pod, sometimes flat and indehiscent, then with hardened seed chambers and sometimes winged. Seeds variously shaped, flat to round in cross-section, with a small to extended hilum, sometimes arillate; testa hard or leathery; embryo with a straight or curved radicle.

A tribe of about 50 genera, mainly tropical.

The following species, belonging to genera not occurring naturally in the Flora area, have been recorded in cultivation. Other introduced and cultivated taxa are treated under the genera below.

Tephrosia, Requienia and *Ptycholobium* by R.K. Brummitt; *Leptoderris, Dalbergiella, Xeroderris, Philenoptera, Derris* by J.M. Lock based on earlier draft by F.A. Mendonça; *Platysepalum, Craibia, Millettia* and *Mundulea* by B. Verdcourt.

Wisteria sinensis (Sims) Sweet. A vigorous woody climber; leaflets 11–13, ovate-elliptic, thinly hairy beneath; flowers 2.2–2.5 cm long, in long pendent racemes, white to mauve or lilac; pods linear-oblong, often somewhat broadened distally, velvety, up to c.8-seeded. Native of China, it is mentioned by Biegel (Check-list Ornam. Pl. Rhod. Parks & Gard.: 108 (19770) as being cultivated in Zimbabwe.

Lonchocarpus sericeus (Poir.) Kunth. A tree with mauve-pink flowers; standard petal appressed-pubescent; leaflets rusty-pubescent on the undersides; fruits indehiscent, rusty-pubescent, long, flat and slightly winged along the upper margin. It is planted in Mozambique: Maputo (*Groenendijk* 2029 (K, SRGH)) and Songo, Tete (*Macêdo* 5364 (LISC)); in Zimbabwe, Harare (*Biegel* 4806 (SRGH)) and probably elsewhere. It occurs naturally in West Africa (including Angola) and South America.

1. Standard pubescent to silky tomentose outside · 2
– Standard glabrous outside or with small scattered hairs mostly towards the apex and margins · 7
2. Lateral nerves of leaflets numerous, closely parallel, steeply ascending, extending to the margin and often united into a horny marginal nerve; herbs, shrubs or rarely small soft-wooded trees · 3
– Lateral nerves of leaflets not closely parallel and steeply ascending; trees, shrubs or lianes with hard wood · 5
3. Pods ± thin-walled, variously contorted, indehiscent; leaves mostly digitately 3-foliolate and 1-foliolate (occasional leaf 5-foliolate with a short rachis); undershrubs · **21. Ptycholobium**
– Pods straight to slightly curved, dehiscent · 4
4. Pods 1-seeded, shortly falcate, up to 10 × 4 mm; leaves 1-foliolate; flowers clustered in the axils · **20. Requienia**
– Pods with several to many seeds; leaves various but if 1-foliolate inflorescences generally extended · **19. Tephrosia**
5. Pods leathery, linear, not exceeding 10 mm wide, velvety, ultimately breaking up irregularly or only tardily dehiscent; filament tips somewhat widened; small tree or shrub · **18. Mundulea**
– Pods dehiscent, sometimes tardily so, or if indehiscent much wider · · · · · · · · 6
6. Fruits dehiscent, not winged; trees, shrubs or lianes · · · · · · · · · · · **17. Millettia**
– Fruits indehiscent, sinuate between the few seeds, narrowly winged along the upper margin; cultivated tree · *Lonchocarpus sericeus*
7. Upper calyx lobes greatly enlarged to form a lip as large as the standard · **13. Platysepalum**
– Upper calyx lobes not so enlarged · 8
8. Panicles lateral, precocious, several together from young shoots which grow out into leafy branches; fruit winged along both margins · · · · · · · · · **12. Xeroderris**
– Inflorescences terminal and/or in axils of fallen leaves; fruits not winged along both margins · 9
9. Flowers in racemes or small panicles, with the flowers inserted singly along the axes · 10
– Flowers in pseudoracemes, the flowers in clusters or crowded on relatively short branches, spurs or nodes along the main axes of the inflorescence · · · · · · · 12
10. Flowers in long pendent racemes, 15–30 cm long, without evident bracts or bracteoles; a cultivated, vigorous woody climber · · · · · · · · · · · *Wisteria sinensis*
– Flowers in short spreading to ascending racemes or panicles, with bracts and bracteoles (if without bracteoles and cultivated, see Robinieae p. 217); native trees and shrubs · 11

11. Leaflets alternate, generally glabrous or sparingly hairy; pods dehiscent; plants evergreen ·· **14. Craibia**
   – Leaflets opposite, subopposite or single, generally conspicuously hairy; fruits flat, indehiscent; plants deciduous ···················· **15. Philenoptera**
12. Margins of the fruit fringed with long densely matted plumose hairs; leaves mostly crowded on spur shoots, usually flowering precociously; trees ·········
   ··· **11. Dalbergiella**
   – Margins of fruits without plumose hairs; leaves not on spur shoots ······· 13
13. Fruits turgid, drupaceous, somewhat compressed laterally, thick-walled, obliquely oblong-ellipsoid in outline; cultivated tree with dark glossy foliage ··
   ··· *Millettia pinnata*
   – Fruits not turgid ·· 14
14. Fruit a dehiscent pod; native trees or subshrubs with stems from a woody rootstock (lianes in this genus generally have hairy standards, but see imperfectly known species) ····························· **17. Millettia**
   – Fruits flat, indehiscent, winged along the upper margin; climbers (cultivated tree species of *Derris* not yet recorded from Flora Zambesiaca area) ······· 15
15. Standard subcircular; flowers clustered severally on very short axes lateral from the otherwise undivided rachis ····························· **16. Derris**
   – Standard narrowly elliptic-oblong; terminal panicles, at least, well branched, with flowers densely crowded on abbreviated ultimate branches ···· **10. Leptoderris**

10. LEPTODERRIS Dunn

by J.M. Lock*

Leptoderris Dunn in Bull. Misc. Inform., Kew **1910**: 387 (1910).

Lianes. Leaves imparipinnate, stipulate; leaflets opposite, stipellate. Inflorescence a terminal or axillary and terminal contracted racemose panicle, the flowers clustered at nodes or on short spurs (pseudoracemes); bracts and bracteoles present. Hypanthium present; disk absent. Calyx campanulate, shortly denticulate. Corolla usually persistent in fruit. Standard oblong-cymbiform; wings adhering slightly to the keel above the claw; keel petals oblong-cucullate, as long as the standard. Stamens connate into a tube, the upper one free at the base but fused to the claw of the standard, the rest fused to the bases of the other petals; anthers versatile. Ovary few-ovulate, stipitate; style filiform, stigma terminal. Pod flat, indehiscent, papyraceous, winged along the upper margin, 1–2-seeded.

A tropical African genus of 20–25 species; 2 in the Flora Zambesiaca area. The genus is much in need of revision on a continent-wide basis.

The leaves of young sterile material of *Leptoderris* may be much larger than those of mature flowering plants, and also different in shape and indumentum. *Edwards* 825 (SRGH) from Mwinilunga, W Zambia, probably represents young material of *L. goetzii*. *Dowsett-Lemaire* 571 (K), from Mzuma Forest, a few kilometres west of Chinteche, N Malawi, has distinctively narrow acuminate leaflets. It has been named as *L. harmsiana* Dunn, a poorly-understood species from Tanzania, but definite determination must await the collection of flowering and/or fruiting material.

Leaflets 2 on each side of the rachis; flowers 12–14 mm long, subsessile; fruits less than three times as long as broad ···························· **1.** *goetzei*

---

* Based on the Flora Zambesiaca account prepared by F.A. Mendonça.

Leaflets 3–4 on each side of the rachis; flowers 8–9 mm long, shortly pedicellate; fruits more than three times as long as broad · · · · · · · · · · · · · · · · · · **2.** *nobilis*

1. **Leptoderris goetzei** (Harms) Dunn in Bull. Misc. Inform., Kew **1910**: 389 (1910). —Harms in Engler, Pflanzenw. Afrikas **3**(1): 643 (1915). —Baker f., Legum. Trop. Africa: 555 (1929). —Hauman in F.C.B. **6**: 35 (1954). —White, F.F.N.R.: 158 (1962). —Sousa in C.F.A. **3**: 370 (1966). —Polhill in F.T.E.A., Legum., Pap.: 77, fig.15 (1971). —White, Dowsett-Lemaire & Chapman, Evergr. For. Fl. Malawi: 328, fig.116 (2001). Type: Tanzania, Rungwe Dist., Likaba foothills near Umuamba, *Goetze* 122 (B† holotype, L, K).

Derris goetzei Harms in Bot. Jahrb. Syst. **30**: 330 (1901).
Leptoderris kirkii Dunn in Bull. Misc. Inform., Kew **1910**: 388 (1910). —Harms in Engler, Pflanzenw. Afrikas **3**(1): 643 (1915). Type: Tanzania, Mafia Is., *Kirk* s.n. (K).

Liane up to 20 m tall. Branchlets fulvous-tomentellous. Leaves 15–25 cm long; petiole 3–8 cm long, rachis 3.5–5 cm long; petiolules 0.5–2 mm long; leaflets 5, 5–15 × 3.5–11.5 cm, leathery, dull green, glabrous above, pubescent beneath, with 5–7 pairs of lateral nerves prominent beneath, reticulation fairly prominent beneath; the proximal pair of leaflets broadly ovate to obovate, somewhat asymmetric, rounded to broadly obovate and mucronulate at the apex, rounded or subtruncate at the base; the distal pair and the terminal leaflet (which appears subsessile between the distal pair) broadly obovate, rounded or truncate or broadly retuse at the apex and apiculate, cuneate to rounded or obtuse at the base; stipules 3–5 mm long, triangular, early-deciduous; stipels 2–3 mm long, subulate. Inflorescence an axillary or terminal racemose panicle 15–40 cm long, fulvous-tomentellous; bracts and bracteoles 1–3 mm long, falling early. Flowers pink, or white with pink markings, 12–15 mm long, subsessile or on pedicels c.1 mm long, clustered on nodes along the axes. Calyx 3.5–5 mm long, tomentellous outside, sericeous inside; upper teeth separated by a small notch, the others 0.6–1 mm long. Standard 12–13 mm long, oblong-obovate, slightly auriculate, claw c.3 mm long; wings c.12 mm long, ciliolate at the apex, auricle acute, claw c.2.5 mm long; keel petals c.13 mm long, obovate, lower margins fused in upper third, claw c.3 mm long. Ovary 2-ovulate, sericeous, stipe c.3 mm long; style 4–5 mm long. Pod 1–2-seeded, 3.5–6 × 2–2.5 cm, elliptic, reticulate, thinly sericeous, wing 2–3.5 mm wide. Ripe seeds not seen.

Zambia. W: Mwinilunga Dist., 6.5 km N of Kalene Hill Mission, Zambezi R., fr. 22.ix.1952, *Angus* 514 (BM, FHO, K). **Malawi**. N: Chitipa Dist., Misuku Hills, Sokora road, c.15 km SE of Misuku Court, fl. 16.ix.1977, *Pawek* 13081 (K, MAL, MO).

Also in Angola, Congo and Tanzania. In evergreen riverine and swamp forest (mushitu); 850–1350 m.

Conservation notes: Within the Flora Zambesiaca area known only from two provinces and in a limited habitat; probably Lower Risk, Near Threatened.

2. **Leptoderris nobilis** (Baker) Dunn in Bull. Misc. Inform., Kew **1910**: 389 (1910). —Harms in Engler, Pflanzenw. Afrikas **3**(1): 643 (1915). —Baker f., Legum. Trop. Africa: 555 (1929). —Hauman in F.C.B. **6**: 39 (1954). —White, F.F.N.R.: 158 (1962). —Sousa in C.F.A. **3**: 371 (1966). Syntypes from Angola (Cuanza Norte). FIGURE 3.3.**14**.

Derris nobilis Baker in F.T.A. **2**: 245 (1871). —R.E. Fries, Wiss. Ergebn. Schwed. Rhod.-Kongo-Exped. 1: 92 (1914).
Leptoderris nobilis var. *latifoliolata* Hauman in Bull. Jard. Bot. État **24**: 227 (1954). Type from Congo.

Scrambling shrub or liane up to 12 m tall. Young parts densely fulvous-pilose or tomentellous. Branchlets fulvous-tomentellous; bark red-brown with paler lenticels. Leaves 12–20 cm long; petiole 2.5–7 cm long, rachis 3–12 cm long, fulvous-tomentellous; petiolules 1–3.5 mm long; leaflets 3–4(5) on each side of the rachis, 3.5–10 × 2.5–5 cm, ovate to elliptic or oblong-obovate,

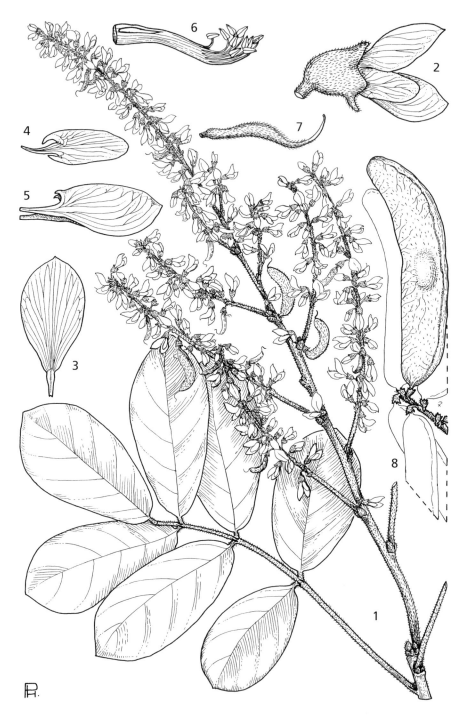

Fig. 3.3.**14**. LEPTODERRIS NOBILIS. 1, flowering branchlet ($\times$ ^{2}/$_3$); 2, flower ($\times$ 4); 3, standard ($\times$ 4); 4, wing ($\times$ 4); 5, keel ($\times$ 4); 6, stamens ($\times$ 4); 7, gynoecium ($\times$ 4), 1–7 from *Fanshawe* 307; 8, pod ($\times$ ^{2}/$_3$), from *Fanshawe* 497. Drawn by Pat Halliday.

broadly obtuse to rounded at the apex and sometimes emarginate, rounded or slightly cordate at the base and occasionally ± asymmetric, dull or light green, glabrescent on upper surface, paler and yellowish subappressed-pilose beneath, lateral nerves in 4–6 pairs, prominent, reticulation conspicuous beneath; stipules c.3 mm long, caducous; stipels small. Inflorescence a terminal panicle 20–65 cm long, fulvous-tomentellous; bracts c.2 mm long and bracteoles very small, falling early. Flowers 8–9 mm long, fasciculate on nodes along the panicle axes, pedicels 1.5–3 mm long. Calyx 2.5–3 mm long, fulvous-pubescent outside, sericeous inside; upper teeth obsolete, the others up to 1.5 mm long, obtuse. Standard 7–9 mm long, oblong-obovate, slightly auriculate, claw c.2 mm long; wings c.6 mm long, shorter than the keel, claw 2 mm long; keel petals c.8 mm long, obovate-hooded, fused from the middle to the apex, claw c.2.5 mm long. Stamens c.8 mm long, the lower 6 mm fused into a sheath. Ovary 2-ovulate, pubescent, stipe c.3 mm long; style pilose below, glabrous above. Pod 1-seeded, 6–9 × 1.6–2.3 cm, usually somewhat falcate, thinly sericeous; wing 2–3 mm wide. Ripe seed not seen. First pair of seedling leaves opposite, later ones alternate, unifoliolate, ovate, apex acute, base cordate to rounded.

Zambia. B: Zambezi Dist., Chavuma, fl. 5.viii.1952, *Gilges* 166 (EA, K, PRE, SRGH). N: Chilubi Dist., L. Bangweulu, Chilubi (Chiluwi) Is., fr. 13.x.1947, *Brenan & Greenway* 8097 (EA, FHO, K, SRGH). W: Ndola, fr. 12.xi.1953, *Fanshawe* 497 (EA, K, LISC, SRGH).

Also in Angola and Congo. In fringing evergreen or mixed forest and mushitu; 900–1700 m.

Conservation notes: On a global scale, probably Lower Risk, Least Concern. The genus as a whole is poorly understood and under-collected.

Two specimens from Zambia (*Holmes* 913, *White* 3103) have been determined as var. *latifoliolata* Hauman. The differences in leaf shape and size used to distinguish the variety seem similar to those often found between crown leaves in full sun and lower leaves in shade; leaflets with acute and rounded apices can be found on the same sheet.

11. **DALBERGIELLA** Baker f.

by J.M. Lock*

Dalbergiella Baker f. in J. Bot. **66**: 127 (1928); Legum. Trop. Africa: 534 (1929).

Trees or (not in the Flora area) lianes. Leaves imparipinnate; leaflets opposite, subopposite or occasionally alternate, without stipels. Inflorescence a contracted racemose panicle, with flowers in clusters along the axes (pseudoracemes). Calyx campanulate, 5-dentate, upper teeth broader than the others. Standard suborbicular, dorsally keeled towards the base; wings oblong; keel petals shorter than the wings. Stamens 10, 9 connate in a dorsally slit sheath, the upper one free; anthers dorsifixed, with a subcircular connective. Ovary few-ovulate, sessile. Pod 1-seeded. Seed compressed; radicle inflexed.

A genus with 3 species in tropical Africa, 1 in the Flora Zambesiaca area.

Geesink (Scala Millettiarum: 75, 1984) refers *Dalbergiella* to Dalbergieae, from which it is easily distinguished by the pseudoracemes, but the flowers do indeed resemble those of *Dalbergia*.

Dalbergiella nyasae Baker f., Legum. Trop. Africa: 535 (1929). —Milne-Redhead in Hooker's Icon. Pl. **32**: fig.3169 (1932). —Burtt Davy & Hoyle, Check-list For. Trees Shrubs, Nyasaland: 59 (1936). —Williamson, Useful Pl. Nyasaland: 47 (1955). —Pardy in Rhod. Agric. J. **53**: 625 (1956). —White, F.F.N.R.: 149, fig.27G

---

* Based on the draft Flora Zambesiaca account prepared by F.A. Mendonça in the 1960s.

(1962). —Gomes e Sousa, Dendrol. Moçamb. Estudo Geral **1**: 285, t.79 (1966). —Sousa in C.F.A. **3**: 362 (1966). —Polhill in F.T.E.A., Legum., Pap.: 93, fig.19 (1971). —K. Coates Palgrave, Trees Sthn. Africa, ed.2: 321 (1988). —M. Coates Palgrave, Trees Sthn. Africa: 388 (2002). Type: Malawi, L. Malawi, Likoma Is. (Lukoma), *Bellingham* s.n. (BM holotype). FIGURE 3.3.**15**.

Deciduous tree up to 10 m tall; crown irregular; bark rough, grey to dark brown, longitudinally fissured. Young parts fulvous or rufous-tomentellous. Leaves 12–28 cm long; petiole 3–5 cm long; rachis 8–20 cm long, tomentellous; petiolules 1–1.5 mm long; leaflets (5)6–10 on each side of the rachis, (1.5)2–6 × 1–2.5 cm, elliptic to narrowly elliptic or oblong-ovate rarely obovate, ± asymmetric at the base, acute to rounded or sometimes obtuse at the apex, recurved at the margin, broadly obtuse to rounded, rarely subcordate at the base, coriaceous, glabrous on upper surface, puberulous beneath; midrib impressed above, prominent beneath; lateral nerves in 4–7 pairs, scarcely visible on upper surface, slightly prominent beneath; stipules 3–7 mm long, linear, falling early. Inflorescence a very dense to lax racemose panicle 6–36 cm long, borne precociously on leafless stems, occasionally also in the axils of last season's leaves, fulvous- or rufous-tomentellous; bracts 2–3.5 mm long, linear, falling early; bracteoles minuscule or absent. Flowers 10–12 mm long, white or cream with the standard at least partly purple, clustered on nodes or on very short pedicels; pedicels 3–7 mm long. Calyx 4–5.5 mm long, puberulous; upper teeth shallow, rounded, the others obtuse or

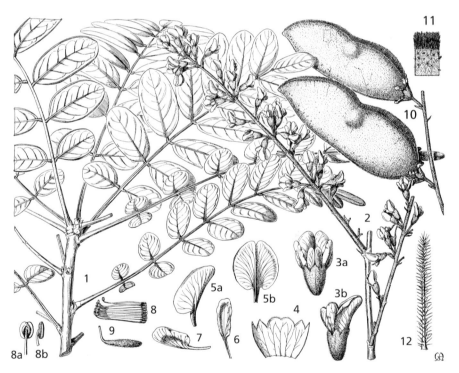

Fig. 3.3.**15**. DALBERGIELLA NYASAE. 1, leafy branch (× ²/₃); 2, flowering branch (× ²/₃); 3, flower, viewed from beneath (a) and side (b) (× 2); 4, calyx, opened out (× 2); 5, standard, viewed from side (a) and front (b) (× 2); 6, wing (× 2); 7, keel petal (× 2); 8, stamens, spread out (× 2); 8a, anthers from back (× 8); 8b, anthers from side (× 8); 9, gynoecium (× 2); 10, pods (× ²/₃); 11, detail of margin of fruit (× 2); 12, marginal hair from fruit (× 14). From Hooker's Icones Plantarum (1932). Drawn by G. Atkinson.

acute, c.1 mm long. Standard 9–10 mm long, suborbicular, somewhat reflexed, broadly emarginate, claw c.3 mm long; wings as long as the standard, auricle ciliolate, claw 4 mm long; keel petals broader than the wings, sparsely puberulous, auricle ciliolate, claw c.5 mm long. Stamens c.8 mm long, 9 connate in the basal 3–4 mm, the upper one free. Ovary 6–8-ovulate, 7–8 mm long, sessile, densely rusty-sericeous; style 2–3 mm long, incurved. Pod 1-seeded, 5–8.5 × 2–2.5 cm, rounded at the apex, broadly obtuse or cuneate at the base, densely fringed with plumose hairs on the margins, particularly the upper, otherwise pubescent-glabrescent, indehiscent. Seed 12–18 × 9–10 × 1.5 mm, reniform; testa reddish-brown, smooth.

Zambia. N: Nchelenge Dist., Chiengi (Chienge), fl. 19.viii.1958, *Fanshawe* 4747 (FHO, K). W: Luanshya, fl. 16.vii.1954, *Fanshawe* 1385 (K, SRGH). C: Lusaka Dist., between Rufunsa R. and Luangwa Beit Bridge (Luangwa R.), fl.& imm.fr. 6.ix.1947, *Brenan & Greenway* 7821 (BM, EA, FHO, K). S: Mazabuka, fr. 14.iv.1952, *White* 2666 (FHO, K). **Zimbabwe**. N: Hurungwe Dist., main road near Vuti, fl. 27.viii.1959, *Goodier* 602 (K, SRGH). W: Hwange (Wankie), st. vii.1974, *Bradfield* in *GHS* 252139 (SRGH). C: Makoni Dist., Chiduku communal land, 16 km W of Rusape, fr. 13.v.1969, *Orpen* in *GHS* 194807 (SRGH). E: Mutare (Umtali), fl. 30.viii.1949, *Chase* 1730 (BM, K, SRGH). S: Bikita Dist., Moodie's Pass, fl. 6.ix.1954, *Greenway* 8801 (FHO, SRGH). **Malawi**. C: Salima Dist., between Chitala and Dowa, fr. 28.x.1941, *Greenway* 6368 (EA, K, PRE). S: Mulanje Dist., Mt. Mulanje massif at foot of Mchese Mt., SE corner above Msomali, Nkhulambe area, fl. 4.ix.1987, *J.D. & E.G. Chapman* 8840 (K, MO). **Mozambique**. N: Montepuez, Balama, fl. 29.viii.1948, *Andrada* 1307 (BM, COI, K, LISC, LMA, LMU, PRE, SRGH). Z: Gurué Dist., between Lioma and Molumbo, fl.& imm.fr. 29.ix.1944, *Mendonça* 2309 (COI, K, LISC, LMU, PRE, SRGH). T: Marávia Dist., between Chicoa and Chiputo (Vila Vasco da Gama), fl. 8.viii.1941, *Torre* 3289 (BR, COI, EA, FHO, LISC, M, WAG). MS: Báruè Dist., Tete road, 40 km from Catandica (Vila Gouveia), fl. 18.ix.1942, *Mendonça* 334 (LISC).

Also in Tanzania. Locally frequent at medium to lower altitudes in woodland, often on rocky ground and escarpments; 600–1400 m.

Conservation notes: A very widespread taxon. Lower Risk, Least Concern.

12. **XERODERRIS** Roberty

by J.M. Lock*

Xeroderris Roberty in Bull. Inst. Fondam. Afrique Noire, sér.A. **16**: 353 (1954).
Ostryoderris sensu Hutchinson, Gen. Fl. Pl. **1**: 381 (1964) in part as regards syn. *Xeroderris* Roberty.

Deciduous tree. Leaves imparipinnate, stipulate; leaflets opposite or subopposite, exstipellate. Inflorescences of lateral panicles borne on new shoots before the leaves. Flowers white. Calyx campanulate, shortly denticulate. Standard suborbicular, abruptly reflexed at the base; wings and keel petals equalling the standard. Stamens in 2 bundles; anthers dorsifixed. Disk intrastaminal, lobulate. Ovary few-ovulate, subsessile. Pod linear, flat, winged on both margins, indehiscent. Seed oblong, radicle inflexed.

A monotypic tropical African genus occurring in seasonally dry regions from Senegal and Guinea eastwards to Kenya, and southwards to Mozambique and South Africa (Limpopo and Mpumalanga Provinces).

Geesink (Scala Millettiarum: 108, 1984) includes *Xeroderris* in *Ostryocarpus* Hook.f., together with *Ostryoderris* Dunn and *Aganope* Miq.

---

* Based on the draft Flora Zambesiaca account prepared by F.A. Mendonça.

Xeroderris stuhlmannii (Taub.) Mendonça & E.P. Sousa in Bol. Soc. Brot., sér.2, **42**: 273 (1968). —Polhill in F.T.E.A., Legum., Pap.: 91, fig.18 (1971). —Palmer & Pitman, Trees Sthn. Africa: 948–951 (1973). —K. Coates Palgrave, Trees Sthn. Africa, ed.2: 327 (1988). —M. Coates Palgrave, Trees Sthn. Africa: 395, fig.110 (2002). Types: Tanzania, Mwanza Dist., Makola, *Stuhlmann* 734; Kilosa Dist., Kidete, *Stuhlmann* 190 (K fragment); no locality, *Fischer* 225; and Malawi, no locality, 1891, *Buchanan* 1043 (B† syntypes). FIGURE 3.3.**16**.

Deguelia stuhlmannii Taub. in Engler, Pflanzenw. Ost-Afrikas **C**: 218 (1895). —Burkill in Append. II in Johnson, Brit. Centr. Africa: 244 (1897).

Ostryoderris stuhlmannii (Taub.) Harms in Engler, Pflanzenw. Afrikas **3**(1): 644 (1915). —Baker f., Legum. Trop. Africa: 563 (1929). —Bremekamp & Obermeyer in Ann. Transvaal Mus. **16**: 420 (1935). —Burtt Davy & Hoyle, Check-list For. Trees Shrubs, Nyasaland: 61 (1936). —Miller in J. S. African Bot. **18**: 34 (1952). —Brenan in Mem. New York Bot. Gard. **8**: 424 (1954). —Hauman in F.C.B. **6**: 51 (1954). —Pardy in Rhod. Agric. J. **50**: 494 (1955). —Williamson, Useful Pl. Nyasaland: 189 (1955). —Hepper in F.W.T.A., ed.2, **1**: 522, fig.163E (1958). —White, F.F.N.R.: 161, fig.27F (1962). —Gomes e Sousa, Dendrol. Moçamb. Estudo Geral **1**: 284, fig.78 (1966).

Tree up to 20 m tall; crown rounded or spreading; bark grey-brown, flaking off in large scales. Young parts fulvous-subsericeous; branchlets fulvous-pubescent, glabrescent; bark greyish, lenticellate. Leaves up to 35 cm long; petiole 2.5–7(8) cm long, rachis 10–20 cm long, fulvous-puberulous, glabrescent; petiolules 2–5 mm long; leaflets 4–7(8) on each side of the rachis, (2)3–8(12) × (2)3–5(8) cm, subcircular or broadly elliptic to elliptic or oblong- ovate, with apex rounded or obtuse or rarely subacuminate, occasionally emarginate, the base rounded or subcordate, often somewhat asymmetric, thinly puberulous-glabrescent on both surfaces; lateral nerves in 4–7 pairs, slightly prominent beneath; stipules 3–8 mm long, linear, falling early. Inflorescence of panicles 3–15(20) cm long, thinly fulvous-tomentellous, glabrescent; bracts and bracteoles falling early. Flowers 14–16 mm long; pedicels 2–5 mm long. Calyx 3–4 mm long, thinly subsericeous; upper teeth very small, the others 0.2–0.4 mm long, depressed-triangular. Corolla white. Standard 12–14 mm in diameter, abruptly reflexed near the base, emarginate, narrowly decurrent into a short claw; wings broadly obovate, auriculate, claw c.3 mm long; keel petals somewhat arcuate, connate in the middle third of the lower margin, auricle c.2 mm long, conduplicate to pouch-shaped, claw c.4 mm long. Stamens 12–15 mm long, 9 connate in the basal 10 mm, the upper one free; anthers 1 mm long, dorsifixed. Ovary 9–10 mm long, 5–7-ovulate, fulvous-appressed-puberulous; style 4–5 mm long, incurved, glabrous; stigma terminal. Pod 1–5-seeded, 6–25 × 2.5–5 cm, obtuse or acute at the apex, cuneate at the base, coriaceous, glabrous, valves adhering between the seeds, thickened over the seeds; wings 5–10 mm wide. The pod is often replaced by a globose gall c.1.5 cm in diameter. Seeds up to 18 × 10 × 4.5 mm; testa brown, glossy. First seedling leaves opposite, very broadly ovate, cordate at base, rounded and mucronate at the apex.

Caprivi Strip. Katima Mulilo, fl.& imm.fr. 21.x.1970, *Vahrmeijer* 2207 (K, PRE, SRGH). **Botswana**. N: Chobe Dist., Kasane, Chobe R., fr. vii.1930, *van Son* in Herb. Mus. Transvl. 28866 (BM, PRE). **Zambia**. B: Sesheke Dist., Katongo Forest Res., fr. 28.i.1952, *White* 1974 (FHO, K). C: Kabwe (Broken Hill) to Lusaka, fr. 27.vii.1930, *Pole Evans* 3062(19) (K, SRGH). E: Chipata Dist., Chikowa Mission, Jumbe, fl.& fr. 13.x.1958, *Robson & Angus* 82 (BM, K, LISC, SRGH). S: Mazabuka, fr. 6.ii.1963, *van Rensburg* 1355 (K, SRGH). **Zimbabwe**. N: Mutoko Dist., road to Mokaka, E of Mutoko

Fig. 3.3.**16**. XERODERRIS STUHLMANNII. 1, leaf (× ¹/₂), from *Dale* in *Kenya Forest Dept.* 3843; 2, flowering branchlet (× ²/₃), from *Michelmore* 828; 3, flower (× 3); 4, flower with petals removed (× 3); 5 & 6, standard (× 3); 7, wing (× 3); 8, keel (× 3); 9, stamens, spread out (× 3); 10, gynoecium, with ovary opened to show ovules (× 3); 11, stigma (× 50), 3–11 from *B.D. Burtt* 5262; 12, pod (× ²/₃), from *Peter* K463; 13, portion of pod, with surface cut away to show seed cavity (× ²/₃); 14, seed (× 1¹/₂), 13 & 14 from *Michelmore* 811. Drawn by Margaret Stones. From F.T.E.A.

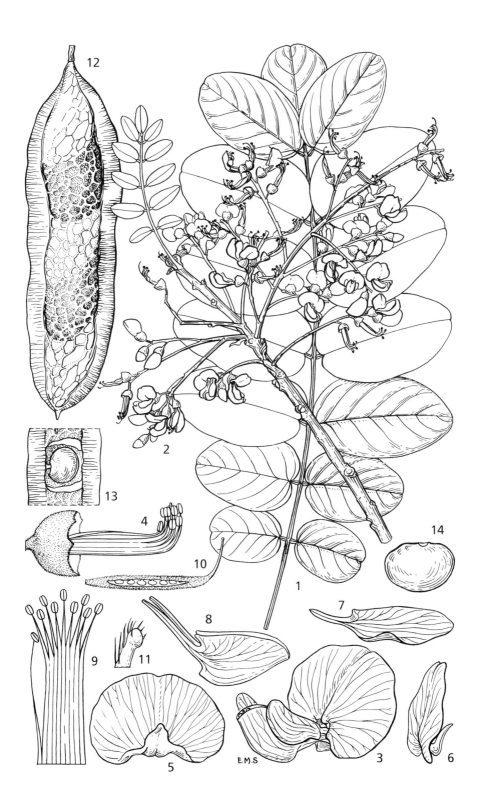

(Mtoko) Reserve, fl. 8.x.1958, *Cleghorn* 411 (SRGH). W: Sebungwe R., fr. 16.v.1956, *Plowes* 1993 (K, SRGH). C: Kadoma (Gatooma), fl. 13.x.1931, *Eyles* 7486 (SRGH). E: Mutare (Umtali), fl.& fr. 6.xi.1949, *Chase* 1814 (BM, COI, K, SRGH). S: Lower Save R. valley, fr. 28.i.1948, *Wild* 2325 (K, SRGH). **Malawi**. N: Nkhata Bay Dist., Chiponde, Likoma Is., fr. 26.viii.1994, *Salubeni & Nachamba* 3894 (K). C: Salima Dist., Dowa, Chitala, fl.& fr. 28.x.1941, *Greenway* 6374 (EA, K, PRE). S: Chikwawa, fl. 5.x.1946, *Brass* 17981 (K). **Mozambique**. N: Palma Dist., Rovuma R. banks, fl. 22.x.1942, *Mendonça* 1052 (LISC, LMU, PRE). Z: Mocuba Dist., near Mugeba, fl. 6.x.1941, *Torre* 3601 (COI, EA, LISC, LMU, M). T: between Chetima and Tete, 41.9 km from Chetima, fr. 1.vii.1949, *Barbosa & Carvalho* 3418 (LISC, LMA). MS: Barué Dist., near Pungué R., Catandica (Vila Gouveia) road, fl.& fr. 16.ix.1942, *Mendonça* 262 (B, FI, G, LISC, LUAU, M, P, SRGH, Z). GI: Guijá Dist., between Mabalane and Caniçado, fl. 5.xi.1944, *Mendonça* 2764 (COI, K, LISC, SRGH, WAG).

Also in Sierra Leone, Guinea and Senegal, extending to Burundi, Kenya, Tanzania, Congo (Katanga) and South Africa (Limpopo and Mpumalanga Provinces). In various mixed deciduous woodlands, including miombo, mopane and *Sterculia*/baobab woodland and in wooded grassland, in granite sandveld and Kalahari sands, often on rocky outcrops and on seasonal river banks; 30–1200 m.

Conservation notes: A very widespread taxon. Lower Risk, Least Concern.

13. PLATYSEPALUM Baker

by B. Verdcourt

Platysepalum Baker in F.T.A. **2**: 131 (1871). —Gillett in Kew Bull. **14**: 464–467 (1960).

Small trees, shrubs or lianes. Leaves imparipinnate; leaflets in 2 or more pairs, opposite, stipellate; stipules very small, falling early. Inflorescences paniculate, axillary, the branches pseudoracemose, many-flowered; bracteoles large, oblong, sometimes persistent. Calyx tube campanulate; upper pair of teeth joined to form a very broad emarginate hood as large as the standard; lower 3 teeth narrowly lanceolate. Standard oblate and emarginate to obcordate, shortly clawed, glabrous. Upper stamen joined only near base of filament sheath. Ovary sessile, pubescent, 5–7-ovuled; style filiform, glabrous, incurved with minute terminal stigma. Pod dehiscent, woody, compressed, 3–5-seeded. Seeds, discoid or compressed-oblong.

A genus of about 12 forest species in tropical Africa, differing from the *Millettia puguensis* group of species only in the hooded form of the upper calyx limb which almost hides the standard.

Platysepalum inopinatum Harms in Notizbl. Bot. Gart. Berlin-Dahlem **12**: 510 (1935). —Brenan, Check-list For. Trees Shrubs Tang. Terr.: 436 (1949). —Gillett in Kew Bull. **14**: 467 (1960); in F.T.E.A., Legum., Pap.: 120, fig.23 (1971). Type: S Tanzania, 45 km W of Lindi, L. Lutamba, *Schlieben* 5392 (B† holotype, EA, K). FIGURE 3.3.17.

Small tree, shrub or liane 4–6 m tall; branchlets grey, pubescent when young. Leaves 5–7-foliolate; petiole 2.5 cm long, brown-yellow pubescent; rachis up to 4.5 cm long; leaflets 3–8 × 2–4 cm, elliptic to oblanceolate or oblong, broadly acute at the apex, narrowed to a rounded base, glabrous on upper surface, appressed-pubescent beneath and sometimes with a golden sheen, lateral nerves 6–7 on each side; stipules 2–5 × 2 mm, triangular, fibrous, deciduous; stipels 1.5–2 mm long, subulate; petiolules 1–1.5 mm long. Inflorescence a narrow terminal many-flowered panicle 7–20 cm long, the lateral branches under 0.5 to 3 cm long, 2–5-flowered;

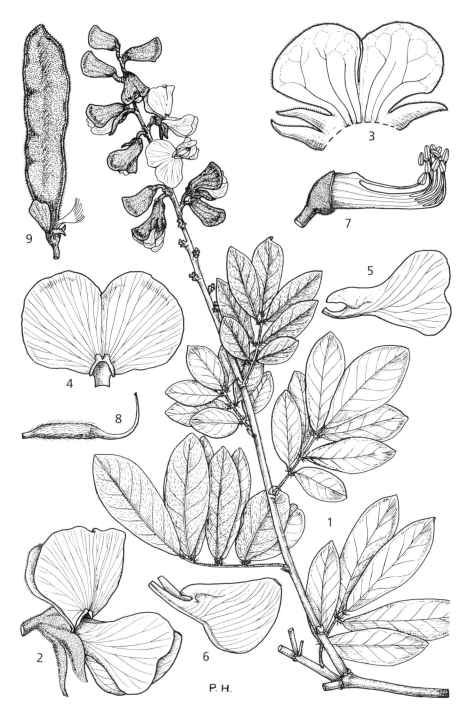

Fig. 3.3.**17**. PLATYSEPALUM INOPINATUM. 1, flowering branchlet ($\times$ ²/₃); 2, flower ($\times$ 2), 1 & 2 from *Mendonça* 993; 3, calyx, opened out ($\times$ 2); 4, standard ($\times$ 2); 5, wing ($\times$ 2); 6, keel ($\times$ 2); 7, stamens ($\times$ 2); 8, gynoecium ($\times$ 2), 3–8 from *Semsei* 1347; 9, pod ($\times$ ²/₃), from *Mendonça* 993. Drawn by Pat Halliday. From F.T.E.A.

bracts c.3 mm long, ovate, pubescent, deciduous; bracteoles at calyx base, c.2 mm long, ovate; pedicels up to 6 mm long. Calyx greenish-brown, velvety; tube c.3 mm long; upper lip longitudinally wrinkled in bud, bilabiate, 15 × 22 mm wide at apex, 8 mm wide at base; the 3 lower teeth c.8 × 3 mm. Corolla lilac, mauve or wine-red, scarcely exserted; standard 14 × 21 mm, oblate, emarginate, glabrous, with transverse callus on each side of base, claw 4 mm long; blade of wing 13 × 11 cm, obtriangular, with auricle 2.5 mm long and claw 3 mm long; blades of keel petals c.12 × 10 mm, ± semicircular, with auricle 15 mm long and a small pocket near its base, claw 6 mm long. Stamen sheath 8–9 mm long with free parts 7–8 mm long; anthers 1.6 mm long; upper filament almost free, widened at the base, 12 mm long. Nectarial disk 10-lobed, 1–1.5 mm long. Ovary 12 mm long, 8-ovuled, densely silky pubescent; style glabrous curved through 90°. Pods 6.5–10 × 1.5–2.3 cm, narrowly oblong, densely rusty pubescent, 4–6-seeded. Seeds blackish, elliptic, 10 mm long (immature).

Mozambique. N: between Nangade and Palma, fl.& fr. 20.x.1942, *Mendonça* 993 (BM, K, LISC).

Also in S Tanzania. Evergreen bushland on orange sandy soil, also on tall trees in forest ('laurisilva'); under 200 m.

Conservation notes: Limited distribution; possibly Vulnerable.

14. CRAIBIA Harms & Dunn

by B. Verdcourt

Craibia Harms & Dunn in Dunn in J. Bot. **49**: 107 (1911). —Gillett in Kew Bull. **14**: 189–197 (1960).

Evergreen trees or shrubs. Leaves usually imparipinnate, the leaflets ± leathery, entire, nearly always alternate (1-foliolate in one West African species); stipels present or absent; pulvinus present; buds well developed, flattened, oval or round with broad scales. Flowers usually fragrant, in axillary or terminal racemes or terminal panicles, the pedicels inserted singly; bracts and bracteoles deciduous. Calyx with broad lobes much shorter than the tube, the upper pair united for at least half their length. Corolla white or speckled and streaked with pink or purple, glabrous or with few hairs; standard without folds or processes, with short distinct claw; wings and keel petals with well marked auricles. Upper stamen almost free, curving away from the sheath at the base and apex but sometimes lightly adhering in the middle; anthers uniform. Disk absent. Ovary glabrous or pubescent, shortly stipitate, 2–6-ovuled; style cylindrical; stigma small, terminal, punctate or capitate. Pod shortly stipitate, flat, tapering at the base, asymmetric, the lower margin evenly curved but the upper ± sinuous, ± concave towards the base and convex in upper half, glabrous or glabrescent, shortly beaked, dehiscing into thin but stiffly woody twisted valves, 1–4-seeded. Seeds black or dark brown, ellipsoid; hilum short, near one end of seed, surrounded by a short white cupular rim aril which is produced on one side into a curved strap-shaped process clasping the funicle.

A genus of 10 species confined to tropical Africa, 4 occurring in the Flora Zambesiaca area.

1. Ovary pubescent all over; corolla 1.8–2.5 cm long (Zambia) · · · · · **4.** *grandiflora*
 – Ovary glabrous or pubescent near base and on sutures; corolla (except in c.*affinis*) less than 1.5 cm long · 2
2. Calyx and pedicels usually glabrous; hairs on inflorescence axis rather sparse and over 0.5 mm long; bracteoles linear or spathulate, c.0.5 mm wide at base; ovary glabrous or with a few hairs on the suture (Mozambique) · · · · · **1.** *zimmermannii*
 – Calyx, pedicels and inflorescence axis densely pubescent with short brown hairs under 0.5 mm long; bracteoles at least 1 mm wide; ovary pubescent near base and on sutures · 3

3. Bracteoles rounded ovate, less than 4 mm long; corolla 1.2–1.9 cm long; filament
   sheath 0.8–1.2 cm long ··· **2. *brevicaudata***
– Bracteoles elliptic-oblong, up to 5.5 mm long; corolla 1.8–2.2 cm long; filament
   sheath 1.3–1.5 cm long ···································· **3. *affinis***

1. **Craibia zimmermannii** (Harms) Dunn in J. Bot. **49**: 108 (1911). —Baker f., Legum.
   Trop. Africa: 249 (1929). —Gillett in F.T.E.A., Legum., Pap.: 149 (1971). —Palmer
   & Pitman, Trees Sthn. Africa: 927–929, (1973). —Ross, Fl. Natal: 203 (1973). —K.
   Coates Palgrave, Trees Sthn. Africa, ed.2: 312 (1988). —Pooley, Trees Natal: 166
   (1993). —M. Coates Palgrave, Trees Sthn. Africa: 379 (2002). Type: NE Tanzania,
   near Amani, Bomole, *Zimmermann* in Herb. Amani 1480 (B† holotype, EA, K).

   Lonchocarpus zimmermannii Harms in Bot. Jahrb. Syst. **45**: 311 (1910).

   Craibia filipes Dunn in J. Bot. **49**: 109 (1911). —Baker f., Legum. Trop. Africa: 250
   (1929). Type: Mozambique, 38 km from Beira, Mafambice (Zimbiti), *P.A. Sheppard* 305 (K
   holotype).

Shrub or small to medium-sized tree 2–15 m tall, with white or pale grey scaly or slightly
flaking bark. Young twigs, leaf rachides, petiolules and leaflet midribs sparsely rusty pubescent
with hairs c.1 mm long, or almost glabrous. Leaves 3–5(6)-foliolate; petiole 2.5 cm long; rachis
up to 6.5 cm long; leaflets ± subcoriaceous, 3–10 × 1.5–5 cm, ovate, elliptic, elliptic-lanceolate
or sometimes round, acuminate at the apex, broadly cuneate to truncate at the base; lateral
nerves 5–6 on each side; stipules falling very early; petiolules 3 mm long. Inflorescences
scented, 15–20 cm long, 5–10-flowered, the rachis with rather long chestnut-brown hairs; bracts
very deciduous; bracteoles 2–3 mm long, under 0.5 mm wide at base, linear to spathulate,
deciduous; pedicels 10 mm long. Calyx reddish, glabrous or nearly so; tube 3–4 mm long; lobes
0.5–1.5 mm long. Corolla white; standard 1.3 × 1.1 cm, with claw 1.5 mm long; wing blades 12
× 4.5 mm, oblong, with claw 4 mm long; blades of keel petals 11.5 × 5 mm, with claw 4.5 mm
long. Filament sheath 10 mm long, with free parts of filaments 4–6 mm long; anthers 0.7 mm
long. Ovary glabrous or with few hairs on the sutures, 2–5-ovuled; style c.7 mm long. Pods
5–9(11) × 1.7–3.5 cm, lanceolate or elliptic-oblong, venose when young. Seeds ± blackish,
1.6–2.3(2.7) × 1.2(1.8) × 0.4–0.9 cm, oblong-ellipsoid, subrhomboid or compressed, obscurely
striate and pitted.

Mozambique. Z: Pebane Dist., Praia de Pebane, fr. 24.x.1942, *Torre* 4669 (BM,
LISC). MS: Inhaminga Dist., Inhansato, sawmill of Cardoso Lopes, fr. 4.iii.1962, *M.F.
de Carvalho* 512 (K, LMU). GI: Vilankulo Dist., Govuro, fl. 9.i.1912, *Dawe* 523 (K). M:
environs of Maputo (Lourenço Marques), between Costa do Sol and P. do Mar, fl.
27.x.1947, *Pedro* 3940 (LMA).

Also in coastal Kenya, Tanzania and South Africa (KwaZulu-Natal). In coastal
evergreen forest and forest margins with *Pterocarpus-Xylia-Newtonia*, and
Erythrophleum-Hirtella, in coastal bush on dunes with *Mimusops, Bridelia, Aloe bainesii,
Euphorbia, Vepris* etc., and on margins of mangroves; also in dry *Afzelia* sand forest;
0–140 m.

Conservation notes: Widely distributed in coastal areas; Lower Risk, Least Concern.

Shantz 333 (K, US) from Maputo (Lourenço Marques), fr. 25.x.1919, has leaflets
much more coriaceous than usual.

2. **Craibia brevicaudata** (Vatke) Dunn in J. Bot. **49**: 108 (1911). —Baker f., Legum.
   Trop. Africa: 247 (1929). —White, F.F.N.R.: 433 (1962). —Gillett in F.T.E.A.,
   Legum., Pap.: 150 (1971). —White, Dowsett-Lemaire & Chapman, Evergreen
   For. Fl. Malawi: 322, fig.114 (2001). Type: Kenya, Mombasa, *Hildebrandt* 1933 (B†
   holotype, BM, K).

   Lonchocarpus brevicaudata (Vatke) Harms in Bot. Jahrb. Syst. **45**: 312 (1910).

Shrub or tree 2–21 m tall*, with a spreading rounded crown and pale grey smooth flaking bark; bole up to 1.2 m diameter, sometimes buttressed. Branchlets pale or dark, sparsely to less often densely brown pilose but becoming glabrescent. Leaves 3–7-foliolate; petiole 1–4 cm long; rachis up to 10 cm long, glabrous or very slightly pilose; leaflets subcoriaceous, (4)5–13.5 × 2–3.5(6.5) cm, ovate, ovate-elliptic or oblong-elliptic, acuminate at the apex, rounded to broadly cuneate at the base, glabrous or with a few hairs on the midrib; 5–8 lateral nerves on each side; stipels round or ovate, soon falling; petiolules 2–5 mm long. Inflorescences showy, sweet-scented, 4–15 cm long, densely brown velvety pubescent; bracts up to 4 mm long, round or ovate, deciduous; pedicels up to 1(1.3) cm long; bracteoles up to 3 mm long, round or ovate-elliptic. Calyx densely brown velvety, tube c.4 mm long, lobes c.2 mm long. Corolla white, very pale blue or pink or sometimes greenish or mauve; standard oblong-elliptic, 1.5 × 1 cm, with claw 2 mm long; wing blades 1.5–3.5 mm, with claw 3 mm long; blades of keel petals 15 × 4 mm, with claw 4.5 mm long. Filament sheath 8–12 mm long, the free parts of filaments 4–6 mm long; anthers 0.5–0.7 mm long. Ovary 1–1.1 cm long, pilose at base and on margins but sides ± glabrous, 3–4-ovuled; style 4–5 mm long. Pods creamy-grey, 6.5–10 × 1.8–5 cm, oblanceolate, flattened, 2–4-seeded. Seeds red-brown, 14–20 × 12–15 × 4–5 mm, ± round or elliptic in outline, discoid.

Subsp. **baptistarum** (Büttner) J.B. Gillett in Kew Bull. **14**: 194 (1960). —Sousa in C.F.A. **3**: 166 (1962). —Gillett in F.T.E.A., Legum., Pap.: 151 (1971). —K. Coates Palgrave, Trees Sthn. Africa, ed.2: 312 (1988). —M. Coates Palgrave, Trees Sthn. Africa: 379 (2002). Type: Angola, Lower Congo region, Tondoa, *Buettner* 222 (B† holotype).

> *Millettia baptistarum* Büttner in Verh. Bot. Vereins Prov. Brandenburg **32**: 50 (1890).
> *Lonchocarpus crassifolius* Harms in Bot. Jahrb. Syst. **26**: 299 (1899). Type: S Malawi, *Buchanan* 101 (B† syntype, BM, K) and *Buchanan* 622 (B† syntype, BM, K, US).
> *Craibia baptistarum* (Büttner) Dunn in J. Bot. **49**: 109 (1911). —Baker f., Legum. Trop. Africa: 248 (1929). —Hauman in F.C.B. **5**: 54 (1954).
> *Schefflerodendron gazense* Baker f. in J. Linn. Soc., Bot. **40**: 55 (1911). Type: Zimbabwe, Chirinda Forest, *Swynnerton* 13 (BM holotype, K, SRGH).
> *Craibia gazensis* (Baker f.) Baker f., Legum. Trop. Africa: 248 (1929) non sensu Brenan (1949).

Leaflets mostly 5–7; inflorescence indumentum golden brown; bracteoles c.2 mm long; filament sheath 1–1.2 cm long.

Zambia. N: Nchelenge Dist., L. Mweru, fl. 13.xi.1957, *Fanshawe* 3958 (K, NDO, SRGH). S: Mumbwa Dist., Nambala Hill, st. 22.vi.1932, *Trapnell* 2079 (K). **Zimbabwe**. C: Hwedza Dist., Hwedza (Wedza) Mt., fr. 14.v.1964, *Wild* 6566 (K, SRGH). E: Mutare, W slope of Murahwa's Hill, fl. 22.xii.1963, *Chase* 8094 (K, SRGH). **Malawi**. N: Chitipa Dist., Misuku Hills, Wilindi Forest, fl. x.1954, *Chapman* 249 (K, PRE). C: Nkhota Kota (Kota Kota) Hills, st. 22.xi.1962, *Chapman* 1759 (K, SRGH). S: Zomba, fl. 1901, *Sharpe* 39 (K). **Mozambique**. Z: Gurué Dist., Lioma, Serra Namuli, fl., *Serramo* 15 (LMA). MS: Manica Dist., Bandula, road to Chicamba (Chicama), fr. 23.iv.1948, *Andrada* 1168 (LISC).

Also in Tanzania, Congo and Angola. Montane evergreen forest and riverine forest where it is often in *Mimusops–Breonadia* associations; also in dense deciduous 'itigi' thicket in N Zambia; 850–1800 m.

Conservation notes: Medium altitude forest; Lower Risk, Least Concern.

Biegel (Check-list Ornam. Pl. Rhod. Parks & Gard.: 43, 1977), mentions it as being cultivated in gardens in Zimbabwe.

According to Swynnerton the seeds are poisonous.

J.B. Gillett (F.T.E.A. 1971) has a subsp. *schliebenii* (Harms) J.B. Gillett (type from Tanzania, Uluguru Mts.) which differs from subsp. *baptistarum* in having bracteoles 3

---

* Chase (387) states 80–100' (24–30 m).

mm long and the inflorescence indumentum dark and often blackish and narrowly acuminate leaflets. He records it from Serra da Gorongosa in Mozambique. However, I have found it difficult to assign some material, and while *Mendonça* 2448 (Serra da Gorongosa) has small bracteoles and is subsp. *baptistarum* sensu stricto, *Mendonça* 2398 (between Gorongosa town and Serra da Gorongosa) has rich chestnut-brown indumentum and bracteoles 2.5 mm long. Material from N Malawi, Viphya Plateau, Chikangawa, *Phillips* 4102, is also intermediate. The original Mozambique material named by Gillett is from Mt Gorongosa, near Marumbozi R. Falls, fl. 23.x.1945, *Pedro* 438 (K, LMA, PRE). I have decided not to formally recognize it in the Flora Zambesiaca area. See Gillett (1971) for information on other subspecies in East Africa.

3. **Craibia affinis** (De Wild.) De Wild. in Ann. Soc. Sci. Bruxelles **38**: 15 (1914). — Baker f., Legum. Trop. Africa: 249 (1929). —Hauman in F.C.B. **5**: 52 (1954). — White, F.F.N.R.: 146 (1962). —Gillett in F.T.E.A., Legum., Pap.: 150 (1971). Type: Congo, Katanga, Lukafu, *Verdick* 115 (BR). FIGURE 3.3.**18**.

> *Lonchocarpus affinis* De Wild. in Ann. Mus. Congo, Bot., sér.4 **1**: 196 (1903).
> *Lonchocarpus dubius* De Wild. in Ann. Mus. Congo, Bot., sér.4 **1**: 196 (1903). Type from Congo.
> *Craibia dubia* (De Wild.) De Wild. in Ann. Soc. Sci. Brux. **38**: 15 (1914).

Shrub or tree, 1.8–9(15) m tall; bark pale grey, thin, smooth; wood yellowish, hard. Branchlets, leaf rachides and petiolules dark brown pilose, glabrescent. Leaves 5–11-foliolate; petiole 2.5–4 cm long; rachis 14.5–26 cm long; petiolules 3.5 mm long; leaflets 3.8–16 × 1.3–5.2 cm, oblong, oblong-elliptic or ovate-lanceolate, acuminate at the apex, rounded to cuneate at the base, nearly always at least 3 times as long as wide, glabrous or with sparse rust-coloured hairs; lateral nerves 6–11 on each side; venation raised and reticulate; stipules up to 10 mm long, ligulate, brown pilose, glabrescent. Inflorescences showy, fragrant, 4.5–9 cm long, the rachis sometimes once- or twice-branched at the base; peduncle 0.5–4 cm long; pedicels 0.6–1.3 cm long; bracts up to 1 cm long, elliptic-oblong, deciduous; bracteoles up to 5.5 × 2 mm, elliptic-obovate or elliptic-oblong, all parts rusty-pubescent. Calyx pinkish, brown-pubescent; tube 4–6 mm long; lobes 1–4 mm long, ovate. Standard white or yellow with a median mark but corolla also described as white or pale purplish or white tipped with pink, 1.7–2.2 × 1.3–1.4 cm, ± round; wing blades 13 × 4 mm, with claw 7-8 mm long; blades of keel petals 14 × 6 mm, with claw 7 mm long. Filament sheath 13–15 mm long; free parts of filaments 4–6 mm long; anthers 0.9 mm long. Ovary 11–13 mm long, glabrous at sides but brown pubescent near the base and on the margins, 3-ovuled. Fruit 6–9.5 × 2.5–3 cm, oblanceolate or elliptic-oblanceolate in outline, 1–2-seeded. Seeds dark violet, 2 × 1.2 × 0.4 cm.

Zambia. N: Mbala Dist., Lunzua Gorge, near waterfall, fl. 11.ix.1966, *Lawton* 1432 (K). W: Mwinilunga, fl. 15.ix.1955, *Holmes* 1189 (K, NDO). C: Mkushi Dist., Chiwefwe, fl. 19.ix.1964, *Mutimushi* 1035 (K, NDO).

Also in S Congo and W Tanzania. Riverine and evergreen fringing forest, *Newtonia* rainforest, gully forest; also in mushitu; 900–1400 m.

Conservation notes: Limited distribution in the Flora area; probably Lower Risk, Near Threatened.

Material from Mbala has darker indumentum. *Mutimushi* 2096 (Ndola, fl. 8.ix.1967 (K, NDO)) has narrow leaves resembling the last species but bracteoles 2 mm long. There is no doubt that intermediates with *Craibia brevicaudata* subsp. *baptistarum* occur but their nature is not evident.

4. **Craibia grandiflora** (Micheli) Baker f., Legum. Trop. Africa: 247 (1929). — Hauman in F.C.B. **5**: 55, pl.3 (1954). —White, F.F.N.R.: 433 (1962). —Gillett in F.T.E.A., Legum., Pap.: 153 (1971). Type: Congo, near Boyoma (Stanley) Falls, *Laurent* s.n. (BR).

Fig. 3.3.**18**. CRAIBIA AFFINIS. 1, flowering branchlet (× ²/₃), from *Angus* 437; 2, standard (× 2); 3, wing petal (× 2); 4, keel (× 2); 5, stamens (× 2); 6, gynoecium (× 2), 2–6 from *Mgaza* 299. Drawn by Pat Halliday.

Pterocarpus grandiflorus Micheli in Bull. Soc. Roy. Bot. Belgique **36**(2): 65 (1897).

Craibia mildbraedii Harms in Mildbraed, Wiss. Ergebn. Deutsch. Zentr.-Afrika Exped., Bot. 2: 257, t.29 (1911). Type: Congo, Kivu Prov., Beni, *Mildbraed* 2441 (B† holotype).

Craibia sp. 8 of Brenan, Check-list For. Trees Shrubs Tang. Terr.: 412 (1949).

Shrub or tree, 9–12(20) m tall; trunk 20–40 cm in diameter; bark finely fissured. Young stems, buds, leaf rachides and petiolules densely ferruginous velvety. Leaves 7–9(11)-foliolate; petiole 1.5–3 cm long; rachis 7.5–11(16) cm long; petiolules 3–4 mm long; leaflets 9–17.5 × 2.5–5 cm, oblong to oblanceolate, acuminate with the narrow acumen up to 2.5 cm long, rounded to cuneate at the base, glabrous or pubescent beneath at least on the nerves, ± subcoriaceous; stipules 6 mm long, oblong, rounded at the apex, deciduous. Inflorescences axillary and terminal, racemose, 5–6(12) cm long, usually much shorter than the subtending leaf; peduncle 3 cm long; bracts up to 1.2 cm long, linear-lanceolate, soon falling; pedicels 6–8 mm long; bracteoles 3–6 × 1 mm, elliptic-oblong or lanceolate, deciduous. Calyx bright rose-coloured, pubescent; tube 4–7 mm long; lobes 2–3 mm long. Standard white or pink with yellow base, 2.3 × 1.3 cm, or 2.5 cm in diameter, oblong or ± round, with claw 3 mm long; wing blades 23 × 5 mm, narrowly oblong, curved, with claw 5 mm long; blades of keel petals curved, 17 × 7 mm with auricle rounded and claw 8 mm long. Filament sheath 1.1–1.3 cm long; free parts of filaments 5–8 mm long; anthers 0.7 mm long. Ovary 1.1–1.2 cm long, densely pubescent, 2–3-ovuled; style 6–7 mm long. Pods purplish, 5–11 × 2–2.5 cm, obovoid, beaked, glossy and venose, 1–3-seeded. Seeds dark brown, 1.5–2.5 × 1.3–1.7 × 0.5 cm, semicircular, elliptic or subquadrate in outline.

Zambia. N: Kawambwa, fl. 28.viii.1957, *Fanshawe* 3628 (K, NDO).

Also in the Central African Republic, Congo, S Sudan and Tanzania. Riparian 'mushitu'; 1300 m.

Conservation notes: Limited distribution in the Flora area; probably Lower Risk, Near Threatened.

15. PHILENOPTERA A. Rich.

by J.M. Lock*

Philenoptera A. Rich., Tent. Fl. Abyss. **1**: 232 (1847). —Schrire in Kew Bull. **55**: 81–94 (2000).

Lonchocarpus sensu auctt. afr., in part, non Kunth (1824).

Capassa Klotzsch in Peters, Naturw. Reise Mossamb. **6**: 27, fig.5 (1861). —Sousa in C.F.A. **3**: 367 (1966).

Trees or shrubs, elsewhere sometimes lianes, deciduous; slash usually producing a blood-red exudate. Leaves imparipinnate or unifoliolate; stipules caducous or persistent; leaflets opposite or subopposite; stipels present. Inflorescence a panicle, terminal or borne precociously in the axils of fallen leaves or at the top of last season's branches, the pedicels inserted separately; bracts and bracteoles caducous. Flowers mauve or violet-purple. Calyx campanulate, 5-dentate, dorsal teeth connate almost to the tip. Standard subcircular to obovate, glabrous, auriculate and/or with distinct calluses at the base; wings oblong, adhering to the keel above the claw; keel petals slightly connate towards the tip. Stamens connate into a tube, the upper one free at the base; anthers versatile. Disk present. Ovary few-ovulate, shortly stipitate; style filiform, incurved, stigma terminal, small. Pod indehiscent, flat, the upper suture thin, thickened or slightly winged, 1–4(8)-seeded. Seed compressed, oblong-reniform, with a small hilum; radicle inflexed.

A genus of 12 species in Africa and Madagascar, recently resegregated from *Lonchocarpus* (Schrire 2000). In early drafts of his account for this Flora, Mendonça

---

*Based on the draft Flora Zambesiaca account prepared by F.A. Mendonça, but taxonomy modified.

separated *Lonchocarpus capassa* into the genus *Capassa* Klotzsch on the basis of its winged pods, but this distinction is not upheld here.

Species distinction within the genus in Africa is often difficult. Mendonça separated the plants here treated as *P. katangensis* and *P. wankieensis* as subspecies, but they are just as distinct (if not more so) than the plants treated as *P. eriocalyx* and *P. bussei*. I treat them all here as good species. It must be recognized that there are intermediates, and not all plants fall definitely within the species accepted here.

1. Leaves unifoliolate (very occasionally trifoliolate); calyx densely white-tomentose
   · **1.** *nelsii*
   – Leaves imparipinnate with 3–15 leaflets; calyx white- or brown-tomentose or
   pubescent · 2
2. Leaves with 3–5 leaflets · 3
   – Leaves with at least 7 leaflets · 4
3. Calyx densely brownish tomentose; pods thin-textured, unwinged · · · · · · · · · ·
   · **2.** *katangensis*
   – Calyx thinly and shortly whitish tomentose or pubescent; pods thicker-textured
   with a wing 1–2 mm wide along the upper margin · · · · · · · · · · · · · **3.** *violacea*
4. Keel petals of flowers distinctly beaked; pod densely pubescent at maturity · · · ·
   · **4.** *wankieensis*
   – Keel petals acute or obtuse but not beaked; pod glabrescent, almost glabrous at
   maturity · 5
5. Lateral teeth of calyx 2–2.5 mm long, obtuse to subacute, densely white-tomentose
   · **5.** *eriocalyx*
   – Lateral teeth of calyx 1–1.5 mm long, rounded to obtuse, appressed fulvous-
   pubescent · **6.** *bussei*

1. **Philenoptera nelsii** (Schinz) Schrire in Kew Bull. **55**: 91 (2000). Type: Namibia, Hereroland, iv.1891, *Fleck* 351 (Z lectotype, chosen by Schrire 2000).

   Dalbergia nelsii Schinz in Bull. Herb. Boissier **6**: 729 (1898).

   Lonchocarpus nelsii (Schinz) Heering & Grimme, Untersuch. Weideverhältn. Deutsch-Südwestafrika: 25 (1911). —Harms in Engler, Pflanzenw. Afrikas **3**(1): 641 (1915). —Baker f., Legum. Trop. Africa: 549 (1929). —Bremekamp & Obermeyer in Ann. Transvaal Mus. **16**: 420 (1935). —Miller in J. S. African Bot. **18**: 34 (1952).—White, F.F.N.R.: 158, 433 (1962). —Mendonça & Sousa in Webbia **19**: 836 (1965). —Sousa in C.F.A. **3**: 366 (1966). —Schreiber in Merxmüller, Prodr. Fl. SW Afrika, fam. 60: 75 (1970). —Palmer & Pitman, Trees Sthn. Africa: 944 (1972). —K. Coates Palgrave, Trees Sthn. Africa, ed.2: 327 (1988). —M. Coates Palgrave, Trees Sthn. Africa: 393 (2002).

   Lonchocarpus laxiflorus var. *sericeus* Baker in F.T.A. **2**: 242 (1871) in part for Angolan material. Type: Angola, Huíla, Ferrão da Sola, fr. iv.1860, *Welwitsch* 1879 (K lectotype, chosen by Schrire 2000).

Tree or shrub up to 12 m tall; crown rounded or irregular; bark rough, grey to dark brown. Branchlets pubescent, glabrescent or tomentose with bark red-brown, becoming blackish. Leaves unifoliolate or occasionally 3-foliolate; lamina leathery or stiff, glabrous on upper surface, ± densely grey-pubescent beneath; lateral nerves in 4–6 pairs, these and the dense reticulation prominent beneath; lamina of unifoliolate leaves up to 16 × 9 cm, broadly ovate to elliptic, petiole 0.8–2 cm long; laminae of leaflets of 3-foliolate leaves smaller, pubescent-glabrescent, apex rounded or obtuse to subacute, base rounded or obtuse; petiole up to 5 cm long; stipules 3–6 mm long, subulate, near-persistent; stipels c.5 mm long, subulate, near-persistent. Inflorescence up to 30 cm long, borne precociously in axils of fallen leaves, or terminal, yellowish-grey-tomentose. Flowers mauve to pinkish, 10–12 mm long. Calyx 4–6 mm long, upper tooth-pair c.2.5 mm long, retuse, the lateral ones 1–2 mm long, obtuse, the lower one 2.5–3 mm long, triangular-acute. Standard patent-reflexed, c.12 mm long, blade subcircular, 10–11 mm in diameter, with prominent calluses at the base, glabrous on the back; wings 12.5–13.5 × 3.5–4 mm, obovate, auricle c.0.8 mm long, acute, claw 1.5 mm long; keel petals 12–12.5 × 5–7 mm, arcuate-cucullate, auricle acute, claw 1.5–3 mm long. Ovary 5–7-ovulate, 6 mm long, shortly stipitate, densely pilose; style 3 mm long, glabrous. Pod 1–4-seeded, linear, up to 10 × 2 cm, rounded at the apex, cuneate at the base, stiff, thinly sericeous; stipe c.3 mm long. Seed 12 × 9 × 2 mm, testa brown, smooth.

Caprivi Strip. 20 km E of Andara, fl. 4.x.1966, *Giess* 9537 (K, WIND). **Botswana**. N: Ngamiland Dist., Moremi Game Res. (Wildlife Res.), N Okavango Swamps, st. vii.1964, *Tinley* 1073 (K, PRE). SW: Kgalagadi Dist., Kang, 300 km W of Gaborone, imm.fr. 20.x.1975, *Mott* 796 (K, UBLS). SE: Kweneng Dist., Lephephe, fl. xi.1968, *Kelaole* in Mahalapye Agric. Herb. 514 (K). **Zambia**. B: Sesheke Dist., Machili, fl. 16.ix.1960, *Fanshawe* 5814 (FHO, K, LISC). S: see note below. **Zimbabwe**. N: Sebungwe, fl. 24.ix.1951, *Lovemore* 113 (SRGH). W: Hwange Nat. Park (Wankie Game Res.), Nkwasha (Ngwashla) road, st. 18.ii.1956, *Wild* 4768 (K, LISC, SRGH). C: Gweru (Gwelo), st. 24.v.1951, *Banks* 6/51 (K, SRGH).

Also in Namibia and SW Angola. In dry open woodland and bushland, generally on Kalahari sands; 800–1200 m.

Conservation notes: A widespread species, often common where it occurs. Lower Risk, Least Concern.

Occasional specimens intermediate between this and the next occur, e.g. Zambia S: Namwala, fr. 20.x.1959, *Fanshawe* F5253 (FHO, K); Zimbabwe W: Hwange Game Reserve, st. 18.ii.1956, *Wild* 4770 (K, SRGH) and in Angola (*Welwitsch* s.n. (K), *Gossweiler* s.n. (K)).

2. **Philenoptera katangensis** (De Wild.) Schrire in Kew Bull. **55**: 90 (2000). Type: Congo, Katanga, Lukafu, *Verdick* 102 (BR holotype).

Lonchocarpus katangensis De Wild. in Ann. Mus. Congo, Bot., sér.4 **1**: 195 (1903). —Baker f., Legum. Trop. Africa: 549 (1929). —Hauman in F.C.B. **6**: 10 (1954). —White, F.F.N.R.: 159 (1962).

Lonchocarpus hockii De Wild. in Repert. Spec. Nov. Regni Veg. **11**: 539 (1913). —Baker f., Legum. Trop. Africa: 550 (1929). Type: Congo, Lubumbashi, *Hock* s.n. (BR holotype).

Lonchocarpus pallescens var. *pubescens* Baker f. in J. Bot. **76**: 21 (1938). Type: Zambia, Mbala Dist., *Gamwell* 244 (BM holotype).

Lonchocarpus nelsii subsp. *katangensis* (De Wild.) Mendonça & E.P. Sousa in Bol. Soc. Brot., sér.2, **42**: 266 (1968). —Polhill in F.T.E.A., Legum., Pap.: 72 (1971).

Tree or shrub up to 10 m tall, crown rounded or spreading; bark rough, dark. Branchlets pubescent, soon glabrescent; bark red-brown, becoming greyish and cracking. Leaves 3–5-foliolate; petiole up to 5 cm long; rachis 4–7 cm long; leaflets pubescent on upper surface, soon glabrescent, whitish-pubescent beneath, especially on the nerves; lateral nerves in 4–6 pairs, brownish and prominent beneath; lateral leaflets 4.5–9 × 2.8–6 cm, elliptic, rounded to acute at the apex, cuneate at the base; terminal leaflet 7.5–14 × 4.5–9 cm, elliptic to obovate, retuse to abruptly acuminate at the apex, cuneate at the base; stipules 3–6 mm long, subulate, early caducous; stipels c.5 mm long, subulate, subpersistent. Inflorescence a panicle up to 30 cm long, usually precocious in the axils of fallen leaves, or occasionally terminal on a leafy shoot, brown-grey-tomentose. Flowers pink to blue, 10–13 mm long. Calyx 4–6 mm long, upper tooth-pair c.3 mm long, retuse, the lateral ones 1.5–3 mm long, obtuse to acute, the lower one 2.5–4 mm long, triangular-acute. Standard patent-reflexed, c.12 mm long, blade subcircular, 10–11 mm in diameter, with calluses at the base, glabrous on the back; wings obovate, 12.5–13.5 × 3.5–4 mm, auricle c.1 mm long, acute, claw 1.5 mm long; keel petals 12–13 × 5–7 mm, arcuate-cucullate, auricle acute, claw 1.5–3 mm long. Ovary 5–7-ovulate, 6 mm long, shortly stipitate, densely pilose; style 3 mm long, glabrous. Pod 1–4-seeded, up to 10 × 2 cm, linear, rounded at the apex, cuneate at the base, chartaceous, thinly pubescent; stipe c.3 mm long. Mature seed not seen.

Zambia. N: Mbala (Abercorn), fl. 30.vii.1951, *Bullock* 4000 (K, SRGH). W: Ndola, fl. 24.ix.1951, *Holmes* 134 (FHO, SRGH).

Also in Congo (Katanga), Burundi and Tanzania (Western Province). In miombo woodland; 1200–1400 m.

Conservation notes: Only a few specimens known from the Flora Zambesiaca area, but fairly widespread outside. Probably Lower Risk, Least Concern.

Mendonça & Sousa (1968) treated this as a subspecies of *L. nelsii*. While there are occasional intermediate specimens at the northern edge of the range of *L. nelsii*, there is otherwise no overlap in distribution and virtually every specimen can be placed in one species or the other. I therefore prefer to treat them as separate species.

3. **Philenoptera violacea** (Klotzsch) Schrire in Kew Bull. **55**: 89 (2000). —M. Coates Palgrave, Trees Sthn. Africa: 394, fig.109 (2002). Type: Mozambique, Sena, *Peters* s.n. (B†, K). FIGURE 3.3.19.

Lonchocarpus philenoptera Benth. in J. Linn. Soc., Bot. **4**: 97 (1860) nom. illegit.*, in part for synonym *Capassa violacea*** and *McCabe* specimen. —Harvey in F.C. **2**: 263 (1862) in part for synonym *Capassa violacea* (sphalm. *Tapassa*). Syntypes include specimens from Mozambique MS: Sena, *Peters* s.n. (K) and Botswana N: Lake Ngami, *McCabe* s.n. (K).

Capassa violacea Klotzsch in Peters, Naturw. Reise Mossamb. **6**: 28 (1861). —Mendonça & Sousa in Webbia **19**: 836 (1965). —Sousa in C.F.A. **3**: 367 (1966).

Lonchocarpus capassa Rolfe in Oates, Matabeleland, ed.2: 397 (1889). —Eyles in Trans. Roy. Soc. S. Africa **5**: 380 (1916). —Baker f., Legum. Trop. Africa: 551 (1929) in part excl. Angolan material. —Burtt Davy & Hoyle, Check-list For. Trees Shrubs, Nyasaland: 60 (1936). —Miller in J. S. African Bot. **18**: 33 (1952). —Pardy in Rhod. Agric. J. **49**: 216 (1952). —Hauman in F.C.B. **6**: 11 (1954). —O. Coates Palgrave, Trees Central Africa: 324 (1957). —White, F.F.N.R.: 159, fig.27H (1962). —Schreiber in Merxmüller, Prodr. Fl. SW Afrika, fam. 60: 75 (1970). —Polhill in F.T.E.A., Legum., Pap.: 66 (1971). —Palmer & Pitman, Trees Sthn. Africa: 946–947 (1972). —Ross in Fl. Pl. Afr. **46**: t.1820 (1980). —K. Coates Palgrave, Trees Sthn. Africa, ed.2 (1988). Type as for *Philenoptera violacea*.

Lonchocarpus laxiflorus sensu Baker in F.T.A. **2**: 242 (1871) in part for synonym *Capassa violacea*, *McCabe* and *Kirk* specimens. —sensu Sim, For. Fl. Port. E. Africa: 45, t.50 (1909). —sensu Gomes e Sousa, Dendrol. Moçamb. Estudo Geral **1**: 287, t.78 (1966) non Guill. & Perr.

Lonchocarpus violaceus (Klotzsch) Baker f. in J. Linn. Soc., Bot. **40**: 60 (1911).

Derris violacea (Klotzsch) Harms in Bot. Jahrb. Syst. **33**: 174 (1902) in part for synonym *Lonchocarpus capassa*; in Engler, Pflanzenw. Afrikas **3**(1): 642 (1915). —Seiner in Mitt. Deutsch. Schutzgeb. **22**: 27 (1909). —R.E. Fries, Wiss. Ergebn. Schwed. Rhod.-Kongo-Exped. **1**: 92 (1914).

Deciduous tree up to 18 m tall; crown rounded; bark greyish, shallow-fissured, flaking off in scales. Young parts pale grey-subsericeous; branchlets slightly striate or terete, bark grey to brownish, thinly puberulous, glabrescent. Leaves up to 28 cm long, imparipinnate; petiole 3.5–9 cm long; rachis 3–10 cm long, puberulous-glabrescent; petiolules 2–6 mm long; leaflets 3–5, the terminal much larger than the laterals; lamina (3)4–13(16.5) × 3–6(9) cm, ovate to elliptic or narrowly elliptic or oblong-obovate, rounded to obtuse or subacuminate-acute at the apex, rounded to broadly obtuse or cuneate at the base, leathery or subcoriaceous, light green on upper surface, paler and thinly tomentellous-glabrescent beneath; lateral nerves in 4–7 pairs, this and the reticulation slightly prominent beneath; stipules 3–6 mm long, falling early; stipels filiform, 1.5–4 mm long, often falling early. Inflorescence a terminal panicle 10–25 cm long, thinly pubescent-glabrescent; bracts c.1.5 mm long; bracteoles smaller, caducous. Flowers 9–11 mm long; pedicels 2–5 mm long. Calyx 4–5 mm long, somewhat asymmetric at the base, thinly pubescent, the upper tooth-pair

---

* Bentham's description included earlier names.
**At that date a manuscript name.

Fig. 3.3.**19**. PHILENOPTERA VIOLACEA. 1, flowering branchlet (× ²/₃); 2, detail of leaflet lower surface (× 2); 3, flower (× 3); 4, standard (× 2); 5, wing (× 2); 6, keel (× 2); 7, stamens (× 2); 8, gynoecium (× 2), 1–8 from *Mendonça* 2749; 9, pod (× ²/₃), from *Balsinhas & Marrime* 475; 10, seed (× 1), from *Torre* 2929. Drawn by Pat Halliday.

obtuse to shallowly lobed, the lateral teeth triangular, c.2 mm long, obtuse or rounded, the lower one narrowly triangular, acute, 2–4 mm long. Standard 10.5–12.5 mm long, broadly obovate, with nectariferous calluses towards the base, narrowly decurrent into the 2–2.5 mm long claw; wings oblong-obovate, 10–12 × 2.5–4 mm, auricle obtuse, claw 2.5–3.5 mm long; keel petals 9–11 × 3.5–6 mm, with subacute auricle, claw 2-3.5 mm long. Stamen sheath 6–9 mm long. Ovary 3–5-ovulate, silky-pubescent; stipe c.1.5 mm long; style glabrous. Pod 1–3-seeded, 8–18 × 1.8–3.5 cm, leathery, thinly velutinous, glabrescent, somewhat reticulate or smooth over the seeds, attenuate-acute at the apex, cuneate to narrowed at the base; stipe 1–4 mm long; wing along upper margin 1–3 mm broad. Seed 12–15 × 6–8 × 2.5–3 mm, reniform; testa reddish-brown, smooth.

Caprivi Strip. Linyanti R. area, fr. 26.xii.1958, *Killick & Leistner* 3170 (BM, K, SRGH). **Botswana**. N: Chobe Dist., Kasane, fr. 25.vi.1930, *van Son* in Herb. Mus. Transvaal 288891 (BM, K, PRE, SRGH). **Zambia**. B: Sesheke, fr. 27.xii.1952, *Angus* 1050 (BM, FHO, K). N: Mbala (Abercorn), fl. 5.x.1956, *Richards* 6356 (K, SRGH). W: Luanshya, fl. 4.x.1955, *Fanshawe* 2487 (EA, K, SRGH). C: Lusaka Dist., Mt. Makulu Research Station, fr. 25.xi.1956, *Angus* 1449 (BM, COI, EA, FHO, K). S: Mazabuka, fl.& fr. 7.x.1930, *Milne-Redhead* 1212 (K). **Zimbabwe**. N: Hurungwe (Urungwe), fl. 17.x.1957, *Goodier* 314 (K, SRGH). W: Hwange (Wankie), fr. 16.v.1956, *Plowes* 1974 (K, SRGH). C: Harare, fl.& fr. xi.1917, *Eyles* 868 (BM, K, SRGH). E: Chipinge Dist., Rupise Hot Springs, fr. 29.i.1948, *Wild* 2375 (K, SRGH). S: Beitbridge, Limpopo/Shashi river junction, fr. v.1959, *Thompson* 24/59 (K, SRGH). **Malawi**. N: Mzimba Dist., above L. Kazuni, fl. 17.x.1973, *Pawek* 7405 (K, MAL, MO, SRGH). C: Dedza, Mua-Livulezi Forest Res., at foot of escarpment, fl. 28.ix.1954, *Adlard* 185 (FHO, K, SRGH). S: Nsanje Dist., 10 km SW of Bangula Station on road to Nchalo, fr. 15.v.1983, *Banda & Balaka* 1942 (K, MAL). **Mozambique**. N: Mandimba, forest fringing the Lugenda R. banks, fl. 13.x.1942, *Mendonça* 807 (BM, COI, K, LISC, LMA, LMU, SRGH). Z: between Ile and Gurué, fl. 26.ix.1941, *Torre* 3519 (B, EBV, FI, LISC). T: Changara Dist., between Boruma (Boroma) and Tete, fr. 26.vi.1941, *Torre* 2929 (COI, G, LISC, LMA, LMU, P). MS: Marrínguè (Marìnguè), banks of R. Inhamapanza, fl. 5.x.1944, *Mendonça* 2353 (FHO, K, LISC, LMA, LMU, M). GI: Chicualacuala Dist., Mapai, banks of R. Limpopo, fl. 4.xi.1944, *Mendonça* 2749 (LISC, M, P, PRE, SRGH, WAG). M: Maputo, between Goba and Catuane, fl. 24.x.1940, *Torre* 1859 (LISC, LISU, LMA, LUAU).

Also in Tanzania, Congo (Katanga), SE Angola, NE Namibia, South Africa (Limpopo, Mpumalanga and KwaZulu-Natal Provinces) and Swaziland. In riverine vegetation and floodplain grassland, in mixed *Acacia* woodland and wooded grasslands, and in plateau miombo woodland, on alluvium and Kalahari sands, often on termite mounds; 30–1400 m.

Conservation notes: A very widespread taxon. Lower Risk, Least Concern.

Biegel (Check-list Ornam. Pl. Rhod. Parks & Gard.: 71, 1977) recorded this species grown as an ornamental in Zimbabwe.

4. **Philenoptera wankieensis** (Mendonça & E.P. Sousa) Lock in Kew Bull. **55**: 95 (2000). Type: Zimbabwe, Hwange (Wankie), *Levy* 67 (PRE holotype, SRGH).

   Lonchocarpus eriocalyx subsp. *wankieensis* Mendonça & E.P. Sousa in Bol. Soc. Brot., sér.2, **42**: 265 (1968). —K. Coates Palgrave, Trees Sthn. Africa, ed.2: 326 (1988).—M. Coates Palgrave, Trees Sthn. Africa: 395 (2002).

Small deciduous tree or bush 3–10 m tall; bark grey, flaking. Branchlets fulvous to white-tomentellous; bark grey, later becoming rough and corky. Leaves imparipinnate with 9–13 leaflets; petiole 3.5–6 cm long, rachis 6–14 cm long, both densely appressed to spreading-pubescent; leaflets 2.5–8 × 2–4 cm, ovate to elliptic, the terminal leaflet often obovate, thin and stiff, reticulation not very prominent beneath, the veins beneath appressed-pubescent with straight hairs. Inflorescence a terminal panicle 13–16 cm long, the axis and branches grey-pubescent; bracts and bracteoles 5–8 mm long, linear, falling early. Flowers 10–12 mm long, pale purple or blue; pedicels 1–2 mm long. Calyx 5–7 mm long, the upper pair of teeth c.1.5 mm long, acute, slightly bilobed at apex, the other teeth 2.5–3 mm long, narrowly triangular, acute or acuminate. Standard strongly reflexed, 11–12 mm long, blade subcircular, 9–11 mm in diameter, with calluses at the base; wings 8–11 × 3–3.5 mm, rounded at apex, auricle acute, claw c.2 mm long; keel petals 10.5–15 × 3.5–4 mm, dorsally slightly connate, apex beaked, auricle rounded. Staminal tube 7 mm long, split dorsally for 1.5 mm at the base, free portion of longer stamens 2 mm long. Ovary 9 mm long, densely appressed-pilose; style c.2 mm long, glabrous. Pod 1(2)-seeded, broadly elliptic, 5–6 × 3 cm, densely velutinous.

Zambia. S: Gwembe Dist., Zambezi escarpment, Lusaka–Chirundu road, fl. 11.xii.1976, and fr. 31.i.1977, *Bingham* s.n. (K). **Zimbabwe**. N: Kariba Dist., S of Nyaodza (Naodsa) R., fl. 26.i.1956, *Phelps* 113 (K, SRGH). W: Hwange (Wankie), hillside with mopane, *Guibourtia* and *Combretum*, fr. xii.1952, *Hodgson* 1/52 (K).

Confined to the mid-Zambezi Valley, on escarpments and rocky hill slopes, usually in mixed woodlands with *Colophospermum mopane* dominant; 700–900 m.

Conservation notes: Endemic to the Flora Zambesiaca area; apparently not uncommon in dry woodlands of the Zambezi Valley. Lower Risk, Least Concern.

The description of the pods is taken from *Bingham* s.n. (K). The leaflets of this are rather larger and more acute than other material. *Fanshawe* 6600 (from S Zambia, Lusitu) was identified as this by Mendonça, but the pods are only sparsely velutinous and much more elongated. There is a need for carefully correlated collections of fruiting and flowering material of this species.

5. **Philenoptera eriocalyx** (Harms) Schrire in Kew Bull. **55**: 86 (2000). Type: Tanzania, Unyanyembe area around Tabora, *Stuhlmann* 509 (B† holotype, K frag.).

> *Lonchocarpus eriocalyx* Harms in Bot. Jahrb. Syst. **30**: 89 (1901); in Engler, Pflanzenw. Afrikas **3**(1): 641 (1915). —Baker f., Legum. Trop. Africa: 549 (1929). —Hauman in F.C.B. **6**: 9 (1954). —White, F.F.N.R.: 159 (1962). —Polhill in F.T.E.A., Legum., Pap.: 70, fig.13/12 (1971). —K. Coates Palgrave, Trees Sthn. Africa, ed.2: 326 (1988).

Tree 4–10 m tall; bark greyish-brown. Branchlets fulvous or grey-tomentellous, bark grey. Leaves 9–20 cm long; petiole 2–5 cm long, rachis 4–10 cm long, fulvous-puberulous; petiolules 1–2 mm long; leaflets 7–9, 2–8 × 1.5–3.5 cm, obovate or oblong-obovate or subelliptic, broadly obtuse or rounded or retuse at the apex, rounded or obtuse at the base, leathery, light green, concolorous, glabrous on upper surface, grey-pubescent beneath; lateral nerves in 4–6 pairs, reticulate venation prominent beneath; stipules 2–5 mm long, acute, persistent; stipels 1–2 mm long or absent. Inflorescence a terminal panicle 15–35 cm long with short or elongate branchlets, grey-tomentose; bracts and bracteoles 2–3 mm long, lanceolate-acute, falling early. Flowers mauve or pale purple, 10–12 mm long; pedicels 1–2 mm long. Calyx 6–7 mm long; tube as long as or longer than the teeth; upper teeth separated by a small notch, the others 2–3 mm long, obtuse or subacute. Standard patent-reflexed, subcircular, 10–12 mm in diameter, with calluses and small inflexed auricles at the base, claw c.2 mm long; wings as long as the standard, rounded at the apex, auricle obtuse, claw c.2 mm long; keel petals broader than the wings, dorsally slightly connate, auricle and claw as in wings. Ovary 3–5-ovulate, shortly stipitate, densely pilose; style c.2 mm long, glabrous. Pod 1–2-seeded, 5–9 × 2–2.5 cm, subsessile, leathery, tomentellous-glabrescent. Seed up to 14 × 8 × 5 mm; testa reddish-brown.

Zambia. N: Mbala Dist., near Mpulungu, fr. 17.xi.1952, *White* 3689 (FHO, K).

Also in Central African Republic, Congo (Katanga), SW Kenya and Tanzania. In open woodland and savanna; c.780 m.

Conservation notes: A widespread taxon, but in only a single province in the Flora Zambesiaca area. At the global scale, Lower Risk, Least Concern.

6. **Philenoptera bussei** (Harms) Schrire in Kew Bull. **55**: 87 (2000). —M. Coates Palgrave, Trees Sthn. Africa: 392 (2002). Type: Tanzania, Lindi Dist., Matapwe, fl. 12.xii.1942, *Gillman* 1124 (K neotype, chosen by Schrire (2000), EA).

Lonchocarpus bussei Harms in Bot. Jahrb. Syst. **33**: 172 (18 Nov. 1902); in Engler, Pflanzenw. Afrikas **3**(1): 641 (1915). —Baker f., Legum. Trop. Africa: 551 (1929). —Polhill in F.T.E.A., Legum., Pap.: 68, fig.13/1–11 (1971). —Drummond in Kirkia **10**: 247 (1975). —Gonçalves in Garcia de Orta, sér.Bot. **5**: 94 (1982). —K. Coates Palgrave, Trees Sthn. Africa, ed.2: 325 (1988).

Lonchocarpus fischeri Harms in Bot. Jahrb. Syst. **33**: 173 (18 Nov. 1902). —Baker f., Legum. Trop. Africa: 550 (1929). —Brenan, Check-list For. Trees Shrubs Tang. Terr.: 431 (1949). —Pedro & Barbosa, Esb. Rec. Ecol.-Agric. Moçambique **2**: 194 (1955). Type: Tanzania, Dodoma Dist., Ugogo, fl. vii.1900, *Busse* 253 (K lectotype).

Lonchocarpus menyharthii Schinz in Bull. Herb. Boiss., sér.2, **2**: 998 (5 Dec. 1902). —Baker f., Legum. Trop. Africa: 551 (1929). —Burtt Davy & Hoyle, Check-list For. Trees Shrubs, Nyasaland: 60 (1936). —White, F.F.N.R.: 159 (1962). Type: Mozambique, Tete, Boroma, *Menyharth* 854 (Z holotype, K photo).

Tree up to 12 m tall; crown spreading; bark brown. Branchlets fulvous-pubescent, hairy or glabrous; bark dark grey to red-brown. Leaves imparipinnate, 12–25 cm long; petiole 2–7(8) cm long, rachis 6–12 cm long, pubescent, hairy or glabrous; petiolules 2–5 mm long; leaflets 7–13(15), opposite or subopposite, 2–7(11.5) × 1–3(5) cm, subcircular to ovate or elliptic or narrowly elliptic to narrowly obovate, rounded to obtuse or acute and sometimes emarginate at the apex, rounded or obtuse or cuneate to narrowed cuneate at the base, stiff, appressed-puberulous and glabrescent on upper surface, pubescent beneath at least on the main nerves; lateral nerves in 4–6 pairs; reticulation somewhat prominent or flat beneath; stipules linear, 3–6 mm long, caducous; stipels 1–3 mm long or absent. Inflorescence a ± flexuose panicle 8–35(40) cm long, usually borne precociously at the top of last season's branches (occasionally 2–3 together), rarely terminal, fulvous-puberulous or glabrous; bracts and bracteoles falling early. Flowers 10–12 mm long; pedicels 2–6 mm long. Calyx 4–5 mm long, thinly fulvous-silky, upper tooth-pair 2.5 mm long, bilobed or entire, the lateral ones 1–1.5 mm long, rounded or obtuse, the lower one 1.5–2 mm long, acute. Standard 9–15 mm long, blade broadly obovate, subcordate with small inflexed auricle, claw 2 mm long; wings 11–11.5 × 4 mm, rounded at the apex, auricle acute, claw 2.5 mm long; keel petals 10–14 × 4.5–5.5 mm, auricle acute, claw 3 mm long. Stamens connate in a tube, the upper stamen free at the base. Ovary 4–6-ovulate, shortly stipitate, puberulous; style 1.5–2 mm long, glabrous. Pod 1–2(4)-seeded, 6–12(14) × 1.2–2.8 cm, 1-seeded pods elliptic, 2-seeded pods linear, glabrous, stiff, apex rounded or obtuse or acute, base obtuse to cuneate; stipe 4–6 mm long. Ripe seed 10–12 × 7–8 mm, flattened, reniform, dark red-brown.

Zambia. W: Ndola, fl. 1.x.1937, *Duff* 234/37 (FHO, K). C: Lusaka, fl. 4.ix.1961, *Lusaka Nat. Hist. Club* 56 (K, SRGH). E: Chipata Dist., Luangwa Valley, NW of Chief Nsefu's village, 750 m, fr. 11.x.1958, *Robson & Angus* 47 (BM, K, LISC, SRGH). S: Mazabuka, fr. 7.x.1930, *Milne-Redhead* 1211 (K, PRE). **Zimbabwe**. N: Mudzi Dist., Mkota C.L., Goromonzi, fl. 20.ix.1951, *Whellan* 561 (K; SRGH). E: Mutare (Umtali), commonage, fl. 28.x.1955, *Chase* 5118 (BM, COI, K, LISC, SRGH). S: Chiredzi Dist., Save R., Puta Camp, fl. 2.x.1958, *Phelps* 272 (K, SRGH). **Malawi**. N: Mzimba Dist., North Viphya, c.6 km from Njakwa toward Uzumara, fl. 13.ix.1976, *Pawek* 11837 (K, MAL, MO). C: Lilongwe Nature Sanctuary, Zone A, fr. 9.xi.1984, *Patel, Salubeni & Seyani* 1681 (K, MAL). S: Mt. Mulanje, Fort Lister Gap, fl.& old fr. 5.ix.1988, *J.D. & E.G. Chapman* 9267 (FHO, K, MO). **Mozambique**. N: Chiúre Dist., road from Chiúre to Mecúfi, fl. 21.viii.1948, *Andrada* 1285 (BM, COI, EA, K, LISC, LMA, LMU, LUAU). Z: Gurué Dist., Mutuáli–Lioma road, 40 km S of Mutuáli, fl. 29.ix.1944, *Mendonça* 2308 (BR, K, LISC, LMU, LUAU, SRGH). T: Cahora Bassa Dist., Chicoa, fl. 8.viii.1941, *Torre* 3224 (BM, COI, K, LISC, LMA). MS: Guro Dist., between Mandié and Mungári, fl. 1.ix.1943, *Torre* 5814 (BM, LISC). GI: Chicualacuala Dist., between Massangena and Mapai, fr. 8.x.1948, *Pedro* 3485 (LMA).

Also in Tanzania, Kenya and South Africa (Mpumalanga). In riverine vegetation and mixed deciduous woodland including miombo, *Combretum* and mopane woodlands; 0–1200 m.

Conservation notes: A very widespread taxon. Lower Risk, Least Concern.

P. bussei is a very variable species throughout its range, from the arid Zambezi Valley to the rainy coastal regions of Tanzania and Mozambique, apparently without any geographical disjunction. It is possible to select some coastal specimens with quite glabrous narrow leaflets and rather broad (more than 2 cm) fruit (sometimes named *Lonchocarpus fischeri*), but with the material actually available it does not seem possible to maintain this species. The distinctions from *P. eriocalyx* are not always clear and some specimens cannot be placed with certainty.

Mendonça's original draft stated 'pod 1–4-seeded'. I have not seen pods with more than 3 seeds.

16. DERRIS Lour.

by J.M. Lock*

Derris Lour., Fl. Cochinch. **2**: 432 (1790) nom. conserv.

Lianes or (not in Africa) trees. Leaves imparipinnate; leaflets opposite, exstipellate. Inflorescence usually a contracted racemose panicle with clusters of flowers on short lateral spurs (pseudoracemes), rarely of single axillary flowers. Flowers pinkish. Calyx campanulate, teeth obsolete. Standard suborbicular; wings oblong; keel petals slightly connate on the upper margins. Stamen-filaments connate into an almost closed tube, the upper filament often free at the base; anthers versatile. Ovary few-ovulate; style filiform, stigma terminal. Pod indehiscent, broad, flattened, winged on dorsal suture. Seed subreniform, radicle inflexed.

A genus of about 50 species in the tropics and subtropics of the Old World; a single species along the east coast of Africa.

Derris trifoliata Lour., Fl. Cochinch. **2**: 433 (1790). —Brenan, Check-list For. Trees Shrubs Tang Terr.: 419 (1949). —Polhill in F.T.E.A., Legum., Pap.: 74, fig.14 (1971). Type: China, Canton, *Loureiro* s.n. (P holotype). FIGURE 3.3.**20**.

> *Robinia uliginosa* Willd., Sp. Pl. **3**, 2: 1133 (1802). Type: E Peninsular India, *Roxburgh* (?B-WILLD holotype).
> *Derris uliginosa* (Willd.) Benth. in Miquel, Pl. Jungh.: 252 (1852); in J. Proc. Linn. Soc., Bot. **4**: 107 (1860). —Baker in F.T.A. **2**: 245 (1871). —Sim, For. Fl. Port. E. Africa: 46 (1909). —Harms in Engler, Pflanzenw. Afrikas **3**(1): 642 (1915). —Baker f., Legum. Trop. Africa: 552 (1929).
> *Derris uliginosa* var. *loureiroi* Benth. in J. Proc. Linn. Soc., Bot. **4**: 108 (1860) in part excl. synonym *D. affinis*, nom. illegit. Type as for *D. trifoliata*.
> *Deguelia trifoliata* (Lour.) Taub. in Bot. Centralbl. **47**: 388 (1891); in Engler, Pflanzenw. Ost-Afrikas **C**: 218 (1895).

Liane with twining stems up to 15 m or more high, glabrous except for the inflorescence. Leaves up to 18 cm long; petiole 2.5–6.5 cm long, rachis 2–6 cm long; petiolules 2–5 mm long; leaflets 1–2(3) on each side of the rachis, 4–7(8) × 2.5–4 cm, ovate to elliptic, the terminal one the largest, thin and stiff to leathery, acuminate with the apex obtuse to subacute, rounded at the base, midrib flat above, prominent beneath; lateral nerves in 5–8-pairs, flat on both surfaces; stipules falling early. Inflorescence 3–15 cm long, axillary, also borne on leafless stems; axis sparsely pubescent; bracts c.1 mm long, ovate-acute, sparsely pubescent. Flowers c.10 mm long, in triads on short lateral branches up to 5 mm long, single towards the axis top; pedicels

---

* Based on the draft Flora Zambesiaca account prepared by F.A. Mendonça.

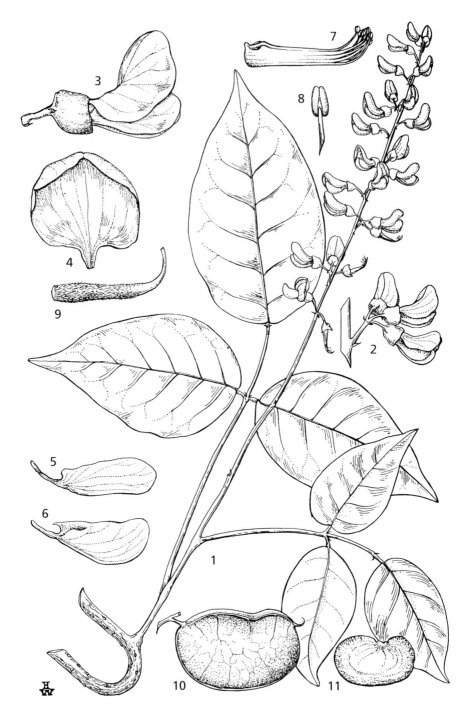

Fig. 3.3.**20**. DERRIS TRIFOLIATA. 1, flowering branchlet (× ²/₃); 2, flower cluster of part of inflorescence (× 1); 3, flower (× 3); 4, standard (× 3); 5, wing (× 3); 6, keel (× 3); 7, stamens (× 3); 8, anther (× 10); 9, gynoecium (× 3); 10, pod (× 1); 11, seed (× 1), 1–11 from *Polhill & Paulo* 686. Drawn by Heather Wood. From F.T.E.A.

2–5 mm long, articulated at or near the base; bracteoles very small, present at anthesis, later deciduous. Calyx 2–2.5 mm long, broadly campanulate, truncate and sparsely hairy above. Corolla pink. Standard c.9 mm in diameter, quadrate-circular, broadly emarginate or retuse, with a 1.5–2 mm long claw at the base; wings obovate, adhering slightly to the keel below the middle, auriculate, with a claw c.2.5 mm long; keel petals sparsely puberulous towards the ciliolate auricle, claw c.3 mm long. Stamen filaments 9–10 mm long, the lower 5–6 mm fused into a tube; anthers c.0.5 mm long, versatile. Ovary 3–6-ovulate, subsessile, puberulous; style 6 mm long; stigma subcapitate. Pod normally 1-seeded, 3–3.5 × 1.5–2 cm, flat, subreniform to subquadrangular, slightly reticulate, wing 1–1.8 mm broad; style persistent. Seed 2.2–2.5 × 1.5 × 0.3–0.4 mm, subreniform with a red-brown testa.

Mozambique. N: Palma Dist., mouth of R. Rovuma, fr. 21.x.1942, *Mendonça* 1009 (BM, COI, K, LISC, LMA, LMU). Z: Marromeu Dist., West Luabo of Kirk (R. Luabo), fl.& fr. 25.v.1858, *Kirk* s.n. (K). MS: Dondo Dist., in mangroves of R. Pungué near bridge on Beira–Nhamatanda (Vila Machado) road, fl. 7.ix.1942, *Mendonça* 171 (BM, COI, K, LISC, LMU, SRGH). GI: Govuro Dist., Nova Mambone, Save R. bank, fl. 8.x.1963, *Leach & Bayliss* 11878 (K, SRGH). M: Maputo R., fl. 25.x.1948, *Gomes e Sousa* 3892 (COI, K, PRE).

In coastal swamp and riverine forests and thickets, including mangrove, of eastern Africa from Kenya to South Africa (KwaZulu-Natal). Also in Asia and Australia; 0–50 m.

Conservation notes: A very widespread taxon; Lower Risk, Least Concern.

17. MILLETTIA Wight & Arn.

by B. Verdcourt*

Millettia Wight & Arn., Prodr. Fl. Ind. Orient. **1**: 263 (1834). —Dunn in J. Linn. Soc., Bot. **41**: 123–243 (1912). —Gillett in Kew Bull. **15**: 19–40 (1961).

Trees, shrubs or lianes, or rarely semi-herbaceous plants with a woody rootstock. Leaves imparipinnate, the leaflets entire, usually opposite; stipels usually present but absent in 2 cultivated species in the Flora area; a pulvinus present at the base of the peduncle. Inflorescence paniculate, but often pseudoracemose by the contraction of the floriferous branches to more knobs. Calyx almost truncate or divided for up to ²/₃ of its length, the upper pair of teeth united for most of their length. Corolla rarely under 1 cm long, white, pink, blue or violet, not speckled, silky or glabrous outside; keel petals lightly united, somewhat pouched but not spurred, their claws and the usually rather longer claws of the wings often ± attached near the base to the filament sheath. Upper stamen usually free in the young bud and remaining so at its base in the mature flower, but usually becoming adherent to its neighbours above; 5 of the 9 united stamens rather longer than the others; anthers uniform, dorsifixed below the middle; filaments not widened at the tip. Disk between stamens and ovary tubular, annular or lobed or flat and undeveloped. Ovary pubescent, sessile or shortly stipitate, with (2)3–11 ovules; style usually pubescent near the base, glabrous above, ± circular in cross-section, straight or incurved at the tip; stigma small, terminal, discoid or capitate. Pod dehiscent, usually flat, rarely subcylindrical, 2- or more seeded; funicle often with a pulpy swelling at the proximal side (towards the pedicel). Seeds usually well separated from one another, sometimes "longitudinal" (long axis parallel to that of pod), more often "transverse" or "obliquely transverse", usually rather flat, usually having around the hilum a short ring-like white or yellow aril prolonged at one side into a strap-shaped process clasping the funicle.

A genus of about 150 species in tropical and subtropical Africa, Asia and Australasia, mainly in forest and woodland.

---

* Use has been made of preliminary work done by Jenny Page who measured all available material and drew floral dissections.

Cultivated species

At least 6 species of *Millettia* have been cultivated in the Flora Zambesiaca area. Biegel (Check-list Ornam. Pl. Rhod.: 76 (1977)) mentions *Millettia grandis* (indigenous) and *Millettia dura. Millettia oblata*, which may be of wider agricultural use, has been included in the sequence of indigenous species, but the other three are mentioned below.

Millettia dura Dunn. A widespread East African species. Tree 4–13 m tall; leaves with 15–19 narrowly oblong leaflets; flowers c.2.5 cm long with mauve corolla, the standard with blade 2 × 1.8 cm; staminal tube 2–2.4 cm long. This will key to *M. eetveldeana* but has larger flowers. **Zimbabwe**. Harare, 11.x.1972, *Biegel* 4002 (K, SRGH); Ewanrigg Aloe Gardens, iii.1958, *Davies* s.n. (K, SRGH); Masvingo (Victoria), Forestry Dept., xi.1953, *Hodgson* in GHS 44363 (SRGH). **Malawi**. Zomba Botanic Garden, 29.ix.1981, *Chapman & Patel* 5899 (K, MAL).

Millettia pinnata (L.) Panigrahi (*Pongamia pinnata* (L.) Pierre). Widespread in tropical Asia and Pacific, usually coastal or lowland. Tree or shrub 15–25 m tall (the one recorded is said to be 15 m tall with pendulous branches); leaves (3)5–7-foliolate; leaflets 5–25 × 2.5–15 cm, ovate, elliptic or oblong, obtuse to acuminate at the apex, rounded to cuneate at the base; inflorescences 6–27 cm long with white or pinkish flowers, purple inside; standard 1.1–1.8 cm long, silky outside. Fruit characteristic, 5–8 × 2.2–3.3 × 1.2 cm, oblong-ellipsoid, indehiscent, glabrous. **Mozambique**. Maputo, Jardim Tunduru (Vasco da Gama), cult., 17.ix.1947, *Pedro & Pedrógão* 1857 (EA, LMA, PRE); vi.1946, *Gomes e Sousa* s.n. (COI).

Millettia thonningii (Schumach. & Thonn.) Baker. Tree 9–12 m tall with several stems; leaves with 7–9 elliptic leaflets 4–7 × 2–4 cm; stipels absent; pedicels with bracteoles in the middle; flowers c.1.5 cm long, the standard silky outside. The long pedicels with bracteoles near the middle, ± glabrous leaflets and lack of stipels distinguish it from all the other species with standard silky outside. **Zimbabwe**. Harare National Botanic Garden, grown from seeds from Ghana, 11.x.1978, *Biegel* 5699 (SRGH).

Millettia xylocarpa Miq. This species was described from Java, based on fruiting material, and left as dubious in Dunn's revision of the genus. Presumably a small tree; leaves 7-foliolate; leaflets c.6 × 3 cm, elliptic to rounded or obtusely acuminate; stipels absent; inflorescence branches slender, up to 15 cm long; pedicels slender, 4–5 mm long; flowers small, 7–9 mm long with standard oblate, 9 × 10 mm, glabrous outside except for a very small tuft at middle of the upper broadly rounded margin; the contrast between the purplish calyx and pale corolla (white?) in dried material is characteristic. These characters immediately separate it from any of the species indigenous to or grown in the Flora Zambesiaca area.

Cultivated material has been seen from Sri Lanka and South Africa. A tree cultivated at Mount Makulu, near Lusaka, Zambia (4.x.1978, *Critchett* s.n. (K)) defied determination when it was first sent to Kew but has now been matched with several specimens of this species annotated by Phan Kê Lôc. Unfortunately Critchett gives no information about habitat or flower colour.

1. Standard glabrous* · 2
– Standard pubescent outside · 3
2. Subshrub or sub-herbaceous, often prostrate, 0.16–1.2 m long or tall; standard 8.5 mm long with claw 2.5 mm long; calyx lobes less than one third the length of the tube; pods 6–7 × 1.5 cm, slightly puberulous · · · · · · · · · · **1.** *makondensis*

---

* Not known in species 11 or 12.

- Tree 6–24 m tall; standard 20–30 mm long with claw 4–5 mm long; calyx lobes much longer; pods 18.5–47 × 3–5.1 cm, puberulous · · · · · · · · · **2.** *stuhlmannii*
3. Lianes; cymule stalks and pedicels short, the inflorescences narrow · · · · · · · · 4
- Trees or shrubs; cymule stalks and pedicels longer, the inflorescences broader, sometimes much-branched · 5
4. Pedicels of mature flowers 0–2 mm long; leaflets glabrous beneath or with short sparse brownish hairs (N Mozambique) · **5.** *impressa*
- Pedicels of mature flowers 2–4 mm long; leaflets velvety appressed silky beneath with longer yellowish to coppery rust-coloured hairs (Malawi and Mozambique) · **6.** *lasiantha*
5. Leaflets 5–9, predominantly rounded at the apex; pods oblanceolate, 9–11.5 × 2.2 –2.5 cm, densely yellow-brown velvety pubescent (N Mozambique) · · · **3.** *bussei*
- Leaflets (5)7–21, predominantly acuminate at the apex · · · · · · · · · · · · · · · 6
6. Leaflets glabrous or nearly so beneath · 7
- Leaflets finely appressed-pubescent to velvety beneath · · · · · · · · · · · · · · · 8
7. Staminal sheath 1–1.3 cm long; leaves often characteristic with lowermost pair of leaflets much the smallest and ± round (but this can occur to some extent in most related species); pods 4–11 × 0.8–1.15 cm, linear-oblong, mostly densely pubescent* · **8.** *usaramensis*
- Staminal sheath 1.5–1.7 cm long; pods 10–16 × 1.4–2.4 cm (but not properly known for variants growing in the Flora area) (Malawi, N Zambia) · · · · · **9.** *eetveldeana*
8. Claws of keel and wing petals more than half as long as blade; leaves 13–21-foliolate (cultivated, N Zambia) · **10.** *oblata*
- Claws of keel and wing petals less than half as long as blade; leaves 5–13(15)-foliolate (wild species) · 9
9. Inflorescences 14–26 cm long and with usually numerous (over 20) lateral branches 1.5–8 cm long; pods oblong or oblanceolate, 5.5–9(15) × 1.5–2.7(3.5) cm, dark brown velvety (Mozambique) · **7.** *grandis*
- Inflorescences usually shorter with fewer branches · · · · · · · · · · · · · · · · · 10
10. Leaflets 3.5–13 × 2.5–7.5 cm with lateral nerves reaching or almost reaching the margins; leaves (5)7(9)-foliolate; calyx lobes 4–5 mm long; pods oblanceolate to linear-oblong, 7.5–14 × 2.3–4 cm, velvety (N Mozambique) · · · **4.** *mossambicensis*
- Leaflets 1.5–7(10) × 1–3.5(4.7) cm, usually small with lateral nerves looping fairly close to but distinctly away from the margin; leaves 7–17-foliolate; calyx lobes less than 2 mm long; pods linear-oblong, less than 1.8 cm wide · return to couplet 7

1. **Millettia makondensis** Harms in Bot. Jahrb. Syst. **33**: 169 (1902). —Dunn in J. Linn. Soc., Bot. **41**: 237 (1912) (adnot.). —Brenan, Check-list For. Trees Shrubs Tang. Terr.: 432 (1949). —Gillett in F.T.E.A., Legum., Pap.: 127 (1971). Type: Tanzania, Newala Dist., Makonde Plateau, *Busse* 1982 (B† holotype, EA, K frag.).

Subshrub 0.16–1.2 m tall or long, from a woody rootstock. Stems ascending or ± prostrate, pubescent, green or drying straw-coloured. Leaves (5)7–9-foliolate; petiole 5–11 cm long; rachis pubescent, 6–10 cm long; leaflets 2–11 × 1.4–7.3 cm (to 10 × 9 cm in Gillett 1971), elliptic to oblong, occasionally almost round or obovate, rounded to shortly bluntly acuminate and often mucronate at the apex, rounded or slightly subcordate at the base, shortly sparsely pubescent on both surfaces; lateral nerves 6–8 on each side, not reaching the margin, ± prominent beneath;

---

* Two imperfectly known species will probably key here but flowers are not known: species 11 with broad leaflets and ± glabrous pods 7.5 × 1 cm, from Zambia N: Mbala; and species 12 with narrow leaflets and ± glabrous pods 6.5–9.8 × 1.2–1.4 cm, from Zambia N: Mansa.

stipules 6–8 mm long, ligulate; stipels 1.2–3 mm long, filiform; petiolules 4–5 mm long, pubescent. Inflorescences (7)10–18 cm long including the 3–5(8.5) cm long peduncle; cymule-stalks c.1 mm long, 1–2-flowered; bracts 1–3 mm long, linear-triangular, pubescent; pedicels c.2 mm long, pubescent; bracteoles c.1 mm long, ± triangular, paired. Calyx reddish, pubescent; tube c.3 mm long; teeth acuminate, c.0.5 mm long, the upper pair triangular, shorter than the lower. Corolla mauve with white keel, glabrous; standard 8.5 × 8 mm, ovate-oblong, emarginate at apex, ± truncate at the base, unappendaged, with claw 2.5 mm long; wing blades 8 × 2 mm, oblong, with claw 3 mm long; keel petals similar. Stamen sheath 7–8 mm long with free parts of filaments 2–3 mm long; upper stamen free at the base; anthers 0.6–1 mm long. Disk absent. Ovary 6 mm long with stipe 1 mm long, 2–3-ovuled, pubescent; style curved, 3 mm long, glabrous except at base; stigma globose. Immature pod flat, 6–7 × 1.5 cm, oblanceolate, 2–3-seeded, sparsely puberulous, later glabrous. Seeds dark brown, 9 × 7 × 4 mm, rounded-oblong; hilum 2.5 mm long.

Mozambique. N: Quissanga Dist., from Mahate towards Macomia, fl. 9.ix.1948, *Pedro & Pedrógão* 5098, 5099 (EA, LMA).

Also in SE Tanzania. Sandy hills; c.200 m.

Conservation notes: Restricted distribution and habitat; probably Vulnerable.

2. **Millettia stuhlmannii** Taub. in Engler, Pflanzenw. Ost-Afrikas **C**: 212 (1895). — Dunn in J. Linn. Soc., Bot. **41**: 203 (1912). —Gomes e Sousa, Dendrol. Moçamb. **1**: 177 (1950). —Gillett in F.T.E.A. Legum., Pap.: 129 (1971). —K. Coates Palgrave, Trees Sthn. Africa, ed.2: 310 (1988). —M. Coates Palgrave, Trees Sthn. Africa: 378 (2002). Types: Mozambique, Isla de Moçambique, *Stuhlmann* 856, and Quelimane, near Puguruni, *Stuhlmann* 868 (both B† syntypes). FIGURE 3.3.**21**.

 Lonchocarpus mossambicensis Sim, For. Fl. Port. E. Africa: 45 (1909). Type: Mozambique, *Sim* 5382 (?PRE holotype).

Tree 6–24 m tall with a spreading crown; bark greenish-yellow, greenish-grey or reddish, smooth, papery and flaky (but has been described as grey and rough in part); slash yellow. Young branchlets with minute appressed white hairs. Leaves 7–9-foliolate; petiole 3–12 cm long; rachis minutely hairy, 7–18 cm long; leaflets 7–24 × 5.5–15.5 cm, oblong to elliptic or obovate, round to emarginate at the apex, broadly cuneate to rounded at the base, almost glabrous above, sparsely pubescent beneath; lateral nerves 7–10 on each side, reaching the margin; stipules 10–14 × 3–6 mm, ± oblong, with an oblong appendage at the base, soon falling; stipels 5–8 mm long, filiform, persistent; petiolules 6–12 mm long. Inflorescences paniculate, pendulous, 20–40(70) cm long, grey-velvety with minute hairs; branches numerous, 1–9 cm long; bracts small, ovate, falling very early; pedicels 8–11 mm long; bracteoles 2–3 mm long, ovate-oblong, up to 1 mm below calyx, soon falling. Flowers scented. Calyx minutely velvety; tube 6–8 mm long; lobes 5–7 mm long, ovate-oblong, rounded, overlapping at the base, laterals 4–5(8) mm wide, the upper pair united to within 1 mm of the tip. Corolla purple or lilac, white or cream at the base, veined, glabrous; standard 2–2.6 × 2.5–2.8 cm, almost round, with auricles 2–3 mm long and 2 shelf-like appendages 3 × 1–2 mm across the top of the 4–5 mm long claw; wing-blades 2.3 × 1.1 cm, with claw 4–7 mm long; blades of keel petals 2.2 × 1.1 cm with similar claws. Stamen sheath 2–2.5 cm long with free parts of filaments 4–8 mm long; upper stamen free and strongly bent at base; anthers 1.3–1.8 mm long. Disk short. Ovary 1–1.1 cm long, 8–10-ovuled, densely silky; stipe 5–6 mm long; style curved, glabrous except at base; stigma globular. Pod yellow-brown, woody, 18–47 × 3–5.1 cm, densely velvety when young, eventually glabrous, lenticellate, rugulose, 6–8-seeded. Seeds chocolate-brown to blackish-purple, transversely positioned, 2–2.4 × 1.5–2 × 0.4 cm, broadly elliptic or ± round, flattened, ± discoid; hilum placed at one end, surrounded by a yellow rim aril prolonged at one side into a ligulate process 1.5–3 mm long pressed against the funicle which is slightly enlarged at both ends.

Zimbabwe. E: Chipinge Dist., near Devon Farm, fl. 9.xii.1949, *Chase* 1855 (BM, K, LISC, SRGH). S: Mwenezi Dist., S of Runde (Lundi) R., fl. 20.xi.1957, *Phelps* 205 (K, SRGH). **Mozambique**. N: Namapa Dist., between Estação Experimental of C.I.C.A. and lands of Cabo Nauacha, on track to Rio Lúrio, fr. 8.iii.1960, *Lemos & Macuácua*

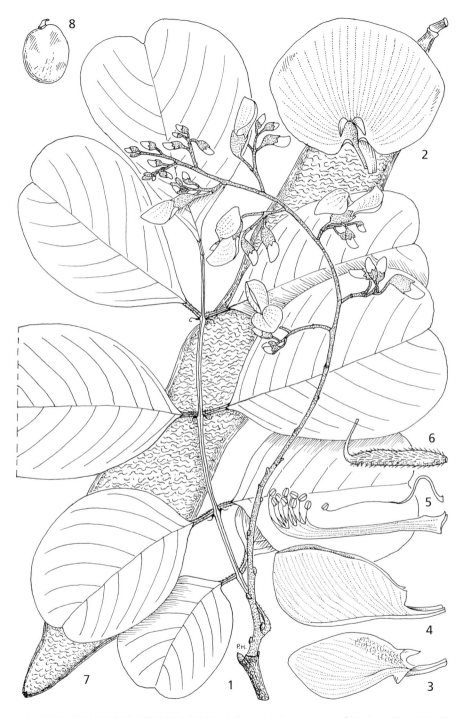

Fig. 3.3.**21**. MILLETTIA STUHLMANNII. 1, flowering branchlet (× ²/₃), from *Dawe* 511; 2, standard (× 2); 3, wing (× 2); 4, keel (× 2); 5, stamens (× 2), 2–5 from *Faulkner* PRE 325; 6, gynoecium (× 2), from *Dawe* 511; 7, pod (× ²/₃), from *Groenendijk, Maite & Dungo* 841; 8, seed (× ²/₃), from *Dawe* 511. Drawn by Pat Halliday.

14 (BM, COI, K, LISC, LMA, LMU, PRE, SRGH). Z: Lugela Dist., Mocuba, Namagoa, fl. ix.1945, *Faulkner* PRE 325 (K, PRE, SRGH). MS: Cheringoma Dist., Inhamitanga, fl. 10.x.1944, *Simão* 157 (LISC). GI: Vilankulo Dist., 4 km W of Mapinhane on Mabote road, fr. 10.v.1971, *Edwards & Vahrmeijer* 4246 (K, PRE). M: Maputo (Lourenço Marques), st. 1908, *Sim* 7128 (K).

Also in S Tanzania. Locally frequent in high rainfall areas at low altitudes, in forest or woodland and beside streams; in *Brachystegia* and *Pterocarpus* woodland, in *Parkia, Pteleopsis, Khaya* gallery forest, and forming an association with *Afzelia, Pteleopsis, Pterocarpus* and *Xylia*, often in riverine and degraded associations, or persisting on cleared grassy plateaux; on white or red sands, often in rocky places; 10–1020 m.

Conservation notes: Widely distributed; Lower Risk, Least Concern.

Also cultivated near Maputo at Matola, e.g. fr. 21.vii.1948, *Gomes e Sousa* 3765 (COI, K, PRE, SRGH).

Yields a good attractive hard durable and termite-proof timber, for which the species has been much exploited commercially.

3. **Millettia bussei** Harms in Bot. Jahrb. Syst. **33**: 170 (1902). —Dunn in J. Linn. Soc., Bot. **41**: 227 (1912). —Brenan, Check-list For. Trees Shrubs Tang. Terr.: 431 (1949). —Gillett in F.T.E.A., Legum., Pap.: 131 (1971). Type: Tanzania, Dodoma Dist., Saranda, *Fischer* 194 (B† holotype, BM sketch, K frag.).

Shrub or tree 3–25 m tall. Branches, foliage and inflorescence axes densely yellow-brown pubescent, the upper surface of the leaves eventually glabrescent but velvety silky all over when very young. Leaves 5–9-foliolate; petiole 2–8 cm long; rachis 4–7 cm long; leaflets 3–16 × 2.8–10 cm, oblong to obovate or ± round, sometimes acute (rarely in Flora Zambesiaca area) to rounded or shallowly emarginate and sometimes very slightly acuminate at the apex, rounded at the base, densely pubescent to glabrescent in mature leaf; lateral nerves 7–11 on each side, reaching the margin; stipules 0.6–1 cm long, oblong; petiolules 5–8 mm long; stipels minute or absent. Inflorescences almost sessile, pseudoracemose, borne in pairs or clusters before the leaves appear or just as they develop, 4–25 cm long on short leafless shoots or at the base of leafy shoots; nodes 1–2 mm long, 1–3-flowered; bracts up to c.1 cm long, linear; pedicels up to 5 mm long; bracteoles 6–8 mm long, linear. Calyx pubescent; tube 4–5 mm long; lowest tooth 3–4 mm long, triangular-acuminate; lateral teeth 2 mm long, broadly triangular, upper teeth united almost to the tip. Corolla mauve; standard 1.7–1.8 × 1.4–1.5 cm, ± round, emarginate at the apex, densely brown-silky outside, glabrous inside, with 2 thickened vertical folds at the base, claw 3–3.5 mm long, pubescent; wing-blades c.11 × 5 mm, ± oblong, rounded at the apex, pubescent at apex and on auricle, the claw 6–7 mm long and slightly hairy; blades of keel petals 10 × 6 mm, mostly silky, the claw 5 mm long and slightly hairy. Stamen sheath 1.2–1.4 cm long with free parts 3–4(7) mm long; upper stamen free at base; anthers 0.8 mm long. Disk annular. Ovary 6–7 mm long, 9–10-ovuled, pubescent; stipe 1–2 mm long; style 1 cm long, curved through 90° near the middle; tip incurved with stigma facing downwards. Pods chestnut coloured, very compressed, 1–4-seeded, 9–11.5 × 2.2–2.5 × 0.4 cm, oblanceolate, widest near the apex, narrowed and without seeds towards the base, densely pale yellow-brown pubescent. Seeds brown-black, placed transversely, 1.2–1.4 × 1.1–1.2 × 0.4 cm, compressed ellipsoid, ± pitted, with a circular aril produced into a ligulate process, appressed against the funicle which is much dilated proximally.

Mozambique. N: Mueda Dist., Negomano, fl.& fr. 2.xii.1960, *Gomes e Sousa* 4595 (COI, K, PRE, SRGH).

Also in SE Tanzania. *Brachystegia* and *Terminalia-Acacia-Sclerocarya* woodland and gallery forest on sandy soils, forest of dense shrubs and tall *Sterculia appendiculata* and *Milicia excelsa* with *Oxytenanthera*, also sisal plantations; 140–410 m.

Conservation notes: Limited distribution; probably Lower Risk, Near Threatened.

Andrada 1280 (Mozambique: Niassa, Chiure, fr. 20.viii.1948) has shortly but distinctly acuminate leaflet apices; J. Page thought it was this species but it resembles the next. The possibility of hybrids needs investigating.

4. **Millettia mossambicensis** J.B. Gillett in Kew Bull. **15**: 23 (1961). Type: Mozambique, Gorongosa Nat. Park (Game Reserve), *Chase* 5078 (K holotype, BM, COI, LISC, PRE, SRGH).

Tree 3–15 m tall; branches grey becoming black, lenticellate; bark longitudinally flaky; slash pale brown, very fibrous. Young branchlets brown-pubescent; buds compressed conic, brown-pubescent; scales boat-shaped. Leaves (5)7(9)-foliolate; petiole 3–6 cm long; rachis brown-pubescent, 6–10 cm long; leaflets 3.5–13 × 2.5–7.5(8.8) cm, smaller towards the base, basal leaflets ovate, upper leaflets elliptic, obovate or oblong-obovate, acuminate at the apex with a deciduous filiform mucro 1–2 mm long, cuneate to truncate at the base, minutely white appressed-pubescent on upper surface, later glabrescent, thinly to densely silky pubescent beneath with hairs c.0.5 mm long, lateral nerves 8–10 on each side ± reaching the margins, brown beneath; stipules very deciduous; petiolules 3–5 mm long, pubescent; stipels 2–4 mm long, filiform. Panicles showy, terminal, 4–13-flowered, 4–10 cm long with branches 1.5 cm long; pedicels 1–2 cm long; bracts and bracteoles very deciduous, not seen. Calyx dark brown silky; tube 3–4 mm long; lobes 4–5 mm long, oblong-triangular, the upper pair almost completely joined. Standard blue, purple or lilac, (1.3)1.5–1.7(1.9) × 1.4–1.75 cm overall, ± round or oblate, without auricles or appendages, white-silky outside; claw 1.5–2.5 mm long; wing blades 13–14 × 6–7 mm, obovate-oblong, slightly hairy at apex, with spur 1.5 mm long and claw 2–4 mm long; blades of keel petals 12–13 × 6–7 mm, rhombic, with spur 2 mm long and claw 4 mm long. Stamen sheath 1–1.2 cm long with free parts 3–5 mm long; upper stamen free at the base and apex; anthers 0.8–1 mm long. Disk minute, angular, crenate. Ovary 5–6 mm long, silky pubescent, 4-ovuled; style 1–1.2 cm long, glabrous except at base, curved through 100° or more at the middle; stigma minute, erect. Pod yellow-brown, (6)7.5–14 × 2.3–4 cm, oblanceolate to linear-oblong, flat, 1–3-seeded, densely velvety, margined when young. Seeds dull black, transversely placed, c.15 × 10 × 4 mm, elliptic-oblong; hilum along the short side; rim aril annular, produced into a ligulate process; funicle not dilated.

Mozambique. N: Meconta Dist., between Namialo and Monapo, fl. 16.x.1948, *Barbosa* 2463 (BM, K, LISC). MS: Gorongosa Dist., Gorongosa Nat. Park (Parque Nacional de Caça), on track No. 3, between to crossings with track No. 1, c.40 m, fl. 4.xi.1963, *Torre & Paiva* 9031 (LISC).

Not known elsewhere. Mixed evergreen and dry forest with *Pterocarpus*, *Xylia* and *Newtonia*, woodland of *Acacia*, *Sterculia*, *Adansonia* on red sand; 40–200 m.

Conservation notes: Only known from N & C Mozambique. Limited distribution and habitat; probably Vulnerable.

5. **Millettia impressa** Harms in Bot. Jahrb. Syst. **26**: 288 (1899). —Dunn in J. Linn. Soc., Bot. **41**: 210 (1912). —Baker f., Legum. Trop. Africa: 234 (1929). —Hauman in F.C.B. **5**: 36 (1954). —Sousa in C.F.A. **3**: 176 (1962). —Gillett in F.T.E.A., Legum., Pap.: 135 (1971). Type: Congo (Brazzaville), Loanga, near Zala, *Soyaux* 99 (K).

Climber to several metres high; stems dark brown pubescent but becoming glabrous. Leaves (5)7(9)-foliolate; petiole to 7 cm long; rachis up to 8 cm long; leaflets 3.5–9.5 × 1.5–5 cm, elliptic-oblong to oblong-obovate, rounded to shortly broadly obscurely acuminate at the apex with a deciduous mucro 0.5–1 mm long, cuneate to rounded at the base, glabrous on upper surface, sparsely brown pubescent or glabrous beneath; main lateral nerves 6–7 on each side, impressed above, not reaching the margin; stipules 4–6 mm long, narrowly triangular; stipels 2–4 mm long, filiform; petiolules 3–6 mm long. Inflorescences terminal and axillary, pseudoracemose, 25–30 cm long, brown pubescent; cymule rachides up to 5 mm long; pedicels 1–2 mm long; bracteoles up to 3.5 mm long, oblong. Calyx appressed brown pubescent; tube 3–4 mm long, the 3 lower teeth 1.5 mm long, triangular; upper pair shorter, almost completely united. Corolla mauve or greenish, striped with purplish-chestnut; standard 7–11 × 8–12 mm, ± round, emarginate at the apex, slightly cordate at the base and with 2 vertical folds, silky brown pubescent outside, the claw 2–4 mm long, glabrous; wing-blades 6–9 × 2–4 mm, slightly auriculate at the base, silky near tip, with glabrous claw 3–5 mm long; blades of keel petals 6–10 × 3.5–4.5 mm, silky pubescent outside with auricle 1–1.5 mm long and glabrous claw 4 mm long. Stamen sheath 7–10 mm long with free parts of filaments 2–3 mm long; upper stamen free at the base, curved; anthers 0.5–0.8 mm long. Disk absent. Ovary 7 mm long, 4–6-ovuled, silky pubescent, with stipe 1 mm long; style curved through 90° near the base, glabrous save at the base; stigma capitate; pod 5–10 × 1–1.9 cm, ± oblong-linear, silky pubescent. Seeds brown, 11–13 × 9–10 × 4.5 mm, oblong or subrhombic in outline.

Subsp. **goetzeana** (Harms) J.B. Gillett in Kew Bull. **15**: 25 (1961); in F.T.E.A., Legum., Pap.: 135 (1971). Type: Tanzania, Uzaramo, c.7°S, 39°E, *Goetze* 11 (B† holotype, K).

> *Millettia goetzeana* Harms in Bot. Jahrb. Syst. **28**: 404 (1900). —Dunn in J. Linn. Soc., Bot. **41**: 209 (1912). —Baker f., Legum. Trop. Africa: 234 (1929). —Brenan, Check-list For. Trees Shrubs Tang. Terr.: 432 (1949).

Leaflets more rounded at the apex; flowers 1–1.4 cm long.

Mozambique. N: Palma Dist., between Nangade and Palma, fl.& fr. 20.x.1942, *Mendonça* 997 (BM, K, LISC).

Also in SE Tanzania. "Hydrophilous forest"; below 200 m.

Conservation notes: Limited distribution and habitat; possibly Vulnerable.

The typical subspecies, with acuminate leaflets and smaller flowers, occurs in NW Angola, western Congo and Congo (Brazzaville).

6. **Millettia lasiantha** Dunn in J. Bot. **49**: 221 (1911); in J. Linn. Soc., Bot. **41**: 226 (1912). —Brenan, Check-list For. Trees Shrubs Tang. Terr.: 432 (1949). —Gillett in F.T.E.A., Legum., Pap.: 136 (1971). —White, Dowsett-Lemaire & Chapman, Evergr. For. Fl. Malawi: 330, fig. 117 (2001). Types: Kenya, Nyika, *Wakefield* s.n. (K syntype); Tanzania, Usambara Mts., *Holst* 2968 & 2215 (K syntypes); Uzaramo Dist., Pugu, *Holtz* 1041 (B† syntype); Hemiafa, *Braun* in Herb. Amani 1790 (B† syntype, EA); Morogoro Dist., Mhonda, *Sacleux* s.n. (P syntype).

> *Millettia leucantha* sensu Taubert in Engler, Pflanzenw. Ost-Afrikas C: 212 (1895) in part, non Vatke.

Liane 4.5 to 40 m tall, reaching the forest canopy layer, becoming massive with lower main stems like vast cables, much-fluted, 7.5–10 cm in diameter. Branchlets velvety with appressed green-brown to chocolate-coloured hairs; stems later glabrescent and white-lenticellate. Leaves 7–9-foliolate; petiole 3.5–6 cm long; rachis 2–10 cm long; leaflets 2.5–10 × 1.7–5 cm, oblong to elliptic or obovate, acute to acuminate at the apex, cuneate to rounded at the base, ± glabrous on upper surface save for the midrib, velvety appressed silky beneath; stipules 8 × 3 mm, oblong-lanceolate, appendaged at the base, hairy like the stems, falling almost at once; stipels 3–5 mm long, subulate; petiolules 2–4 mm long. Inflorescences branched or unbranched, (2)10–30 cm long, appressed brown pubescent; peduncle (0)4 cm long; cymule-stalks nodular, c.1 mm long, 3–5-flowered; bracts 5 mm long, lanceolate, very deciduous; bracteoles 1.2–1.5 mm long, oblong; pedicels 2–4 mm long. Calyx silvery-brown velvety; tube 2.5 mm long; lateral lobes 1.5–2 mm long, lower lobe 2.5–3 mm long. Corolla mauve-blue or violet; standard sometimes with white spots at the base, 10–15 × 7–12 mm overall, rounded oblong, emarginate at the apex, densely silvery brown silky outside; without appendages but with slight folds at the base; auricles 2 mm long; claw 3–4 mm long, broadly obtriangular; wing blades 9 × 3–4 mm, oblong, with claw 3 mm long; blades of keel petals 9–11.5 × 4.5–6 mm, oblong-elliptic, with claws 2.5 mm long. Stamen sheath 8.5 mm long with free parts (2)3–4 mm long; anthers 0.8 mm long; upper filament free at the base. Ovary 6–8 mm long, densely silky pubescent with white and brown hairs, 4–6-ovuled; style glabrous, 4–6 mm long, curved through 90°; stigma small, erect. Pods flat, 5–10.5 × 2–3 cm, ± oblong, rounded at the apex with a sharply deflexed beak, (1)2–5-seeded, densely yellow-brown or brown appressed silky. Seeds dark brown, transversely placed, 14 × 10 × 3.5 mm, flattened ellipsoid or oblong; hilum with an incomplete rim aril produced into a ligulate process appressed to the non-dilated funicle.

Malawi. N: Nkhata Bay Dist., Rose Falls (Lady Roseveare's cottage), 8 km E of Mzuzu, fl. 14.xi.1970, *Pawek* 3976 (K); fr. 6.iii.1971, *Pawek* 4479 (K). S: Mulanje Mt., Great Ruo Gorge, Lufiri, fl. 18.vi.1962, *Richards* 16768 (K). **Mozambique**. Z: Gurué, stream margins, fl. 6.iv.1943, *Torre* 5084 (BM, K, LISC).

Also in E Kenya and E Tanzania. *Newtonia* and other montane forest, usually riverine, dense undergrowth at forest edges, also secondary scrub; 770–1500 m.

Conservation notes: Fairly widely distributed; probably Lower Risk, Near Threatened.

J.B. Gillett has pointed out on herbarium covers that material from SE Tanzania and the Flora Zambesiaca area had the pseudoracemes unbranched and that a subspecies should perhaps be recognized. This has, however, proved illusory. *Chapman* 472 (Mt. Mulanje, along upper Lukulezi R., 15.x.1957 (FHO, K)) has them unbranched, but *Chapman* 6672 (Mt. Mulanje, lower slopes above Esperanza Estate, 27.x.1985 (K, MO)) has them branched.

7. **Millettia grandis** (E. Mey.) Skeels in U.S. Dept. Agric. Bur. Pl. Industr. Bull. **248**: 55 (1912). —Palmer & Pitman, Trees Sthn. Africa: 923 (1972). —K. Coates Palgrave, Trees Sthn. Africa, ed.2: 310 (1988). —M. Coates Palgrave, Trees Sthn. Africa: 377, fig.101 (2002). Types from South Africa (KwaZulu-Natal).

  Virgilia grandis E. Mey., Comment. Pl. Afr. Austr.: 1 (1836).
  Millettia caffra Meisn. in Hooker, London J. Bot. **2**: 99 (1843). —Harvey in F.C. **2**: 211 (1862). —Dunn in J. Linn. Soc., Bot. **41**: 206 (1912). —Ross, Fl. Natal: 203 (1973). Type from South Africa (KwaZulu-Natal).
  Fornasinia ebenifera Bertol. in Novi Comment. Acad. Sci. Inst. Bononiensis **9**: 589 (1849). Types: Mozambique, opposite Inhambane Is., *Fornasini* s.n. (BOLO syntypes, K photo).

Tree or shrub, sometimes stunted or gnarled, 2–13 m tall* ; bark grey or brown, smooth to ± flaky. Leaves 7–13(15)-foliolate; petiole 2–4.5(6) cm long; rachis 4.5–10.5 cm long; leaflets 2–9 × 1.5–4 cm, elliptic or oblong to oblong-elliptic or ovate-elliptic, rounded to acute and apiculate at the apex, rounded to cuneate at the base, glabrous on upper surface, finely appressed-pubescent beneath; lateral nerves in 9–12(16) pairs; stipules 4–6 mm long, deciduous; stipels 2–4 mm long, persistent; petiolules 3–4 mm long; characteristic curved striate buds in the leaf-axils. Inflorescences branched panicles, 14–26 cm long; branches 1.5–8 cm long; bracts 1.7 mm long; pedicels 2.5–4 mm long; bracteoles 1.5–2 mm long, ovate, deciduous. Calyx with silvery grey to dark brown velvety hairs; tube 2–4 mm long; lobes 2–2.5 mm long, rounded ovate, the upper two united except for the apical 0.3 mm. Corolla violet, standard 1.3–1.6 × 1.2–1.3 cm, ± round, with transverse callus, silvery grey shiny silky pilose outside, with claw 2.5–3 mm long; wing-blades 13–14 × 4.5 mm, oblong-elliptic, with spur 1 mm long and claw 3 mm long; blades of keel petals 9–12 × 6 mm, elliptic, toothed at the base on upper side and with claw 3 mm long. Stamen sheath 1–1.3 cm long; free parts of filaments 4–5 mm long; anthers 0.8–1 mm long; upper filament free at the base. Ovary 8 mm long, densely silky with light brown hairs, 5-ovuled; style 4.5 mm long, glabrous, bent through 40–45°; stigma small. Pods woody, 5.5–9(15) × 1.5–2.7(3.5) cm, oblong or oblanceolate, margined, densely dark brown velvety when young, later glabrescent, with short beak, 1–3-seeded. Seeds brownish-black, 12(20) × 9 mm, oblong, compressed.

Mozambique. GI: Massinga, fl. xii.1935, *Gomes e Sousa* 1688 (COI, K, LISC); between Xai-Xai (Vila João Belo) and Inchobane (Bilene), fl. 10.i.1943, *Torre* 4771 (K, LISC); between Quissico (Zavala) and Inharrime, fr. 28.i.1941, *Torre* 2569 (BM, LISC).

Also in South Africa (KwaZulu-Natal and NE Cape). Open dry coastal forest and secondary forest on sandy soil and on shale; c.100–150 m.

---

* Dunn gives 40 m, but this is presumably incorrect – error for feet?

Conservation notes: Widespread; Lower Risk, Least Concern.

The calyx of Mozambique material is pale silvery grey hairy whereas that of South African material is dark brown save for a cultivated sheet of unknown origin; if a varietal or subspecific name were required Bertoloni's epithet would be available.

The wood is very hard, close-grained and durable, and is usually considered one of the best South African timbers.

8. **Millettia usaramensis** Taub. in Engler, Pflanzenw. Ost-Afrikas **C**: 212 (1895). — Dunn in J. Linn. Soc., Bot. **41**: 224 (1912). —Gillett in F.T.E.A., Legum., Pap.: 137, fig.24 (1971). —K. Coates Palgrave, Trees Sthn. Africa, ed.2: 311 (1988). — White, Dowsett-Lemaire & Chapman, Evergr. For. Fl. Malawi: 330 (2001). —M. Coates Palgrave, Trees Sthn. Africa: 378 (2002). Types: Tanzania, Uzaramo, *Stuhlmann* 6358, 6395 & 7044 (B† syntypes).

Shrub or small tree (2)3–7(10) m tall, sometimes many-stemmed and thicket-forming; bark light grey or mottled light grey with darker round flakes, smooth; buds small, rounded. Leaves 7–17-foliolate; petiole 2–3(5.5) cm long; rachis 8.5–16(20.5) cm long; leaflets (1)1.5–7(10) × 0.8–3.5(4.7) cm, upper leaflets oblong, obovate or elliptic, but the basal ones relatively broadly ovate, broadly elliptic, oblate or ± round, shortly acuminate at the apex with the tip rounded, truncate or emarginate, asymmetrically cuneate at the base, glabrous to reddish pubescent; lateral nerves c.5–7 on each side, not reaching the margin; stipules very deciduous; stipels 0.5–2 mm long, filiform; petiolules 2–4 mm long, pubescent. Inflorescences axillary, 12–14(16) cm long; the individual cymules 0.1–2.5 cm long, 1–3-flowered; bracts up to 1 mm long, ligulate, deciduous; pedicels 0.4–1.3 cm long; bracteoles c.0.5 mm long, ovate. Flowers scented. Calyx glabrous save at margins, or pubescent; tube 3 mm long; teeth up to 0.5 mm long, broadly triangular. Corolla purple-blue or rose-coloured; standard 1.1 × 1.3 cm, cordate, not thickened or folded at the base, thinly white silky outside; claw 2–3 mm long; wing-blades 8–10 × 3.5–4 mm, narrowly oblong, pubescent at the tips with claw 6.5 mm long; blades of keel petals 10 × 5.5–7 mm, oblong-elliptic, pubescent at the tips, with claw 5.5 mm long. Stamen sheath 1–1.3 cm long; free parts of filaments 2–4 mm long; upper filament free at base; anthers 0.7 mm long. Disk absent. Ovary 8 mm long, silky with long appressed hairs save at tip, 5–8-ovuled; style glabrous, c.8 mm long, curved through 90° near the middle but often curved back through 180° after flowering. Pod 4–11 × 0.8–1.15 cm, linear-oblong, densely silky pubescent or sometimes ± glabrescent at maturity, glabrous and deflexed at the tip. Seeds ± black, 5–6 × 2 mm, discoid; funicle dilated proximally.

Subsp. **usaramensis** —Gillett in F.T.E.A., Legum., Pap.: 138, fig.24 (1971).

Leaflets lower surface and inflorescences almost glabrous; calyx ± glabrous except at margins; pods mostly silvery hairy.

Zimbabwe. N: Shamva, fr. 11.vi.1922, *Eyles* 3503 (K, SRGH). S: Chiredzi Dist., Chuhonja Range, Majidando, fr. iii.1959, *Farrell* 64 (SRGH). **Mozambique**. N: between Angoche and Corrane, fl. 1.xi.1942, *Torre* 4746 (BM, K, LISC). Z: Milange Dist., Milange road, 450 m, fl. 27.xi.1949, *Faulkner* Kew 476 (K).

Also in Kenya and Tanzania. In miombo and mixed *Acacia* woodland, and in wooded grassland, sometimes in riverine vegetation, on sandy soil and also on termite mounds; 10–900 m.

Conservation notes: Widespread; Lower Risk, Least Concern.

All Flora Zambesiaca material seen of this subspecies is var. *usaramensis*. Var. *parvifolia* Dunn, distinguished from the typical variety by leaflets only 2–3 cm long, occurs in SE Tanzania and may occur in N Mozambique.

Subsp. **australis** J.B. Gillett in Kew Bull. **15**: 31 (1961); in F.T.E.A., Legum., Pap.: 138 (1971). Type: Mozambique, lower Búzi R., Chironda, *Swynnerton* 1425 (K holotype, BM, SRGH).

> *Millettia usaramensis* sensu Dunn in J. Linn. Soc., Bot. **41**: 225 (1912) in part, non Taub. sensu stricto.

All parts with conspicuous reddish pubescence.

Zimbabwe. N: Mutoko Dist., Nyadire R., between Nyagogo and Tembo rivers, fl. 27.x.1955, *Lovemore* 460 (K, LISC, PRE, SRGH). E: Chimanimani Dist., Save Valley, Nyanyadzi R., on road to Birchenough Bridge, fl. 14.xi.1948, *Chase* 1317 (BM, COI, K, SRGH). **Malawi**. S: Mangochi Dist., Manjawira, fl. 15.xii.1954, *Jackson* 1409 (BM, FHO, K). **Mozambique**. N: Mecanhelas Dist., road to Tchamba, along Rio Lúrio, fl. 19.x.1948, *Andrada* 1426 (COI, LISC). MS: Chibabava Dist., Madanda Forest, fl. ix–x.1911, *Dawe* 446 (K). GI: about 48 km SW of Massangena, fl. 16.ix.1959, *Goodier* 614 (K, SRGH).

Not known elsewhere. Rocky outcrops and hillsides in *Brachystegia tamarindoides* woodland and mixed deciduous woodlands (including *Terminalia-Combretum* woodland); also beside rivers in thicket vegetation; 60–1095 m.

Conservation notes: Subspecies endemic to the Flora area, but widespread; Lower Risk, Least Concern.

The subspecies are usually easily separated, but material from Zimbabwe named as subsp. *usaramensis* is suspect – for example, both subspecies occur in Zaka Dist. (Zim: S). *Balaka, Seyani & Kaunda* 828 (K, MAL) from S Malawi (Mulanje, Macheaba Hill, st. 17.xi.1984) is intermediate. *Goldsmith* 138/62 from Chipinge (fr. v.1962 (K, PRE, SRGH)) is also intermediate and the Kew sheet has been referred to subsp. *usaramensis* by Gillett. It appears to have lost much indumentum through maturity.

A specimen *Chapman* 6136 (K) from S Malawi (Thyolo Dist., Zoa Falls, fr. 18.ix.1984, rocky river bank at 450 m) must be closely allied to *M. usaramensis*, but has larger leaves and pods. Without further flowering material the status of this specimen will remain uncertain. It is a small tree or shrub c.4 m tall; branchlets with ridged bark, glabrous. Leaves 9–13-foliolate; leaf rachis 27 cm long including petiole of 5.5 cm, sparsely pubescent; terminal leaflet obovate; intermediate lateral leaflets rhombic-elliptic with asymmetric midrib, up to 9 × 4.5 mm, acuminate but with minutely emarginate apex, cuneate to rounded at the base, the midrib ± pubescent beneath; lowermost leaflets almost round, up to 3 × 2.8 cm. Inflorescence rachis and petiole c.10 cm long; fruit stalks c.1 cm long; pods linear, up to 12 × 1.5 cm, the margins somewhat raised, sparsely appressed pilose, up to 5-seeded.

9. **Millettia eetveldeana** (Micheli) Hauman in F.C.B. **5**: 28, fig. 2A (1954). —Gillett in Kew Bull. **15**: 32 (1961); in F.T.E.A., Legum., Pap.: 141 (1971). Type: Western Congo, *Laurent* (BR holotype).

> *Lonchocarpus eetveldeanus* Micheli in Bull. Soc. Roy. Bot. Belg. **36** (2): 67 (1897); in Ann. Mus. Congo, sér.1, **1**: 17 (1898).
> *Millettia leptocarpa* Dunn in J. Bot. **49**: 221 (1911); in J. Linn. Soc., Bot. **41**: 225 (1912). Type: Congo, Mukenge, *Pogge* 337 & 841 (B†, K drawing & frag.).

Shrub or small tree, 1.8–17(25) m tall, sometimes many-stemmed. Young branches dark grey, sparsely rusty pubescent; older branches grey with elliptic lenticels, noduled and ridged. Leaves (11)13–17-foliolate; petiole 1.5–6 cm long, sparsely to densely rusty pubescent; rachis 5.5–19(24) cm long; leaflets thin, 1–9 × 1–3 cm, elliptic-oblong or elliptic with the basal pair rounded-ovate, subacute to acuminate at the apex, asymmetrically cuneate or rounded at the base, glabrous on upper surface and practically so beneath or appressed silky ferruginous on both surfaces, later glabrescent; lateral nerves 8–9 on each side, not reaching the margin; reticulate venulation prominent beneath; stipules 3–7 mm long, ligulate, very deciduous; petiolules 3–4 mm long; stipels 1–3 mm long, filiform, or absent. Inflorescence (8.5)12–25 cm long including peduncle 1.5–13 cm long, sparsely rusty pubescent; cymule rachis stout, 4–7 mm long, 2–4(5)-flowered; pedicels (2)6–10 mm long, with red pubescence; bracts 2 mm long, triangular subulate; bracteoles 0.7–1 mm long. Calyx sparsely rusty-pubescent; tube c.4 mm long; teeth 1–1.5 mm long, broadly triangular, obtuse. Corolla showy, purple or blue with yellow patches at base of standard; standard 1.4–1.9 × 1.2–1.6 cm, almost round, white silky outside, tapering into the 3–5 mm long claw, folds and appendages lacking; wing blades 12–13 × 4–6 mm, oblong, glabrous except at apex, with claw 6 mm long; blades of keel petals similar, 12–13 × 6 mm, glabrous, with claw 5–6 mm long. Stamen sheath 1.5–1.7 cm long; free parts of filaments (2)3–4 mm long; upper filament free or adhering near the middle; anthers 0.6–0.7 mm long. Disk absent. Ovary 1 cm long, linear, with short spreading yellowish hairs, (6)8–9-ovuled; style 1.1 cm long, pubescent at the base, curved through 90° at the middle. Pods brown, 10–16 × 1.4–2.4 cm, linear-oblong, deflexed at the apex, 2–6-seeded, glabrescent. Seeds blackish, 9 × 7 mm, compressed oblong-ellipsoid.

Var. **brevistipellata** (De Wild.) Hauman in F.C.B. **5**: 30 (1954). Type from Congo (Katanga).

> *Millettia brevistipellata* De Wild. in Ann. Mus. Congo, Bot., sér.4 **1**: 193 (1903). —Dunn in J. Linn. Soc., Bot. **41**: 221 (1912). —Baker f., Legum. Trop. Africa: 237 (1929).

Stipels less than 1 mm long; leaflets more leathery, with venation raised and apparent on both surfaces; corolla 2–2.2 cm long.

Zambia. N: Kaputa Dist., Mweru-Wantipa, road near Kanjiri (Kangiri), fl. 17.xii.1960, *Richards* 13738 (K). **Mozambique**. N: Mueda Dist., Mueda area, fl. 23.ix.1948, *Pedro & Pedrógão* 5307 (EA, LMA).

Also in Congo (Brazzaville), Congo, Uganda, Tanzania and Angola. In chipya woodland (fire-degraded dry evergreen forest) often on termite mounds, in itigi thickets on sandy soil, on mushitu margins, and in high-water table wooded grassland with *Parinari*; 650–1000 m.

Conservation notes: Probably Lower Risk, Least Concern.

J.B. Gillett (notes on sheets) suggested that all the Zambian material is var. *brevistipellata* (De Wild.) Hauman, not var. *eetveldeana*.

Var. ?

Large forest tree, 10.5 m to first branch; bark shedding in flakes; stripped branches smell like freshly stripped Mountain Ash (*Sorbus*) bark. Lower leaflets 3.5 × 2.5 cm, ovate to elliptic, the rest 10.5 × 2.5 cm, oblong, acuminate at apex, somewhat glaucous beneath. Inflorescences to 23 cm long including peduncle 8 cm long, very floriferous; calyx brown with teeth to 2 mm long; corolla blue. Good timber.

Malawi. N: Chitipa Dist., Misuku Hills, Mugesse Forest Res., fl. ix.1953, *Chapman* 157 (FHO, K).

Conservation notes: If this is a separate taxon, it probably has a very limited distribution and should be provisionally considered as Vulnerable, especially given its good timber.

There is a fruit in a packet (mounted on the K sheet of *Chapman* 157) just like that of *M. bussei*, 9.5 × 2.5 cm. If it belongs to *Chapman* 157 then, as J.B. Gillett has written on the sheet, the specimen belongs to an undescribed species; if not he suggests it is probably a glabrescent form of *M. eetveldeana*. *M. bussei* is not known to occur in N

Malawi and if the fruit does not belong the error must have resulted in the herbarium. Only further material from the same forest with undoubted fruits will solve the problem. White, Dowsett-Lemaire & Chapman (Evergr. For. Fl. Malawi: 330 (2001)) attribute this specimen to *Millettia dura* Dunn.

10. **Millettia oblata** Dunn in J. Bot. **49**: 221 (1911); in J. Linn. Soc., Bot. **41**: 223 (1912). —Brenan, Check-list For. Trees Shrubs Tang. Terr.: 432 (1949). —Gillett in F.T.E.A., Legum., Pap.: 142 (1971). Type: Tanzania, E Usambara Mts., Amani, *Warnecke* 46 (K lectotype, EA).

Shrub or tree 2–30 m tall. Young branches dark rusty pubescent. Leaves 9–23-foliolate; petiole up to 9 cm long, rusty pubescent; rachis 3–21 cm long; leaflets 2–11(14) × 1.7–4 cm, oblong-lanceolate, or the lowest 3–6 × 2.5–4 cm, ovate, acuminate at the apex, rounded or even slightly subcordate at the base, sparsely pubescent or glabrous on upper surface except the main venation, rusty pubescent to velvety beneath; stipules 5–9 mm long, ligulate; petiolules 2–5 mm long, pubescent; stipels 1–5 mm long, filiform. Inflorescences axillary, 21–35 cm long, including peduncle 4–10 cm long, rusty pubescent; cymule stalks up to 1.5 cm long or reduced to knobs, 1–4-flowered; pedicels 0.5–1 cm long, pubescent; bracteoles 1–2 mm long, oblong, usually well below the calyx base and often near the middle of the pedicels. Calyx rusty pubescent; tube 4–5 mm long; teeth 1–2 mm long. Corolla blue or purplish-blue. Standard 1.2–1.5 × 1.4–1.8 cm, rounded oblate, cordate at the base, without folds or appendages, white silky outside, with claw 3 mm long; blades of wings 10–11 × 5–7 mm, rounded oblong, with claw 7 mm long; blades of keel petals similar, 10–11 × 6–7 mm, with claw 5–6 mm long; both keel and wings pubescent only at the tip. Stamen sheath 1.2–1.6 cm long; free parts of filaments 2–4 mm long; upper filament adhering at the middle only; anthers 8–10 mm long. Disk absent. Ovary 6–9-ovuled, rusty pubescent; style 8–9 mm long, curved through 110° below the middle. Pod brown, 10–12(16) × 1.8–2.5 cm, linear-oblong, glabrescent. Seeds brown, c.1 cm in diameter, ellipsoid or round in outline, with white annular aril produced into a ligulate process appressed to the apically dilated funicle.

Subsp. **stolzii** J.B. Gillett in Kew Bull. **15**: 33 (1961); in F.T.E.A., Legum., Pap.: 143 (1971). Type: Tanzania, Rungwe, *Stolz* 183 (K holotype).

Millettia oblata sensu White, F.F.N.R.: 159 (1962) non Dunn sensu stricto.

Tree up to 10 m tall. Leaflets 13–21, with dense indumentum beneath. Inflorescences up to 35 cm long, with lower floriferous branches up to 15 mm long. Blade of standard up to 15 mm long. Filament sheath c.13 mm long. Ovules c.9. Pod up to 2.5 cm wide.

Zambia. N: Mbala Dist., Sunzu Farm, fl. 18.ix.1952, *White* 3723 (FHO, K). C: Lusaka Dist., Chilanga, Mt. Makulu Res. Station, fl. 23.ix.1962, *Angus* 3328 (FHO, K).

Native to the Southern Highlands of Tanzania. Cultivated in Zambia as an ornamental and as a coffee shade tree.

Conservation notes: Introduced in the Flora area.

Several other subspecies are recognized in East Africa, see Gillett in F.T.E.A. (1971).

11. **Millettia** sp.

Millettia sp. 1 of White, F.F.N.R.: 160 (1962).

Scrambling or spreading bush. Young shoots glabrous, with ridged bark. Leaves up to 9–?11-foliolate; rachis plus petiole up to 14.5 cm long, glabrous; leaflets 5–6.5 × 3–4 cm, ovate to elliptic-oblong, shortly narrowed to a very slightly emarginate apex, rounded at the base, glabrous. Flowers unknown. Pods 7.5 × 1 cm, linear, with a deep central longitudinal groove (according to the collector, less evident in dried fruit), glabrous (a few very scattered hairs visible at high magnification), 4-seeded.

Zambia. N: Mbala (Abercorn), lower Lufu Valley, fr. iii.1937, *Trapnell* 1734 (K). "A chief component of dense bush on sand"; 750 m.

The fruits indicate affinity with *M. usaramensis*; despite the collector's comment no more seems to have been collected. The native name is given as Mupando (Lungu).

12. **Millettia** sp.

Multi-stemmed shrub to 1.8 m tall. Branchlets glabrous with ridged bark. Leaves 13-foliolate; rachis plus petiole 13 cm long, glabrous; leaflets 2.3–5.5 × 1.2–2 cm, elliptic to narrowly oblong, acuminate at the apex, the actual tip rounded, cuneate at the base, glabrous; petiolules 2 mm long; stipels persistent, 2–2.5 mm long. Flowers not known, ultimate cymule stalks 5 mm long in fruit. Pods brown, 6.5–9.8 × 1.2–1.4 cm, linear-oblong, slightly margined, narrowed to the base, glabrous save for few scattered hairs towards the base, 2–3-seeded. Seeds red-brown, 8.5 × 8 × 2.5 mm, discoid, narrowed at margin.

Zambia. N: Mansa (Fort Rosebery), fr. 5.v.1964, *Fanshawe* 8569 (K, NDO). Miombo woodland on rocky scarp; c.1150 m.

The pods would indicate an affinity with *M. usaramensis*, but the material is not adequate for determining its true relationship. It could be a form of what is here treated as a variant of *M. eetveldeana* from the Mweru-Wantipa area.

STERILE MATERIAL

Seymour-Hall 59/51 in SRGH 35046 (K, SRGH): Zimbabwe, Chiredzi Dist., Chipinda Pools, 13.x.1951.

Tree to 7.8 m tall with white bark. Stems with nodular petiole bases and subpersistent stipules. Leaves 9–11-foliolate; petiole 3.5 cm long, densely yellow-brown spreading pubescent; rachis slender, c.9 cm long; terminal leaflet c.4 × 2 cm, narrowly obovate, lowest leaflets c.1.5 × 1.2 cm, rounded elliptic, the intermediate ones elliptic to oblong-elliptic, all rounded and very slightly emarginate at the apex; some of leaflets not quite opposite; leaflets dry pale grey-green, appressed-pubescent beneath, with close raised reticulum on both surfaces; stipels subulate.

Determined as *Millettia* sp.

Chase 6068 (BM, K, SRGH): Zimbabwe, Mutare (Umtali), rocky mountain sides at 960 m, 8.iv.1956.

Shrub or small tree to 3 m. Terminal leaflet up to 7.5 × 3 cm; leaflet-margin recurved; leaflets sometimes distinctly not quite opposite.

Same species as *Seymour-Hall* 59/51.

Chase 6073 (BM, K, SRGH): Zimbabwe, 18 miles S of Mutare (Umtali), 18.xii.1955.

Tree or shrub 2.7 m tall; all leaflets opposite, some ± acuminate but tip emarginate.

On rocky mountain sides among large trees at 930 m; seen over a number of years but never in flower. It appears to be the same species as *Seymour-Hall* 59/51.

Chase 7418 (K, SRGH): Mozambique, Amatongas Forest, grown at Mutare (Umtali), La Rochelle, Imbeza Valley, 1200 m, 26.i.1961.

Climber; young stems ferruginous pubescent. Leaves 9–11-foliolate; petiole 8.5 cm long; rachis 11.5 cm long; leaflets up to 11 × 4.5 cm, oblong-elliptic, appressed-pubescent beneath; petiolules 6 mm long; stipels broadly triangular.

This might be a cultivated *Derris* allied to *D. ferruginea* (Roxb.) Benth., but that has lateral nerves impressed above. It seems more likely to be a new indigenous plant, but flowers and fruits are needed for confirmation.

18. MUNDULEA (DC.) Benth.

by B. Verdcourt

Mundulea (DC.) Benth. in Miquel, Pl. Jungh.: 248 (1852).
Tephrosia Pers. sect. *Mundulea* DC., Prodr. **2**: 249 (1825).

Small, silky-pubescent trees and shrubs. Leaves imparipinnate; leaflets reticulately veined (rarely with numerous parallel nerves as in *Tephrosia* spp.), the blades mostly widest below the middle; stipules small; stipels absent. Flowers sometimes scented, blue-purple, red or rarely white, in terminal pseudoracemes; bracts small; bracteoles absent. Calyx teeth short, ± unequal, the upper 2 ± connate. Standard silky outside, with transverse basal callus, the claw short but sharply defined from the blade; wings falcate-oblong and keel incurved, obtuse, both pubescent at the margins near the base. Upper filament free and sharply bent near to base, often connate with sheath above; tips of free parts of filaments sometimes ± widened. Ovary sessile, many-ovuled; style hardened, tapering, glabrous; stigma terminal, minute. Pod linear, usually under 1 cm wide, densely hairy, pubescent (or glabrous in some Madagascan species), several-seeded, not or very tardily dehiscent or breaking irregularly (in some Madagascan species splitting into flat or spiralling valves), margins thickened. Seeds reniform, with aril.

A genus of 12 species all in Madagascar and one widespread in Africa and cultivated throughout the tropics. The genus links *Tephrosia* with *Millettia*, but to combine it with either would lead to problems of definition. Certainly the three cannot be united, but *Mundulea* could perhaps be considered a subgenus of *Tephrosia*.

Mundulea sericea (Willd.) A. Chev. in Compt. Rend. Hebd. Acad. Sci. Paris **180**: 1521 (1925). —Hepper in F.W.T.A., ed.2, **1**: 527 (1958). —White, F.F.N.R.: 160 (1962). —Sousa in C.F.A. **3**: 169 (1962). —Schreiber in Merxmüller, Prodr. Fl. SW Afrika, fam. 60: 90 (1970). —Gillett in F.T.E.A., Legum., Pap.: 155, fig.28 (1971). —Palmer & Pitman, Trees Sthn. Africa: 919–923 (1972). —K. Coates Palgrave, Trees Sthn. Africa, ed.2: 309 (1988). —M. Coates Palgrave, Trees Sthn. Africa: 376 (2002). Type: S India, Tranquebar, unknown collector (B holotype).

Cytisus sericeus Willd., Sp. Pl. **3**: 1121 (1802).

Tephrosia suberosa DC., Prodr. **2**: 249 (1825). —Klotzsch in Peters, Naturw. Reise Mossamb. **6**(1). 46 (1861). Type: plant cultivated in Calcutta, India, *Wallich* cat.no. 5628 (K).

Mundulea suberosa (DC.) Benth. in Miquel, Pl. Jungh.: 248 (1852). —Baker in F.T.A. **2**: 126 (1871). —Harms in Engler, Pflanzenw. Afrikas **3**(1): 591 (1915). —Baker f., Legum. Trop. Africa: 216 (1929).

Tephrosia petersiana Klotzsch in Peters, Naturw. Reise Mossamb. **6**(1): 584 (1864) & t.9 (1861) as "*T. suberosa*" in text. Type: Ilha de Mocambique, *Peters* s.n. (B† holotype, BM, K).

Small tree or many-stemmed deciduous shrub 0.5–7.5(12) m tall. Young stems velvety silky; bark yellow-brown or pale grey, or striate with ridges grey and fissures white, corky, smooth or fissured. Leaves grey or silvery, (5)9–21(29)-foliolate; petiole 1–2 cm long, silky-velvet; rachis 5–8 cm long, leaflets subcoriaceous, ± opposite, 1–6.5(9) × 0.4–3.3 cm, ovate, ovate-elliptic or elliptic to lanceolate, narrowly rounded at the apex, rounded at the base, minutely appressed hairy on both surfaces or glabrescent on upper surface, with 8–10 pairs of lateral nerves; stipules 1–3 mm long, triangular-lanceolate, velvety; petiolules 2 mm long. Flowers lilac, purple, mauve or pink, in pairs at each node of dense terminal velvety pseudoracemes 5–14 cm long; bracts c.0.5 mm long, triangular, persistent; pedicels 0.8–1.8 cm long. Calyx tube c.3 mm long, the 3 lower teeth c.3 mm, triangular, the upper pair of teeth almost completely united. Standard (1)1.8–2.2 × (1)1.5-1.6 cm, elliptic, emarginate, claw curved, c.3 mm long, inrolled; wing blades 15 × 5 mm, narrowly elliptic, the auricle and claw 2 mm long; blades of keel petals 14 × 5 mm, elliptic-oblanceolate, the auricle 1 mm long and claw 4 mm long. Stamen sheath (1)1.5 cm long, the free parts 3 mm long; upper filament free; anthers 0.5 mm long. Ovary linear, c.8-ovuled, velvety; style gently curved. Pod yellow-brown, (4)7–9(12.5) × 0.7–1.1 cm,

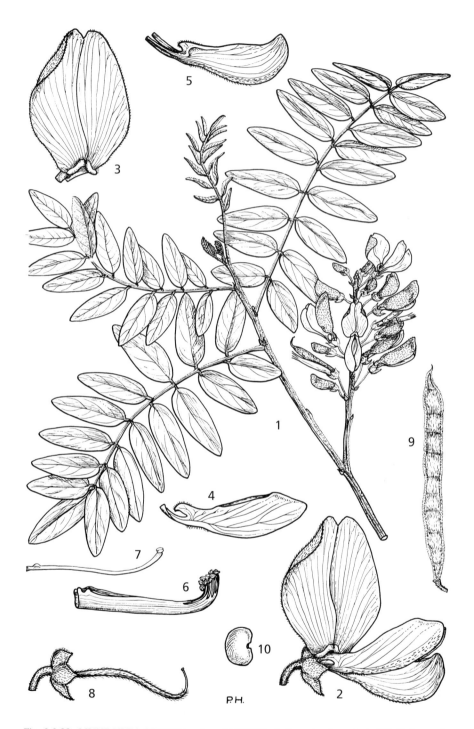

Fig. 3.3.**22**. MUNDULEA SERICEA subsp. SERICEA. 1, flowering branch (× ²⁄₃); 2, flower (× 2); 3, standard (× 2); 4, wing (× 2); 5, keel (× 2); 6, stamens (× 2); 7, upper stamen (× 2); 8, gynoecium (× 2), 1–8 from *Bally* 12181; 9, pod (× ²⁄₃); 10, seed (× 2), 9 & 10 from *Greenway* 4246. Drawn by Pat Halliday. From F.T.E.A.

linear, often constricted between the seeds, velvety, (1)4–10-seeded, indehiscent or eventually dehiscent after a long period. Seeds dark green, 4(5) × 3 mm, reniform.

Subsp. **sericea**. FIGURE 3.3.**22**.

Flowers large, 1.6–2.1 cm long.

Botswana. N: Ngamiland Dist., Toteng–Tsao road at 20°15'S, 22°54'E, fl.& fr. 21.x.1974, *P.A. Smith* 1123 (K, SRGH). SW: Ghanzi Dist., between D'Kar and Kuke, fr. 7.iii.1987, *Long & Rae* 136 (K). SE: Kweneng Dist., E of Molepolole, fl.& fr. 21.ix.1977, *Hansen* 3193 (C, GAB, K, PRE, SRGH). **Zambia**. C: Lusaka Dist., Kafue, river banks near Kafue R. road bridge, fr. 20.i.1964, *van Rensburg* 2804 (K). E: Chipata (Fort Jameson), fr. 31.v.1958, *Fanshawe* 4478 (K). S: Namwala Dist., Kafue Nat. Park, c.6.4 km N of Ngoma, fr. 23.iii.1964, *Mitchell* 25/10 (K). **Zimbabwe**. N: Hurungwe Dist., Kariba road at Savory's Folly, fl. 24.vii.1959, *Goodier* 568 (K, LISC, PRE, SRGH). W: Matobo Dist., Matopos Hills, Gladstone Farm, fl. 19.x.1958, *Darbyshire* 2716 (LISC, SRGH). C: Seke Dist., Mupfure (Umfuli) R., E of Beatrice, fl. 7.x.1956, *Robinson* 1807 (K, LISC, NDO, SRGH). E: Mutare Dist., 20.8 km S of Mutare (Umtali), fl. 21.xi.1948, *Chase* 1261 (BM, COI, K, LISC, SRGH). S: Chiredzi Dist., 3.2 km N of Chipinda Pools, fl. 17.xii.1959, *Goodier* 731 (K, PRE, SRGH). **Malawi**. C: Ntcheu Dist., Mphepozinai (Mphepo Zinai), fl. 23.xi.1967, *Salubeni* 904 (K, SRGH). S: Machinga Dist., W side of Chikala Hills, fr. 17.ii.1975, *Brummitt, Banda, Seyani & Patel* 14370 (K). **Mozambique**. N: Macomia Dist., road from Quiterajo to Mocímboa da Praia, fl. 12.ix.1948, *Andrada* 1343 (COI, LISC, LMA). Z: Gurué Dist., 6 km from Nintulo towards Lioma, fr. 10.ii.1964, *Torre & Paiva* 10526 (LISC). T: Zobué, fr. 28.vi.1947, *Hornby* 2769 (K, PRE, SRGH). MS: between R. Revué and Manica (Macequece), fl.& fr. 10.iii.1948, *Barbosa* 1157 (LISC). GI: Vilankulo Dist., 14.4 km S of Cheline on Maxixe–Nova Mambone road, fl.& fr. 5.x.1963, *Leach & Bayliss* 11835 (K, SRGH). M: Marracuene Dist., between Marracuene and Bobole, fl. 31.iii.1959, *Barbosa & Lemos* 8406 (COI, K, LISC, LMA, PRE).

Widespread in tropical Africa from Central African Republic to Sudan and Somalia, East Africa, Angola, Namibia and South Africa (North-West, Limpopo, Gauteng, Mpumalanga and KwaZulu-Natal Provinces); also Madagascar, India and Sri Lanka. Widely cultivated elsewhere in the tropics. Frequent at medium to low altitudes, in open woodland (miombo and *Combretum–Pteleopsis–Xeroderris, Acacia–Commiphora–Dichrostachys, Colophospermum–Combretum*, etc.), in *Baikiaea* forest and deciduous thickets on Kalahari sands, and in wooded grasslands on sandy soils; also on rocky hillsides, along streams and on wooded dunes; 0–1440 m.

Conservation notes: Widespread; Lower Risk, Least Concern.

A well known fish poison, the active principle being a rotenone; wood also used for tool handles. According to Gillett in F.T.E.A., its natural range has been greatly extended owing to cultivation as a fish poison.

A subspecies with flowers 1.1–1.4 cm long occurs in Madagascar.

19. TEPHROSIA Pers.

by R.K. Brummitt

Tephrosia Pers., Syn. Pl. **2**: 328 (1807) nom. conserv. —Brummitt in Bol. Soc. Brot., sér.2, **41**: 219–393 (1968); in Kew Bull. **35**: 459–473 (1980).
> *Caulocarpus* Baker f., Legum. Trop. Africa: 169 (1926).
> *Lupinophyllum* Hutch., Gen. Fl. Pl. **2**: 626 (1967).

Annual or perennial herbs, or softly woody shrubs, rarely small trees. Leaves usually imparipinnate, less often 1-foliolate, rarely palmately 3–7-foliolate; leaflets usually narrowest at the base and widest above the middle, with the lateral nerves parallel, running through to the margin and often united there to form a well-developed marginal nerve; stipels absent except in palmately-leaved species. Inflorecences usually in terminal and leaf-opposed or axillary pseudoracemes, with flowers clustered at the nodes, the inflorescences sometimes paniculate or contracted and dense, or the flowers clustered (elsewhere rarely singly) in the axils of upper leaves; bracteoles generally absent. Calyx 5-lobed, the upper pair of lobes joined higher, the lowest lobe often longest. Petals usually reddish-purple, sometimes salmon-pink or red or yellowish-orange; standard pubescent or silky outside, with a well-defined claw; keel petals oblong-elliptic or oblong-falcate, with a distinct claw and auriculate at the base of the blade, joined along the lower margin, slightly adhering to the wings. Upper filament lightly attached to the others or less often free, widened and often arched at the base; free parts relatively short, not widened at the tip; anthers dorsifixed. Disk usually present around the base of the ovary. Ovary usually sessile, 1–22-ovulate; style sharply or gradually upcurved, linear or tapering, sometimes twisted, glabrous or pubescent; stigma a transverse line, punctate or minutely capitate, in the last case often pencil-like. Pod usually linear to oblong, variously beaked, dehiscent. Seeds oblong-reniform, with a small hilum; aril developed to varying degrees; radicle incurved.

A large genus of about 400–500 species widespread in the tropics and subtropics.

Two subgenera are recognized, distinguished most easily by the pubescence of the style (see Gillett in Kew Bull. **13**: 414–419 (1959) and Brummitt in Kew Bull. **35**: 460, 1980). Subgenus *Tephrosia* (species 1–30) has the style glabrous except for an apical penicillate tuft of hairs, and has the wings and keel curving upwards (the keel often appearing L-shaped). Subgenus *Barbistyla* Brummitt (species 32–70) has the style pubescent throughout at least one of its surfaces, the wings and keel more forwardly directed. Species 31 is of uncertain position. The style character is not always easy to see, but identification will usually be much more confidently made if it can be ascertained. Other characters are often more obvious, and the following list will assist in rapid identification of some species.

While the pattern of variation in this genus makes a key somewhat difficult to construct and use, there are some characters which are easily observed and which can be found in just a few species. The more obvious characters are detailed below with an indication of which species may show them. Where these are not constant in a species, this is indicated by brackets. A dichotomous key to the species follows.

Native status

— introduced probably as a fish poison: 4 *noctiflora*, 34 *elata* subsp. *elata*, 49 *candida*, 50 *vogelii*

Habit

— suffrutex with erect unbranched stems from rootstock: (9 *elongata*), (24 *aurantiaca*), 25 *hockii*, (39 *longipes* var. *lurida*), 48 *laxiflora*, 65 *zambiana*, 66 *dasyphylla*, 67 *muenzneri*

— plant tufted, with several much branched stems ascending from near the base of the plant: 10 *dregeana*, (13 *malvina*), (14 *micrantha*), 44 *euprepes*, 45 *ringoetii*, (46 *stormsii*)

— bushy shrub 2 m high to small tree: (31 *miranda*), 36 *interrupta*, (41 *faulknerae*), (49 *candida*), (50 *vogelii*), (51 *aequilata*), 54 *montana*, 55 *praecana*, 56 *festina*, (57 *chimanimaniana*), 58 *grandibracteata*

Young stems and inflorescence axes

— villous, with dense ascending or spreading hairs: 5 *villosa*, 6 *rhodesica*,17 *coronilloides*, 21 *linearis* var. *discolor*, (22 *richardsiae*), (24 *aurantiaca*), (35 *nyikensis*), (37 *bracteolata*), 38 *nana*, (42 *reptans*), (43 *caerulea*), 50 *vogelii*, (51 *aequilata*), 59 *rupicola*, 65 *zambiana*, 66 *dasyphylla*

Stipules
— more than 2 mm broad: 34 *elata* var. *abercornensis*, (37 *bracteolata*), (38 *nana*), (41 *faulknerae*), 50 *vogelii*, 51 *aequilata*, 52 *robinsoniana*, 53 *whyteana*, 55 *praecana*, 59 *rupicola*, 60 *meisneri*, 61 *brummittii*, 62 *cordata*, 64 *radicans*
— ovate, often conspicuously auriculate: 59 *rupicola*, 60 *meisneri*, 61 *brummittii*, 62 *cordata*, 64 *radicans*

Leaves
— unifoliolate: (26 *paniculata*), 27 *acaciifolia*, 28 *kindu*, 29 *forbesii*, (39 *longipes* var. *drummondii*), 66 *dasyphylla* subsp. *amplissima*, 67 *muenzneri*
— digitately 3-5-foliolate (leaflets all arising from same point): 1 *lupinifolia*, 48 *laxiflora*, 70 *chisumpae*
— with 1–2 pairs of lateral leaflets: 1 *lupinifolia*, 2 *pentaphylla*, (3 *uniflora*), 9 *elongata*, (10 *dregeana*), 11 *limpopoensis*, (26 *paniculata*), (39 *longipes* var. *drummondii*), (44 *euprepes*), (45 *ringoetii*), (46 *stormsii*), (47 *paradoxa*), (48 *laxiflora*), (63 *gobensis*), 65 *zambiana*, 66 *dasyphylla*, 68 *cephalantha*, 69 *tanganicensis*, 70 *chisumpae*
— with more than 10 pairs of leaflets: (6 *rhodesica*), (34 *elata*), 37 *bracteolata*, (49 *candida*), (50 *vogelii*), (51 *aequilata*), (54 *montana*), (57 *chimanimaniana*), (58 *grandibracteata*)
— subsessile: 27 *acaciifolia*, 28 *kindu*, 29 *forbesii*, 67 *muenzneri*, 70 *chisumpae*

Leaflets
— more or less linear, more than 10 times as long as broad: (1 *lupinifolia*), 9 *elongata*, (10 *dregeana*), 21 *linearis*, 28 *kindu*, (29 *forbesii*), (37 *bracteolata*), 39 *longipes*, (44 *euprepes*), 45 *ringoetii*, (46 *stormsii*), (47 *paradoxa*), 48 *laxiflora*
— less than 3 mm broad: (9 *elongata*), 10 *dregeana*, 11 *limpopoensis*, 21 *linearis*, (23 *curvata*), (39 *longipes*), (46 *stormsii*), (47 *paradoxa*)
— markedly obtriangular: 16 *burchellii*, (19 *argyrotricha*), 38 *nana*, 63 *gobensis*, 64 *radicans*, 70 *tanganicensis*
— silvery-silky beneath at maturity: 2 *pentaphylla*, (3 *uniflora*), 12 *purpurea* subsp. *canescens* and *dunensis*, (14 *micrantha*), (17 *coronilloides*), 18 *lepida*, 19 *argyrotricha*, 20 *decora*, 21 *linearis*, 22 *richardsiae*, 24 *aurantiaca*, 28 *kindu*, (51 *aequilata*), 52 *robinsoniana*, 65 *zambiana*, 66 *dasyphylla*

Inflorescences
— some subterranean: 1 *lupinifolia*
— flowers all in axils of foliage leaves: 2. *pentaphylla*, 3. *uniflora*, (28 *kindu*), 29 *forbesii*, (31 *miranda*)
— more or less sessile terminal clusters surrounded by leaves: (24 *aurantiaca*), (25 *hockii*), 41 *faulknerae*, (50 *vogelii*), 51 *robinsoniana*, (53 *whyteana*), (54 *montana*), 55 *praecox*, 56 *festina*, 57 *chimanimaniana*, 58 *grandibracteata*, 59 *rupicola*, 63 *gobensis*, 64 *radicans*, 66 *dasyphylla*, 67 *muenzneri*, 68 *cephalantha*, 69 *tanganicensis*, 70 *chisumpae*
— compact clusters or heads on distinct peduncles: 36 *interrupta*, 39 *longipes* var. *swynnertonii*, (45 *ringoetii*), 60 *meisneri*, 61 *brummittii*
— with bracts more than 2 mm broad: 50 *vogelii*, (54 *montana*), 55 *praecana*, 56 *festina*, 57 *chimanimaniana*, 58 *grandibracteata*, 59 *rupicola*, 60 *meisneri*, 61 *brummittii*, 62 *cordata*, 64 *radicans*

Standard petals
— white: (4 *noctiflora*), (23 *curvata*), (50 *vogelii*), 68 *cephalantha*
— yellow to orange or apricot and 16-24 mm long: 24 *aurantiaca*, 25 *hockii*, 28 *kindu*, 69 *tanganicensis*

— varying from yellowish or orange to salmon or pinkish-purple, usually less than 18 mm long: 9 *elongata*, 17 *coronilloides*, 18 *lepida*, 19 *argyrotricha*, 20 *decora*, 21 *linearis*, 22 *richardsiae*, (23 *curvata*), (26 *paniculata*), 27 *acaciifolia*

— mostly 20–34 mm long: 25 *hockii*, (36. *interrupta*), 49 *candida*, 50 *vogelii*, (51 *aequilata*), (53 *whyteana*), (54 *montana*), 55 *praecana*, 58 *grandibracteata*, (59 *rupicola*), (61 *brummittii*), 66 *dasyphylla*

— less than 10 mm long: 1 *lupinifolia*, 2 *pentaphylla*, 3 *uniflora*, (6 *rhodesica*), 7 *polystachya*, 8 *multijuga*, 10 *dregeana*, 11 *limpopoensis*, 13 *malvina*, 14 *micrantha*, 15 *pumila*, (17 *coronilloides*), 19 *argyrotricha*, 20 *decora*, 21 *linearis*, 27 *acaciifolia*, (29 *forbesii*), (32 *punctata*)

Style

— glabrous except for a penicillate apical tuft; spp. 1-30

— pubescent along at least one surface, without penicillate apical tuft: spp. 31- 70

— twisted through 180° near base: 4 *noctiflora*, 5. *villosa*, (27 *acaciifolia*), (28 *kindu*)

— deflexed at right angles: 15 *pumila*

— often reflexed above the pod: (18 *lepida*), 24 *aurantiaca*, 30 *zoutpansbergensis*, 31 *miranda*, (50 *vogelii*), 51 *aequilata*, 59 *rupicola*, 62 *cordata*, 66 *dasyphylla*, 68 *cephalantha*, 70 *chisumpae*

Pods

— papery, contorted: see *Ptycholobium*

— mostly more than 9 cm long: (25 *hockii*), (38 *interrupta*), (49 *candida*), 50 *vogelii*

— mostly less than 3.5 cm long: 1 *lupinifolia*, 8 *multijuga*, 10 *dregeana*, 14 *micrantha*, 23 *curvata*, (29 *forbesii*), 30 *zoutpansbergensis*, (51 *aequilata*), (52 *robinsoniana*), 64 *radicans*

— with surfaces appressed-pubescent to villous and the margins with markedly darker hairs: 4 *noctiflora*, 18 *lepida*, 19 *argyrotricha*, (21 *linearis*), 22 *richardsiae* subsp *erucifera*, (26 *paniculata*), 27 *acaciifolia*, 28 *kindu*, (32 *punctata*), 66 *dasyphylla*, (68 *cephalantha*), (70 *tanganicensis*)

— more than 7 mm broad: (4 *noctiflora*), 25 *hockii*, 49 *candida*, 50 *vogelii*, (51 *aequilata*), 52 *robinsoniana*, (54 *montana*), 55 *praecana*, 56 *festina*, (57 *chimanimaniana*), (58 *grandibracteata*), 59 *rupicola*, 60 *meisneri*, 61 *brummittii*, 62 *cordata*, 63 *gobensis*, 66 *dasyphylla* 67 *muenzneri*, 69 *tanganicensis*, 70 *chisumpae*

Seeds

— 1 per pod: (23 *curvata*), *zoutpansbergensis*, see also *Requienia*

— 2–3 per pod: 10 *dregeana*, (23 *curvata*), (51 *aequilata*), (52 *robinsoniana*)

— more than 16 per pod: (37 *bracteolata*), (38 *nana*), (39 *longipes*), (47 *paradoxa*), 50 *vogelii*

— usually 12–15 per pod: 3 *uniflora*, (35 *nyikensis*), (36 *interrupta*), 37 *bracteolata*, 39 *longipes*, 41 *faulknerae*, 42 *reptans*, 43 *caerulea*, 44 *euprepes*, 45 *ringoetii*, 46 *stormsii*, 60 *meisneri*, 61 *brummittii*

— elongated transversely in the pod: (52 *robinsoniana*), 59 *rupicola*, 60 *meisneri*, 61 *brummittii*, 62 *cordata*, 63 *gobensis*, 64 *radicans*, 65 *dasyphylla*, 68 *cephalantha*, 69 *tanganicensis*, 70 *chisumpae*

— more or less obliquely elongated: 37 *bracteolata*, 38 *nana*, (52 *robinsoniana*), 67 *muenzneri*

Key to species

1. Leaves all 1-foliolate ··· 2
  – Leaves mostly 3-foliolate to pinnate or digitate ····················· 8
2. Leaves more than 2.5 cm broad; unbranched erect suffrutices with flowers in subsessile clusters ··· 3
  – Leaves less than 2.5 cm broad; or if more then stems branched and most flowers in elongate racemes ··· 4
3. Leaves petiolate (W Zambia) ··············· **66.** *dasyphylla* subsp. *amplissima*
  – Leaves sessile (E Zambia) ···················· **67.** *muenzneri* subsp. *pedalis*
4. Style pubescent; petiole more than 2 cm (E Zimbabwe) ·················
    ·· **39.** *longipes* var. *drummondii*
  – Style glabrous, penicillate; petiole less than 2 cm ····················· 5
5. Petiole 3–15 mm; leaves usually more than 1.5 cm broad ······ **26.** *paniculata*
  – Petiole 1–2 mm; leaves rarely more than 1.5 cm broad ·················· 6
6. Leaves up to 6 mm broad, linear, the lower surface with dense silky hairs obscuring the veins ··· **28.** *kindu*
  – Leaves 4–15 mm broad, linear to linear-oblong, venation clearly visible beneath ·· 7
7. Flowers mostly in lax terminal racemes, only the lowermost sometimes in leaf axils ·· **27.** *acaciifolia*
  – Flowers all in leaf axils ································· **29.** *forbesii*
8. Leaves digitately 3–5-foliolate (all leaflets arising from same point) ········ 9
  – Leaves 3-foliolate to pinnate, not digitate ························· 11
9. Branching shrub to 1.5 m; leaves sessile ················· **70.** *chisumpae*
  – Procumbent herb or erect suffrutex; petioles 3–13 mm long ············ 10
10. Procumbent herb, often producing underground inflorescences ·· **1.** *lupinifolia*
  – Suffrutex with simple erect stems, without underground inflorescences ·······
    ·· **48.** *laxiflora*
11. Pods with 1 or rarely 2 seeds; shrub 1–2 m high (Botswana and Limpopo Valley)
    ··· **30.** *zoutpansbergensis*
    Pods with 2–many seeds; habit various ························· 12
12. Slender annual with usually 2–4 pairs of linear or linear-oblong leaflets; pods mostly 20–25 mm long with 2–6 seeds (Botswana) ············ **10.** *dregeana*
  – Not as above ··· 13
13. Stems virgate, robust; plant annual to 2 m or shrub or small tree to 4 m; flowers 11–28 mm long; pods mostly 4.5–10 cm long with 8–14 seeds (montane areas)
    ·· 14
  – Not as above ··· 15
14. Annual to 2 m high; flowers 11–15(17) mm long ·············· **35.** *nyikensis*
  – Woody shrub or small tree to 4m; flowers 17–28 mm long ······· **36.** *interrupta*
15. Shrub to 1.7 m with leaflets silver-silky on both surfaces; pods c.35 × 9 mm, with 3–5 seeds elongated transversely or obliquely (C Zambia) ···· **52.** *robinsoniana*
  – Not as above ··· 16
16. Bushy shrubs to small trees 2–5 m high, if less than 2 m high then inflorescence a dense terminal cluster surrounded by leaves (often in high montane areas or introduced) ··· 17
  – Annual to perennial herbs, suffrutices and subshrubs ················· 28
17. Standard white or tinged pinkish, occasionally purple in *T. vogelii* but then pods 13–17 mm broad (fish poisons, cultivated and naturalised) ············· 18
  – Standard blue to purple or pink ···································· 20

18. Stems, leaf rachides and pods villous; inflorescence bracts up to 16 × 13 mm · **50.** *vogelii*
–   Stems, leaf rachides and pods shortly pubescent to tomentose · · · · · · · · · · · 19
19. Style glabrous · **4.** *noctiflora*
–   Style pubescent · **49.** *candida*
20. Shrubs with numerous regularly alternate, axillary elongate inflorescences below the terminal one (N Mozambique) · 21
–   Shrubs to small trees; inflorescences terminal, subsessile and scarcely exceeding leaves (montane areas) · 22
21. Pods c.35 × 7.5 mm, with 3–5 seeds · **31.** *miranda*
–   Pods 50–64 × 4–5 mm, with 12–14 seeds · · · · · · · · · · · · · · · · · **41.** *faulknerae*
22. Bracts to 3(4) mm broad, linear-lanceolate to ovate · · · · · · · · · · · · · · · · · · · 23
–   Bracts 4–13 mm broad, obtriangular or broadly elliptic to suborbicular-acuminate or sometimes ovate · 25
23. Young stems and leaf rachides shortly brown-tomentose; pods 70–80 × 7 mm, brown-tomentose · **54.** *montana*
–   Young stems and leaf rachides not shortly brown-tomentose, though sometimes with long villous hairs; pods 20–60 × 5–7 mm, appressed-pubescent to villous · 24
24. Inflorescences dense, the nodes not clearly visible; bracts and stipules persistent; pods 20–40 × 5–7 mm, with 3–5 seeds · **51.** *aequilata*
–   Inflorescences fairly lax with several nodes clearly visible; bracts and stipules falling early; pods 40–60 × 6 mm, with c.8 seeds · · · · · · · · · · · · · · **53.** *whyteana*
25. Young stems, leaf rachides and bracts clothed with short fine closely-appressed golden-brown hairs; pods closely appressed-pubescent · · · · · **58.** *grandibracteata*
–   Young stems, leaf rachides and bracts variously pubescent but not as above; pods tomentose to villous · 26
26. Bracts and calyces villous with long spreading hairs (Chimanimani Mts) · **57.** *chimanimaniana*
–   Bracts and calyces shortly tomentose · 27
27. Bracts broadly obtriangular and slightly apiculate; calyx 6–9 mm; petals 21–22 mm; pods 72–78 × 7–8 mm, shortly brown-tomentose · · · · · · · · · · **55.** *praecana*
–   Bracts suborbicular-acuminate; calyx 9–12 mm; petals 13–17 mm; pods 52–60 × 7–8 mm, lanate-tomentose · **56.** *festina*
28. Stipules conspicuous, 3–14 mm broad, broadly ovate and ± cordate at base, usually dark purple; inflorescence compact on a long peduncle; pods with 8–16 transversely elongate seeds · 29
–   Stipules not markedly conspicuous, up to 4 mm broad, linear to triangular, not cordate at base, usually not dark purple; inflorescence various; pods with various seeds but if these transverse then fewer than 8 · 31
29. Perennial herb with prostrate stems from crown of a woody rootstock · **61.** *brummittii*
–   Subshrubs to shrubs 0.5–3 mm high · 30
30. Compact woody shrub; stipules 9–17 × 7–14 mm · · · · · · · · · · · · · · · **62.** *cordata*
–   Lax subshrub; stipules 7–13 × 3–8 mm · **60.** *meisneri*
31. Standard yellow to orange or apricot, 16–24 mm long; leaves mostly with 2–4 pairs of leaflets · 32
–   Standard blue, purple, red, pink or white, if yellowish to orange then less than 16 mm long; leaves various · 34
32. Annual; flowers in terminal clusters surrounded by the upper leaves; pods densely villous · **69.** *tanganicensis*

– Perennial with woody rootstock; flowers in elongate racemes; pods pubescent to villous · 33

33. Leaflets strongly discolorous, densely pubescent to silky beneath; petiole 10–16 mm; flowers 16–21 mm · **24.** *aurantiaca*
– Leaflets not or weakly discolorous, sparsely pubescent beneath; petiole 18–42 mm; flowers mostly 20–24 mm · **25.** *hockii*

34. Leaves 3-foliolate; leaflets ± linear, up to 12 mm broad, if silky beneath then less than 5 mm broad (Zimbabwe and Zomba Plateau) · · · · · · · · · · · · · · · · · · 35
– Leaves pinnate, or if 3-foliolate then at least the terminal leaflets 10–23 mm broad and usually silky beneath · 36

35. Style glabrous · **9.** *elongata*
– Style pubescent · **39.** *longipes*

36. Leaves 3-foliolate or some 5-foliolate, leaflets mostly 10–23 mm broad and usually silky beneath · 37
– Leaves pinnate; leaflets various · 40

37. Annual or perennial; stems branched above; standard white or pinkish to salmon or orange · 38
– Suffrutex with simple erect stems from a woody stock; standard blue to purple or pinkish · 39

38. Branching perennial herb to subshrub with erect, decumbent or scrambling stems up to 3.3 m; standard orange, salmon, red or purplish · · · **26.** *paniculata*
– Erect annual; standard white or sometimes pinkish · · · · · · · · · · **68.** *cephalantha*

39. Flowers in ± elongate terminal raceme, or only the lowermost in axil of a foliage leaf; leaves densely silky beneath (NE Zambia) · · · · · · · · · · · · · · **65.** *zambiana*
– Flowers in a dense terminal cluster surrounded by the upper foliage leaves; leaves moderately silky beneath · **66.** *dasyphylla*

40. Pods 12–35 × 6–9 mm, with transversely elongate seeds; distance between middle of adjacent seeds less than pod width; leaflets obovate or obtriangular, usually in 3–4 pairs (in Zimbabwe more than 4) · 41
– Not as above · 43

41. Herb with several stems prostrate up to 1.6 m from rootstock; pods with 2–4(5) seeds (widespread, weedy) · **64.** *radicans*
– Erect subshrub to 1.5 m; pods with 3–9 seeds · 42

42. Stems with brownish hairs; pods with 7–9 seeds (Zimbabwe) · · · · · **59.** *rupicola*
– Stems with grey appressed pubescence; pods with 3–5 seeds (S Mozambique) · **63.** *gobensis*

43. Flowers in axils of foliage leaves; leaves usually with 2–4 pairs of markedly discolorous leaflets (under surface much lighter than upper) · · · · · · · · · · · 44
– Flowers in racemes or terminal clusters; leaves various · · · · · · · · · · · · · · · · · 45

44. Stems and leaf rachides appressed-pubescent; rachis and petiole together 3–14 mm; seeds 6–9 per pod · **2.** *pentaphylla*
– Stems and leaf rachides villous with ascending hairs; rachis and petiole together 10–30(35) mm; seeds 9–14 per pod · **3.** *uniflora*

45. Low bushy shrub to 50 cm high, young stems covered with dense appressed whitish hairs (Limpopo Valley) · 46
– Not as above · 47

46. Pods c.3.2 cm long with 7–8 seeds; style glabrous · · · · · · · · · · · **11.** *limpopoensis*
– Pods mostly 6–6.5 cm long with 11–13 seeds; style pubescent · · · · · **40.** *euchroa*

47. Style pubescent, not terminated by a tuft of penicillate hairs · · · · · · · · · · · · · 48
– Style glabrous except for a terminal tuft of penicillate hairs · · · · · · · · · · · · · 64

48. Leaflets in 4–15 pairs, elliptic to linear-oblong, the larger ones more than 8 mm broad, or if narrower then in more than 7 pairs · 49

–   Leaflets in 1–7 pairs, linear-oblong to linear, the larger ones 2–9(10) mm broad
 · 54

49.   Flowers 8–12 mm long; seeds 6–10 per pod · 50

–   Flowers (10)13–19 mm long; seeds 10–19(21) per pod · · · · · · · · · · · · · · · 51

50.   Calyx 2–3 mm long, the ventral part with pellucid gland dots; pods with 8–10 seeds · **32.** *punctata*

–   Calyx 3.5–6 mm long, without pellucid dots; pods with 7–8 seeds · · **33.** *kasikiensis*

51.   Leaflets 3–7(11) mm broad, linear to linear-oblong; bracts 1–3 mm broad, ovate, purplish, fairly conspicuous though often falling early · · · · · · · · **37.** *bracteolata*

–   Leaflets 5–18(25) mm broad, elliptic to linear-oblong; bracts up to 1 mm broad, linear-triangular, not conspicuous   · 52

52.   Perennial with vertical rootstock; leaflets up to 12(14) mm broad, linear-oblong to linear · **43.** *caerulea*

–   Annual with tap root; leaflets 5–18(25) mm broad, elliptic or obovate to linear-oblanceolate · 53

53.   Calyx shortly appressed-pubescent, the tube 2.5 mm long; pods ± pale straw-coloured with 9–12(13) seeds · **34.** *elata*

–   Calyx ± villous, the tube 1–2 mm long; pods ± brown with 12–19 seeds · · **38.** *nana*

54.   Stems with spreading hairs at least in the lower parts · · · · · · · · · · · · · · · · · 55

–   Stems glabrous or appressed- to ascending-pubescent · · · · · · · · · · · · · · · 59

55.   Annual, tufted, stems often much branched just above base, with persistent spreading stipules there and more lax branches above   · · · · · · · · · · · · · · · 56

–   Perennial with a woody stock, if annual then with robust strongly decumbent stems up to 1 m long · 58

56.   Inflorescence axis terminated by a ± dense head of flowers with numerous persistent bracts (often also with flowers at 1–3 nodes below head) · · **45.** *ringoetii*

–   Inflorescence elongate, without a terminal head · 57

57.   Plant usually 20–50 cm high, branching near base only; stem hairs usually brown; pods with long appressed hairs as well as short ones (Kalahari and lowveld areas) · **44.** *euprepes*

–   Plant usually 60–100 cm high, stems branching irregularly for most of length; stem hairs grey; pods shortly pubescent with irregular geniculate hairs · **46.** *stormsii* var. *pilosa*

58.   Stems decumbent; hairs c.1 mm; petiole 0.2–2.4(4.4) cm · · · · · · · · · **42.** *reptans*

–   Stems erect; stem hairs up to 2.5 mm; petiole (1.2)2.5–5 cm (Kalahari) · **43.** *caerulea*

59.   Suffrutex with stems simple or 1(2)-branched from a woody rootstock; leaves with (0)1–3(4) leaflet pairs with veins prominent on both surfaces, the lower surface subglabrous to sparsely and shortly appressed-pubescent · · **48.** *laxiflora*

–   Stems regularly branched and up to 1.8 m high, or decumbent, if simple to 1(2)-branched then stems and leaflet undersurface appressed-pubescent, veins not prominent on both surfaces, variously pubescent beneath · · · · · · · · · · · · · · 60

60.   Leaflets to 3.2(4) cm long, oblanceolate to linear-oblong   · · · · · · · · **42.** *reptans*

–   Leaflets mostly 4–14 cm long, linear or linear-oblong · · · · · · · · · · · · · · · · 61

61.   Leaflets linear-oblong, mostly 4–7.5 × 0.5–1.2 cm; petiole and rachis together usually 12–18 cm (Kalahari) · **43.** *caerulea*

–   Leaflets linear, mostly 4–14 × 0.2–0.9 cm, if linear-oblong then petiole and rachis together 2–6(8) cm   · 62

62.   Racemes terminal and leaf-opposed; stems strongly appressed- or ascending-pubescent · **39.** *longipes*

–   Racemes terminal and axillary; stems glabrous to sparsely appressed-pubescent (inflorescence sometimes densely pubescent) · 63

63. Lateral racemes slender (axis 0.2–1 mm diameter) and curving upwards, the axes glabrous to shortly pubescent; calyx 3–5.5 mm, appressed-pubescent; stipules (3)4–12 × 0.3–0.8 mm · **46.** *stormsii*
   – Lateral racemes stout (axis 1–2 mm diameter) and ± stiffly erect, the axes densely pubescent to villous; calyx (4)5–10 mm, densely pubescent to villous; stipules (8)13–24 × c.1.2 mm · **47.** *paradoxa*
64. Pods with 1–2 seeds, tapered towards the seedless base; standard usually white, rarely tinged pinkish or yellowish · **23.** *curvata*
   – Pods with more than 2 seeds, not tapered to the base; standard not white · · 65
65. Pods with style bent sharply downwards at 90°; distance between centres of adjacent seeds less than breadth of pod; leaflets mostly in 3–5 pairs · · **15.** *pumila*
   – Pods with style not or only slightly bent downwards; distance between centres of adjacent seeds equalling or exceeding breadth of pod; leaflets in 3–11 pairs · 66
66. Calyx teeth 2–4 times as long as tube; leaflets obtriangular to oblanceolate (Kalahari or rare in C Zimbabwe) · 67
   – Calyx teeth up to 1¹⁄₂ times as long as tube; leaflets various · · · · · · · · · · · · 68
67. Annual or short-lived perennial with decumbent or ascending stems; stems, leaf rachides and calyx grey-pubescent or glabrescent · · · · · · · · · · · · ·**16.** *burchellii*
   – Perennial with woody base and stems erect to 25–45 cm; stems, leaf rachides and calyx brown-pubescent · **17.** *coronilloides*
68. Pod surface thinly pubescent with short appressed or spreading hairs, hairs at margins not markedly darker; stems subglabrous to appressed-pubescent · · 69
   – Pod surface densely appressed-pubescent to villous, hairs often matted together, the margins often markedly darker; upper stems usually tomentose to villous with ascending to spreading hairs, if pubescence intermediate then leaflets silver beneath · 73
69. Much-branched trailing to erect perennial herb up to 1.5(2) m high; leaves with 5–11 pairs of leaflets (mountains of S Mozambique) · · · · · · · · · · · · · · · · · 70
   – Stems various but seldom exceeding 60 cm; leaves with 3–10 pairs of leaflets 71
70. Leaflets mostly 2–10 mm broad, space between them about equalling leaflet width, calyx 2.5–7 mm long · **7.** *polystachya*
   – Leaflets mostly 1–4 mm broad, space between them about twice leaflet width; calyx c.2 mm long · **8.** *multijuga*
71. Stems usually tufted with several much-branched stems ascending from plant base, occasionally more elongate or mat-forming; leaflets often grey or silver beneath; petals 4–5(6) mm long · **14.** *micrantha*
   – Stems usually procumbent, occasionally erect, but not tufted; leaflets not grey or silver beneath; petals (5)6–10 mm long · 72
72. Pods 3–4.5(5) mm broad, dark or dull brown, with 6–10 seeds; inflorescence axis slender, not winged; petiole usually shorter than lower pair of leaflets · **12.** *purpurea*
   – Pods 4–5 mm broad, ± straw-coloured, with (6)7 seeds; inflorescence axis relatively stout and often narrowly winged; petiole usually longer than lower pair of leaflets · **13.** *malvina*
73. Pods 4.5–7 mm broad; style twisted through 180° near base; petals 9–17 mm long · 74
   – Pods 3–5 mm broad; style not twisted; petals 6–12 mm or if longer then either with flowers in a dense terminal cluster (*T. richardsiae*) or pods with silver hairs on surface and very dark hairs on margins (*T. lepida*) · · · · · · · · · · · · · · · · · 75
74. Pods with light brown pubescence on surface and darker brown hairs on the margins (probably introduced, occasionally naturalised) · · · · · · · · · **4.** *noctiflora*

– Pods uniformly grey-villous · **5.** *villosa*
75. Flowers 12–18 mm long, mostly in dense terminal clusters scarcely exceeding leaves (some also in axils of upper leaves); pods villous with densely matted hairs · **22.** *richardsiae*
– Flowers 6–15(16) mm long, in lax racemes exceeding the leaves; pods appressed-pubescent to loosely villous · 76
76. Leaflets 1–4 mm broad, linear, 9–15 times as long as broad · · · · · · · **21.** *linearis*
– Leaflets 2–10 mm broad, narrowly elliptic to oblanceolate, 3–6 times as long as broad · 77
77. Upper stems usually villous with brown or grey hairs; pods with uniform brown tomentum; seeds 5–7 per pod · **6.** *rhodesica*
– Upper stems appressed-pubescent with grey or silver hairs; pods with silver, grey or brown hairs on the surface, usually much darker hairs on margins; seeds 8–15 per pod · 78
78. Petals (10)12–16 mm long · **18.** *lepida*
– Petals 6–9 mm long · 79
79. Pods 3–3.5 mm broad, silvery-silky on surface and dark brown or blackish-pubescent at margins; petiole 1–3 mm (N Mozambique) · · · · · **19.** *argyrotricha*
– Pods 3.5–4.5 mm broad, only thinly pubescent or sometimes densely so at margins only; petiole (2)3–22 mm · **20.** *decora*

1. **Tephrosia lupinifolia** DC., Prodr. **2**: 255 (1825). —Harvey in F.C. **2**: 204 (1862). —Baker in F.T.A. **2**: 107 (1871). —Eyles in Trans. Roy. Soc. S. Africa **5**: 375 (1916). —Baker f., Leg. Trop. Africa **1**: 183 (1926). —Young in Ann. Transv. Mus. **14**: 398 (1932). —Cronquist in F.C.B. **5**: 90 (1954). —Torre in C.F.A. **3**: 148 (1962). Type: South Africa, Northern Cape, near source of Kuruman R., between Little Klobbokhonni & Fig Tree Rock , *Burchell* 2488 (G holotype, K). FIGURE 3.3.**23**
    Tephrosia digitata DC., Prodr. **2**: 255 (1825). Type from West Africa (Senegal).
    Tephrosia lupinifolia var. *digitata* (DC.) Baker in F.T.A. **2**: 107 (1871). —Eyles in Trans. Roy. Soc. S. Africa **5**: 375 (1916). —Baker f., Legum Trop. Africa **1**: 183 (1926).
    Tephrosia laevigata Baker in Oliver, F.T.A. **2**: 107 (1871). —R.E. Fries, Wiss. Ergebn. Schwed. Rhod.-Kongo-Exped. **1**: 81 (1914). —Harms in Engler, Pflanzenw. Afrika **3**(1): 587 (1915). —Baker f., Legum Trop. Africa **1**: 183 (1926). —Torre in C.F.A. **3**: 148 (1962). Type from Angola (Huila).

Perennial with prostrate branches up to 100 cm long arising from a woody vertical stock, often with subterranean inflorescences. Stems puberulous to densely spreading-pubescent. Leaves digitate with (1)3–7 leaflets and a pair of stipellae (very rarely an occasional leaf produced with a short rachis between a lower pair of leaflets and 3 digitate terminal leaflets); petiole 3–100(130) mm long (see note below), variously pubescent; leaflets (25)32–70(95) × (3)5–15 mm, obovate to linear, from 2.5 to 16 times as long as broad, rounded to subacute at the apex, the upper surface glabrous or very rarely appressed-pubescent, the lower surface appressed or spreading-pubescent; stipellae 2–5 mm long, subulate; stipules 3–5(7) × 1–1.5 mm, narrowly triangular. Flowers in aerial terminal racemes up to 15(25) cm long, and also in shorter, geotropic, whitish, subterranean racemes (see note below); aerial inflorescences with lanceolate-acuminate bracts 2–4 × 1 mm, and pedicels 3–7 mm long. Calyx 3–4(5) mm long, appressed or spreading-pubescent, with the teeth subequal and about equalling the tube, the 2 upper ones joined for × to × their length. Petals pink to purple, 5–9 mm long; standard limb subtruncate to subcuneate; keel considerably shorter than wing petals. Stamen tube 6–7 mm. Ovary pubescent; style glabrous, subcapitate, caducous; aerial pods 22–35 × 3–4 mm, linear-oblong, pubescent, brownish, (4)5–7-seeded; subterranean pods 7–10 × 3–4 mm, pubescent, whitish, 1(2)-seeded.

Botswana. N: Ngamiland Dist., Chobe Nat. Park, between Serondella and Ngwezumba, fl.& fr. 17.x.1972, *Pope, Biegel & Russell* 805 (K, SRGH). SE: Kweneng

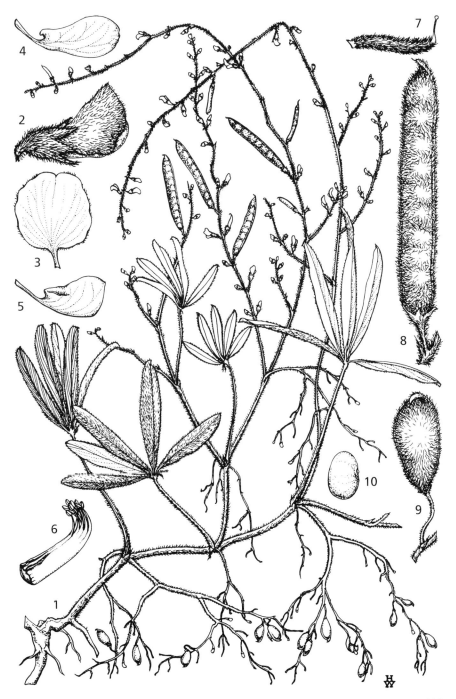

Fig. 3.3.**23**. TEPHROSIA LUPINIFOLIA. 1, habit (× ¹/₂), compiled from *Verboom* 889, *Buchanan* s.n. and *E.A. Robinson* 1099; 2, flower (× 3¹/₂); 3, standard (× 3¹/₂); 4, wing (× 3¹/₂); 5, keel (× 3¹/₂); 6, stamens (× 3¹/₂); 7, gynoecium (× 3¹/₂), 2–7 from *Milne-Redhead* 1189; 8, pod (× 2), from *Buchanan* s.n.; 9, geocarpic fruit (× 3), from *Verboom* 889; 10, seed (× 3), from *Buchanan* s.n. Drawn by Heather Wood. From Kew Bulletin (1980).

Dist., Matlolakgang Ranch, fl. 17.iii.1977, *Hansen* 3087 (GAB, K, PRE, SRGH). **Zambia**. B: Senanga Dist., Shangombo, fl.& subterr.fr. 16.viii.1952, *Codd* 7566 (K, PRE). N: Mbala Dist., Kasaba (Cassawa) sand dunes, Lake Tanganyika, fl.& fr. 16.ii.1959, *Richards* 10906 (K). W: Luakela (Luakera) Falls, N of Mwinilunga, fl. 25.i.1938, *Milne-Redhead* 4337 (K). C: Lusaka Dist., near Joubert's Farm, c.28 km E of Lusaka, fl.& imm.fr. 12.ii.1975, *Brummitt & Lewis* 14344 (K). S: Namwala Dist., Kabulamwanda, fl.& fr. 13.ii.1955, *Robinson* 1099 (K). **Zimbabwe**. N: Gokwe, fl.& fr. 12.iii.1962, *Bingham* 159 (K, SRGH). W: Nkayi Dist., Lukampa Valley, fl.& subterr.fr. iii.1954, *Davies* 742 (SRGH). C: Goromonzi Dist., beside Harare–Marondera (Marandellas) road near Nora Siding, fl. 13.iii.1958, *Corby* 860 (K, SRGH). S: Masvingo Dist., Makaholi Expt. Farm, fr. 23.iii.1948, *D.A. Robinson* 310 (K, SRGH) **Malawi**. N: Nkhata Bay Dist., Old Bandawe Graveyard, fl. & fr. 25.vi.1977, *Pawek* 12813 (K, MAL, MO, SRGH). C: Salima Dist., near Lake Nyasa Hotel, Salima, fl. 14.ii.1959, *Robson & Steele* 1601 (K, LISC, SRGH). ?S: no precise locality, fl. 1891, *Buchanan* s.n. (K).

Also in West Africa (Senegal and N Nigeria), Congo, Angola, Namibia and South Africa. In grassy and open places, especially on sandy ground; 480–1400 m.

The palmate leaves with a pair of stipellae below the leaflets, and the production of cleistogamous flowers, were cited by Hutchinson as grounds for referring this species to the segregate genus *Lupinophyllum*, following an earlier suggestion by J.B. Gillett which Gillett later retracted (see Hutchinson in Gen. Fl. Pl. **1**: 398 (1964); **2**: 626 (1967) and Gillett in F.T.E.A., Legum., Pap.: 169 (1971)). See also Brummitt in Kew Bull. **35**: 467–469 (1980).

The species is very variable, particularly in petiole length, leaflet number and pubescence, and apparently shows sporadic differentiation into local races. An extreme variant, described originally from Angola, having trifoliate leaves with short obovate leaflets (up to 3.5 cm) and very short petioles, has often been regarded as a distinct species, *Tephrosia laevigata*, which has been recorded from the Flora area (see synonymy above), but there appears to be no clear-cut distinction between this and plants having leaves with longer leaflets and petioles. Plants with trifoliate leaves occur frequently throughout the range of the species. Three collections from Solwezi in W Zambia – *Milne-Redhead* 1189 (K), *Fanshawe* 235 (K) and *Holmes* 1507 (K, SRGH) – differ from typical *T. laevigata* only in having somewhat narrower, less broadly rounded leaflets. A number of collections from Matopos in W Zimbabwe have short petioles but five leaflets. Although extreme plants may look very dissimilar, it does not seem possible at the moment to give formal taxonomic recognition to any of these local variants.

There appears to be little information available about the phenology and reproductive biology of this interesting species and further investigation in the field is highly desirable.

2. **Tephrosia pentaphylla** (Roxb.) G. Don in Sweet, Hort. Brit., ed.3: 170 (1839). — Brummitt in Bol. Soc. Brot., sér.2, **41**: 271 (1968). —Gillett in F.T.E.A., Legum., Pap.: 169 (1971). Type: India, Mysore, *Roxburgh*, Icones, pl.1628.

    Galega pentaphylla Roxb. [Hort. Bengal.: 57 (1814) nom. nud.], Fl. Ind. **3**: 384 (1832).

    Tephrosia senticosa sensu J.G. Baker in Hooker, Fl. Brit. India **2**: 112 (1876); sensu Baker f., Legum. Trop. Africa: 186 (1926), non (L.) Pers.

Annual or perennial, 30–40 cm high, bushy. Young stems densely covered with silvery-grey or rarely brown, appressed to somewhat ascending hairs. Leaves with (1)2(3) pairs of lateral leaflets; petiole 0.1–0.4 cm, petiole and rachis together 0.3–1(1.4) cm long; leaflets 1–3.2(4.5) × 0.3–0.8(1.1) cm, strongly oblanceolate, cuneate at the base, broadly rounded to slightly emarginate at the apex; upper surface glabrous, lower surface densely covered with usually silvery-grey hairs, stipules 3–7 × 0.2–0.4 mm, subulate. Flowers all in clusters in the axils of foliage leaves; pedicels 3–7 mm long. Calyx 3.5–5 mm long, covered with grey appressed hairs; teeth usually approximately equalling the tube, the two upper ones connate for only ¹/₄ their length. Petals 7–8 mm long, salmon-pink to scarlet or purple. Staminal tube joined above. Ovary pubescent; style glabrous. Pod 32–45(52 in India) × 3.5–4.5 mm, densely covered with grey appressed hairs. Seeds 7–9, or frequently 6 in Asia, c.3 × 2 × 1.3 mm, ± oblongoid, the hilum and very small aril towards one end of one of the longer sides.

Mozambique. N: Macomia Dist., between Mucojo and Macomia, fl. 29.ix.1948, *Barbosa* 2265 (LISC).

Also in Tanzania, Kenya, Uganda, Sudan, Ethiopia, Somalia, Arabia, Iran and S India; in the Flora area known only from the one cited collection. Recorded on black soils and gravel, but more ecological data required.

Tephrosia elegans Schumach. is recorded from Zambia by Gillett in F.T.E.A., Legum., Pap.: 169 (1971) and Lock, Leg. Afr. Check-list: 372 (1989). No material has been seen, but the species does occur in the Ufipa District of Tanzania, as well as being widespread from Guinea-Bissau to Sudan and south to Angola and Congo, and may well occur in the Flora Zambesiaca area. It has very short pedicels, 1–2 mm long, a densely silvery hairy calyx, with the lobes narrowly triangular rather than drawn out into fine points, pods densely silvery pubescent on the sides and brown along the sutures, and seeds with a prominent aril.

3. **Tephrosia uniflora** Pers., Syn. Pl. **2**: 329 (1807). —Schreiber in Merxmüller, Prodr. Fl. SW Afrika, fam. 60: 120 (1970). —Gillett in F.T.E.A., Legum., Pap.: 171, fig.30 (1971). Type: Senegal, *Roussillon* s.n. (Herb. Lamark, P).

Subsp. **uniflora** —Gillett in Kew Bull. **13**: 114 (1958). —Torre in C.F.A. **3**: 149 (1962). FIGURE 3.3.**24**.

    Tephrosia anthylloides Webb in Hooker's J. Bot. Kew Gard. Misc. **2**: 345 (1850). —Baker in F.T.A. **2**: 118 (1871). Types: Ethiopia, 1841, *Schimper* 1614; Sudan, *Kotschy* 87 & 111 (K syntypes).

    Tephrosia mossambicensis Schinz in Bull. Herb. Boiss., sér.2, **2**: 948 (1902). —Baker f., Legum. Trop. Africa: 205 (1926). Type: Mozambique, Boroma, *Menyhart* 632 (Z holotype).

    Tephrosia uniflora sensu Baker f., Legum. Trop. Africa: 184 (1926). —sensu Hepper in F.W.T.A., ed.2, **1**: 529 (1958).

    Tephrosia uniflora var. *parvifolia* Baker f., Legum. Trop. Africa: 184 (1926). Type from Ethiopia.

Annual or short-lived perennial with a taproot. Stems much branched from the base upwards, up to 80(130) cm high, the lower parts usually somewhat woody; young stems appressed- to spreading-pubescent or villous. Leaves with 2–4(usually 3) pairs of leaflets; petiole and rachis together 1–3.5(5) cm long, appressed or ascending-pubescent; leaflets decreasing in size towards the leaf base, the terminal one 15–35 × 4–14 mm, the lower ones down to 7–3 mm, obovate to oblanceolate or sometimes elliptic, rounded to emarginate at the apex, cuneate at the base, the upper surface glabrous or sometimes (in northern tropical Africa only) appressed-pubescent, the lower surface greyish appressed-villous; stipules (3)5–7(8) × c.0.5 mm, linear-subulate. Flowers solitary or paired in the axils of the foliage leaves; pedicels 2–4 mm long, ascending- to spreading-tomentose. Calyx 4–8 mm long, ascending- grey villous, the teeth c.4–5 times as long as the tube, long-acute, subequal or the lower one slightly longer than the others, the two upper ones only very slightly connate at the base. Petals 7–10 mm long, pink to purple; limb of standard suborbicular, rounded to cuneate at base. Stamen tube open above, upper stamen free. Ovary pubescent; style glabrous, with penicillate tip. Pod 46–64 × 3.5–5 mm, linear-oblong with the distal end curved upwards, rather loosely appressed- to ascending-pubescent. Seeds 9–14, 2–3 × 1.5–2 mm, oblong, flattened, brown, the funicle attached at the middle of one of the longer edges.

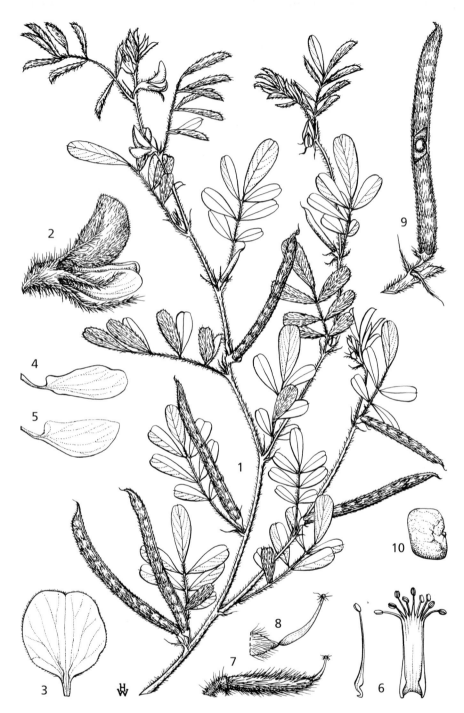

Fig. 3.3.**24**. TEPHROSIA UNIFLORA subsp. UNIFLORA. 1, flowering and fruiting branch (×
²/₃); 2, flower (× 3); 3, standard (× 3); 4, wing (× 3); 5, keel (× 3); 6, stamens (× 3); 7, gynoecium
(× 3); 8, style (× 6); 9, pod, with one seed exposed (× 1); 10, seed (× 4), 1–10 from *Lemos &*
Macuácua 123. Drawn by Heather Wood.

Botswana. SE: Central Dist., Orapa, Mine Manager's house, fl.& fr. 21.ii.1975, *Allen* 297 (K, PRE). **Zimbabwe**. N: Lake Kariba, Binga, DC's harbour, fr. 10.iii.1966, *Jarman* 406 (K, SRGH). W: Hwange Dist., Deka R., fr. 21.vi.1934, *Eyles* 7963 (K, SRGH). E: Chipinge Dist., lower Save Valley, fl.& fr. i.1957, *R.M. Davies* 2400 (K, SRGH). S: Chiredzi Dist., between Sibonja Hills and Chipinda Pools, fr. 1.v.1962, *Drummond* 7857 (K, PRE, SRGH). **Mozambique**. T: Cahora Bassa Dist., between Chicoa and Chetima, 19.2 km from Chicoa, fr. 30.vi.1949, *Barbosa & Carvalho* 3400 (K, LMA). MS: Chemba Dist., Chiou, Estação Experimental do C.I.C.A., fl.& fr. 20.iv.1960, *Lemos & Macuácua* 123 (COI, K, LISC, SRGH). GI: Massingir Dist., Limpopo R., 70 km from Chókwe (Guijá) towards Massingir (Massingire), fr. 22.v.1948, *Torre* 7912 (K, LISC).

Also from Cape Verde Islands, Senegal to N Nigeria, Sudan, Ethiopia and Eritrea, Somalia, Kenya, Tanzania, Angola, Namibia and South Africa (Limpopo Province). In mopane and mixed woodlands, rocky outcrops and open disturbed ground; up to 900 m.

Subsp. *petrosa* (Blatt. & Hall.) J.B. Gillett & Ali, distinguished by its shorter pods with fewer seeds, thinner pubescence in most parts, and leaflets in 1–2 pairs, is from the Arabian peninsula, W Pakistan and NW India.

4. **Tephrosia noctiflora** Baker in F.T.A. **2**: 112 (1871). —De Wildeman in Bull. Soc. Roy. Bot. Belg. **57**: 123 (1925). —Baker f., Legum. Trop. Africa: 193 (1926). —Forbes in Bothalia **4**: 965 (1948). —Cronquist in F.C.B. **5**: 102 (1954). —Hepper in F.W.T.A., ed.2, **1**: 529, 530 (1958). —Brummitt in Bol. Soc. Brot., sér.2, **41**: 228 (1968). —Gillett in F.T.E.A., Legum., Pap.: 182 (1971). Type: Zanzibar, 1830, *Bojer* s.n. (K holotype).

   Tephrosia noctiflora Bojer, Hortus Maurit.: 93 (1837) nom. nud.
   Tephrosia villosa sensu Klotzsch in Peters, Naturw. Reise Mossamb. **6**(1): 47 (1861). —sensu Baker in F.T.A. **2**: 112 (1871) in part.

Usually a shrub, but at least sometimes annual, 0.6–1.3(2) m high, the branches often widely spreading, sometimes almost prostrate. Young stems densely pubescent with appressed to spreading, grey to brown hairs (appressed and greyish in all material so far seen from the Flora Zambesiaca area). Leaves pinnate with (4)6–9(11) pairs of leaflets; petiole 0.5–1.4(2) cm long, petiole and rachis together (2)4–10(12) cm long, pubescent like the stem; leaflets (1)2–3.8(4.5) × (0.4)0.5–0.9(1.2) cm, narrowly obovate or sometimes narrowly elliptic, cuneate at the base, rounded to subacute at the apex; upper surface glabrous, lower surface appressed-pubescent; stipules 4–10 × 0.5–1 (2 in some Kenya specimens) mm, ± linear. Flowers in terminal (often appearing leaf-opposed), fairly lax racemes up to 25(30) cm long; bracts 2–6 × 0.5–1 mm, linear or linear-triangular; pedicels (3)4–5(6) mm long. Calyx (4)5–7(8) mm long, covered with appressed to ascending, brown or grey hairs, the lower tooth usually about twice as long; upper teeth short, connate for $^1/_2$–$^3/_4$ their length. Petals 9–13 mm long, variously described as purplish-, bluish- or pinkish-white, reddish-purple, yellowish or yellow and brown. Staminal tube joined above. Ovary pubescent; style glabrous, twisted near the base so that the lower surface lies uppermost, with penicillate tip. Pods (42)45–55(60) × 4.5–7 mm, densely pubescent to villous, with brown or grey, loosely appressed or ascending, often geniculate hairs, often conspicuously darker at the margins. Seeds (6)8–10, 3.2–4.2 × 2–2.8 × 1–1.5 mm, oblongoid, brown, rugose, with the hilum and small aril more or less at the middle of one of the longer sides.

Zimbabwe. N: Mutoko Dist., Special Native Area 'C', Morussi Camp, fl.& fr. 13.ii.1962, *Wild* 5642 (COI, K, SRGH). E: Chipinge Dist., Save R., East Makesa near Mahenya's Kraal, fl.& fr. 18.v.1959, *Savory* 471 (SRGH). S: Gwanda Dist., Doddieburn Ranch, fr. 27.iv.1924, *Davison* 51 (PRE). **Malawi**. S: Blantyre Dist., Shire R., Mpatamanga Gorge, fl.& fr. 28.ii.1961, *Richards* 14481 (K). **Mozambique**. N: near

Palma, garden, fl. 4.v.1917, *Pires de Lima* (PO). T: Tete, fl.& imm.fr. ii.1859, *Kirk* (K). M: Namaacha Dist., Goba mountains, fr. 15.xi.1940, *Torre* 2009 (LISC).

The natural distribution of this species has been obscured through its cultivation in many parts of the world. In Africa it is probably native in eastern parts, from Kenya and Uganda to South Africa (Mpumalanga Prov.), but is introduced in W Africa and Congo. It occurs either native or naturalized also in the Mascarene Is., Seychelles, Sri Lanka, India (Madras, Bombay, Assam), Java and the Philippines; in the West Indies and S America it is certainly introduced. In the Flora Zambesiaca area recorded from river banks, stony and rocky places and woodland; below 1000 m.

5. **Tephrosia villosa** (L.) Pers., Syn. Pl. **2**: 329, no. 23 (1807).—Klotzsch in Peters, Naturw. Reise Mossamb. **6**(1): 47 (1861) for name, excl. specimen. —Baker in F.T.A. **2**: 122 (1871) for name, excl. specimen. —Harms in Engler, Pflanzenw. Afrikas **3**(1): 589 (1915). —Gillett in Kew Bull. **13**: 121 (1958). —Torre in C.F.A. **3**: 161 (1962). —Brummitt in Bol. Soc. Brot., sér.2, **41**: 224 (1968). —Gillett in F.T.E.A., Legum., Pap.: 190 (1971). Type: Sri Lanka, *Herb. Hermann* 1 f.31 (BM holotype).

    Cracca villosa L., Sp. Pl.: 752 (1753). —Kuntze, Rev. Gen. Pl. **1**: 173 (1891) in part excl. *b purpurea* etc. —Hiern, Cat. Afr. Pl. Welw. **1**: 223 (1896).

    Cracca incana Roxb. [Hort. Beng.: 57 (1814), nom. nud.] Fl. Ind. ed.2, **3**: 385 (1832). Type: India, no locality, *Roxburgh* (K lectotype, selected by Brummitt 1968; BM).

    Tephrosia incana (Roxb.) [Sweet, Hort. Brit., ed. 2: 142 (1830) nom. nud.; Wall., Numer. List no. 5644 (1831–32) nom. nud.] Wight, Cat. Ind. Pl.: 57 (1833). —Baker f. in F.T.A. **2**: 123 (1871). —Harms in Engler, Pflanzenw. Afrikas **3**(1): 589 (1915).

    Tephrosia villosa var. *incana* (Roxb.) Baker in Hooker, Fl. Brit. India **2**: 113 (1876). — Torre in C.F.A. **3**: 161 (1962).

    Cracca villosa var. *incana* (Roxb.) Hiern, Cat. Afr. Pl. Welw. **1**: 223 (1896).

Subsp. **ehrenbergiana** (Schweinf.) Brummitt in Bol. Soc. Brot., sér.2, **41**: 225 (1968). —Schreiber in Merxmüller, Prodr. Fl. SW Afrika, fam. 60: 120 (1970). —Gillett in F.T.E.A., Legum., Pap.: 190, fig.33 (1971). Type: Eritrea, *Ehrenberg* s.n. (B† lectotype); Dongolla, *Pappi* 65 (K neotype, selected by Brumitt 1968).

    Tephrosia apollinia sensu Klotzsch in Peters, Naturw. Reise Mossamb. **6**(1): 47 (1861). — sensu Eyles in Trans. Roy. Soc. S. Africa **5**: 375 (1916).

    Tephrosia ehrenbergiana Schweinf. [Klotzsch in Peters, Naturw. Reise Mossambique **6**, part 2: 576 (1864), nom. nud.], Beitr. Fl. Aethiop.: 18 (1867). —Baker f., Legum. Trop. Africa: 209 (1926). —Forbes in Bothalia **4**: 972 (1948) in part, excl. *T. rhodesica*. —Brenan, Check-list For. Trees Shrubs Tang. Terr.: 446 (1949). —Cronquist in F.C.B. **5**: 100 (1954).

Annual or short-lived perennial 30–140 cm high, bushy and often becoming woody in the lower parts. Stems densely clothed with appressed or spreading to deflexed, white to brown hairs. Leaves pinnate with (4)5–9(10) pairs of leaflets; petiole (0.2)0.5–1.6(2.2) cm long, petiole and rachis together (2)3–11(15) cm long, pubescent like the stem; leaflets (0.7)1.2–3.5(4.2) × (0.3)0.6–1.1(1.4) cm, oblanceolate to subelliptic, cuneate at the base, obtuse to rounded at the apex; upper surface thinly to fairly densely appressed-pubescent; lower surface densely appressed-pubescent with usually greyish, sometimes subsericeous hairs; stipules 4–9(11) × 0.5–2(2.3) mm, linear-triangular. Flowers in usually fairly lax terminal and leaf-opposed racemes up to 16(20) cm long, or the lowermost flowers sometimes in the axils of the uppermost 1–3 foliage leaves; bracts 3–5 × c.1 mm, densely pubescent; pedicels (1)2–4 mm long, densely pubescent. Calyx (8)9–12(15) mm long, ± villous with brownish hairs, the teeth much longer than the tube, the lowermost being (3)4–6 times as long, the tube 1–2(3) mm long; teeth linear-triangular, the two uppermost connate at their base only. Petals (10)12–15(17) mm long, ± purple. Staminal tube connate above. Ovary villous; style glabrous, twisted near the base so that the lower surface lies uppermost,

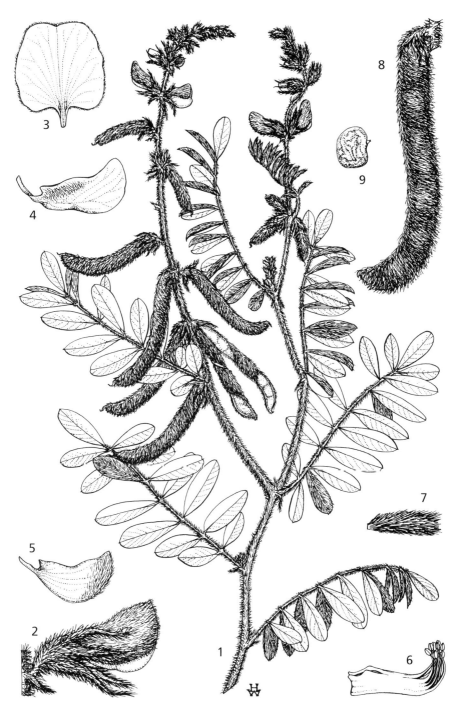

Fig. 3.3.**25**. TEPHROSIA VILLOSA var. EHRENBERGIANA. 1, flowering and fruiting branch
(× ²/₃), from *Monteiro* s.n. and *Barbosa & Correia* 9131; 2, flower (× 2); 3, standard (× 2); 4, wing
(× 2); 5, keel (× 2); 6, stamens (× 3); 7, gynoecium (× 3), 2–7 from *Barbosa & Correia* 9131; 8,
pod (× 1¹/₂), from *Monteiro* s.n.; 9, seed (× 3), from *Davies* 2401. Drawn by Heather Wood.

with penicillate tip. Pod (30)35–46(54) × 4.5–6.5(7) mm, markedly curved upwards in the distal third, densely villous with lightish brown to greyish hairs. Seeds 7–9, 2.8–4 × 2 × 1.3 mm, oblongoid to subreniform, brown, rugose, with the hilum and small rim aril at the middle of one of the longer sides.

Var. **ehrenbergiana** —Brummitt in Bol. Soc. Brot., sér.2, **41**: 226 (1968). FIGURE 3.3.**25**.

> Stems and leaf rachides with spreading to deflexed brown or brownish-grey hairs.

Zimbabwe. E: Chipinge Dist., Save (Sabi) Valley, fl. & fr. i.1957, *R.M. Davies* 2401 (K; SRGH). S: Gwanda Dist., Doddieburn Ranch, Mzingwane (Umzingwane) R., fr. 8.v.1972, *Pope* 676 (K; SRGH). **Malawi**. S. Mangoche Dist., Monkey Bay, by Fisheries Research Station, fl. & fr. 2.iii.1970, *Brummitt & Eccles* 8845 (K). **Mozambique**. N: Mossuril Dist., Lumbo, near the aerodrome, fl. & fr. 5.v.1948, *Pedro & Pedrógão* 3117 (K; LMA). T: road to Tete, fl. 28.ii.1961, *Richards* 14496A (K). MS: Tambara Dist., 19 km from Nhacolo (Tambara) towards Lupata, fr. 16.v.1971, *Torre & Correia* 18476 (K; LISC).

From Ethiopia to South Africa (Limpopo Province), Angola and Namibia, also in Madagascar. Usually recorded from sandy soils, often in grassland or disturbed habitats; 50–750 m.

Var. **daviesii** Brummitt in Bol. Soc. Brot., sér.2, **41**: 227 (1968). Type: Zimbabwe, Gwanda Dist., Special Native Area 'G', Chimaneli Camp, fl.& imm.fr. 15.xii.1956, *Davies* 2285 (K holotype, SRGH).

> Stems and leaf rachides with closely appressed silvery-white hairs, usually subsericeous.

Zimbabwe. N: Lake Kariba, Ruaru Is., fl. 9.i.1963, *Whellan* 2007 (K, SRGH). W: Hwange Dist., Deka R., fl.& fr. 21.vi.1934, *Eyles* 7962 (K; SRGH). S: Beitbridge Dist., Pioneer Memorial, Shashi Drift, Tuli, fl.& fr. 23.iii.1959, *Drummond* 5960 (K, SRGH). **Mozambique**. T: Tete, fl.& fr. ii.1859, *Kirk* (K).

Also in Namibia. In the Flora Zambesiaca area apparently confined to river valleys in the drier lowland parts of the Zambezi and Limpopo River basins; it has been recorded from basaltic soils in *Colophospermum mopane* woodland, but more information is required; 550–1100 m.

Subsp. *villosa* occurs in India and has much smaller flowers and fruits; it is usually smaller in vegetative parts, with a shorter and rather twiggy habit, and appressed-pubescent with silvery-white hairs, occasionally with spreading hairs. Subsp. *ehrenbergiana* var. *ehrenbergiana*, widespread in tropical Africa, differs from the typical subspecies markedly both in size of flowers, fruit and vegetative parts and indumentum, but var. *daviesii*, restricted to Africa south of the Zambezi valley, may be regarded as somewhat intermediate, having the indumentum as in the commoner Indian plants and a tendency to slightly smaller flowers and fruits than in var. *ehrenbergiana*.

6. **Tephrosia rhodesica** Baker f., Legum. Trop. Africa: 208 (1926). —Cronquist in F.C.B. **5**: 100 (1954). —Brummitt in Bol. Soc. Brot., sér.2, **41**: 229 (1968). — Schreiber in Merxmüller, Prodr. Fl. SW Afrika, fam. 60: 119 (1970). —Gillett in F.T.E.A., Legum., Pap.: 189 (1971). Type: Zimbabwe, Bulawayo, ii.1902, *Eyles* 1062 (BM holotype, K, SRGH).

> *Tephrosia ehrenbergiana* sensu Forbes in Bothalia **4**: 972 (1948) in part.

A usually short-lived perennial, or possibly sometimes annual, 60–160(200) cm high, rather bushy, often woody in the lower parts. Young stems ± tomentose with usually rusty-brown, occasionally greyish, spreading or ascending or deflexed hairs. Leaves pinnate with (4)6–11(14) pairs of leaflets; petiole 0.1–1.2(1.5) mm long, petiole and rachis together (2)4–12(16) cm long; leaflets (0.7)1.2–3.2(4.2) × (0.3)0.5–1.2(1.4) cm, oblanceolate or sometimes ± obovate or elliptic, cuneate at the base, rounded or sometimes emarginate at the apex; upper surface rather thinly pubescent or very rarely ± glabrous, lower surface appressed-pubescent; stipules (3)5–9(12) × 0.5–1.5 mm, linear-triangular. Flowers in fairly lax, terminal (?very rarely axillary) racemes up to 30 cm long; bracts 4–9 × 0.5–1 mm, linear-triangular; pedicels 2–5 mm long. Calyx 5–9(11) mm long, with spreading greyish hairs; lateral teeth linear-triangular, equalling or up to 3 times as long as the tube; upper teeth connate for $^1\!/_4$–$^1\!/_2$($^2\!/_3$) their length. Petals 8–12(14) mm long, pink or crimson to purple. Staminal tube joined above. Ovary pubescent; style glabrous, not twisted, with penicillate tip. Pods 28–46(55) × 3.5–5.5(6), markedly curved upwards in the distal part, densely pubescent to conspicuously villous (see varieties) with pale or rusty-brown or rarely greyish hairs. Seeds (4)5–7, 4–4.5 × 3–3.5 × 1.5–2, oblongoid, brown to blackish, with a small aril in the middle of one of the longer sides.

1. Pods fairly densely shortly pubescent, pale straw-coloured beneath the indumentum · **ii)** var. *polystachyoides*
– Pods tomentose to villous · 2
2. Indumentum of stem, leaf rachis and pods densely tomentose to villous, light to rusty-brown · **i)** var. *rhodesica*
– Indumentum of stem, leaf rachis and pods grey; pods rather thinly tomentose with appressed hairs · **iii)** var. *evansii*

i) Var. **rhodesica** —Brummitt in Bol. Soc. Brot., sér.2, **41**: 230 (1968).
    Tephrosia polystachyoides sensu Burtt Davy, Fl. Pl. Ferns Transv., pt.2: 376, 378 (1932) for S. African specimens. —sensu Forbes in Bothalia **4**: 971 (1948) for S. African specimens.
    Tephrosia ehrenbergiana sensu Forbes in Bothalia **4**: 972 (1948) in part.

Indumentum of stem, leaf rachis and pods light to rusty-brown. Calyx 6–9(11) mm long (in South Africa 5–8 mm). Pods (28)34–46(55) × (4)4.5–5.5(6) mm (in South Africa 28–42 × 3.5–4 mm), densely tomentose to villous, with matted hairs.

Botswana. SE: South East Dist., Kgale Mt., c.6.5 km SW of Gaborone, fl.& fr. 23.iii.1974, *Mott* 199b (K, GAB). **Zambia**. B: Mongu, fl.& fr. 2.iv.1964, *Verboom* 1024 (K). N: Samfya, fl.& fr. 12.vii.1955, *Fanshawe* 2385 (K, SRGH). W: Kitwe, fl. 20.i.1963, *Fanshawe* 7654 (K). C: Lusaka, fl. & yng.fr. 5.ii.1965, *Fanshawe* 9148 (K). E: Chipata Dist., Ngoni Reserve (Area), fl.& fr. ii.1962, *Verboom* 425 (K, SRGH). S: Choma Dist., Kanchomba Agric. Station, fl.& fr. 17.ii.1960, *White* 7155 (K, SRGH). **Zimbabwe**. N: Mutoko Dist., Nyamahere Hill, 17°23'S, 32°13'E, fl.& fr. 26.ii.1978, *Pope* 1615 (K). W: Matobo Dist., Farm Besna Kobila, fl.& fr. ii.1959, *Miller* 5765 (K, PRE, SRGH). C: Harare, fl.& fr. 17.i.1937, *Eyles* in Herb. QVM 8936 (K). E: Chipinge Dist., Sabi Valley Expt. Station, fl.& fr. ii.1960, *Soane* 287 (K, LISC, SRGH). S: Masvingo Dist., Makoholi Expt. Station, fl.& fr. 23.iii.1948, *D.A. Robinson* 302 (SRGH). **Malawi**. N: Nkhata Bay Dist., Old Bandawe, fr. 25.vi.1977, *Pawek* 12814 (K, MAL, MO, SRGH). C: Lilongwe, near capital site, fl.& fr. 29.iii.1970, *Brummitt & Little* 9488 (K). S: Mwanza Dist., road to Tete, Mozambique border, fl.& fr. 28.ii.1961, *Richards* 14496 (K, SRGH). **Mozambique**. N: Nampula Dist., Serra de Mesa, fl.& fr. 13.iv.1961, *Balsinhas & Marrime* 387 (COI, K, LISC, LMA, SRGH). MS: Gondola Dist., between river Revué and Munhinga, fr. 27.iv.1948, *Barbosa* 1598 (LISC).
Also in Congo, Burundi, Uganda, Kenya, Tanzania, South Africa (North-West, Limpopo, Mpumalanga and Gauteng Provinces) and Namibia. In grassland, thicket, woodland, roadsides, etc., often on sandy soils; 350–1550 m in the Flora area.

Gonçalves (in Garcia de Orta, sér.Bot. **5**: 117 (1982)), recorded this variety from Tete Province of Mozambique.

Plants from South Africa referred to this variety tend to have smaller flowers and rather markedly narrower fruit, and might perhaps be separable taxonomically.

ii) Var. **polystachyoides** (Baker f.) Brummitt in Bol. Soc. Brot., sér.2, **41**: 230 (1968). Type: Zimbabwe, Mutare Dist., Odzani River Valley, *Teague* 14 (K holotype).

> *Tephrosia polystachyoides* Baker f., Legum. Trop. Africa: 193 (1926). —Burtt Davy, Fl. Pl. Ferns Transvaal, pt.2: 376, 378 (1932) in part for type, excl. S. African specimens. —Forbes in Bothalia **4**: 971 (1948) in part for type, excl. S. African specimens. —Miller in J. S. Afr. Bot. **18**: 37 (1952). —Martineau, Rhod. Wild Fl.: 37 (1953).

Indumentum of stem, leaf rachis and pods light to rusty-brown. Calyx 4–6(8) mm long. Pods (28)30–45 × (3.5)4–5(5.5) mm, fairly densely shortly pubescent but not tomentose or villous, pale straw-coloured beneath the indumentum.

Zambia. S: Mazabuka Dist., Mochipapa to Sinazongwe, mile 24, fl.& fr. 2.iii.1960, *White* 7548 (K). **Zimbabwe**. N: Murehwa (Mrewa), cattle breeding station, fl.& fr. ii.1954, *Davies* 778 (SRGH). W: Matobo Dist., Matopos, on road to Pasture Res. Station, fl.& fr. 4.iii.1930, *Rattray* 95 (SRGH). C: Chegutu Dist., Poole Farm, fr. 6.iii.1954, *Hornby* 3335 (K, LISC, SRGH). E: Mutare Dist., Marange Communal Land (Maranke Reserve), fl.& fr. 10.ii.1953, *Chase* 4763 (SRGH). S: Buhera Dist., near Buhera, fl.& imm.fr. 7.ii.1954, *Masterson* 36 (SRGH). **Malawi**. S: Shire Valley 16 km W of Zomba, slopes of Lulanga Hill overlooking Chingale, fl.& imm.fr. 12.iii.1977, *Brummitt & Patel* 14825 (K, MAL). **Mozambique**. Z: Montes de Ile, fl.& fr. 2.iv.1943, *Torre* 5050 (LISC).

Known only from the Flora Zambesiaca area, between latitudes 15° and 21°S. Usually in grassland on sandy soil; 660–1495 m.

Gonçalves (in Garcia de Orta, sér.Bot. **5**: 117, 1982) recorded this variety from Tete Province of Mozambique.

iii) Var. **evansii** (Hutch. & Burtt Davy) Brummitt in Bol. Soc. Brot., sér.2, **41**: 232 (1968). Type: South Africa, Mpumalanga, Komatipoort, *Pole-Evans* H16853 (K).

> *Tephrosia evansii* Hutch. & Burtt Davy, Fl. Pl. Ferns Transv., pt.2: 376, 378 (1932). — Forbes in Bothalia **4**: 975 (1948).

Indumentum of stem, leaf rachis and pods grey. Calyx 5–8 mm long. Pods 30–44 × 4–4.5 mm, rather thinly tomentose with appressed hairs.

Botswana. SE: Southern Dist., on road from Khakhea (Kakia) to Kanye, fl. 24.ii.1960, *de Winter* 7503 (K, SRGH). **Zimbabwe**. S: Mwenezi Dist., Malangwe R., SW Mateke Hills, fl.& fr. 6.v.1958, *Drummond* 5640 (COI, K, SRGH).

Known only from Mpumalanga Province in South Africa and from two collections from Mwenezi Dist. of Zimbabwe and one from Botswana. Rocky and sandy places, often by rivers; 300–625 m.

The varieties recognized differ markedly in their indumentum, particularly of the pods, but apparently not significantly in other characters. Further investigation of their relationships and taxonomic status is required.

7. **Tephrosia polystachya** E. Mey., Comment. Pl. Afr. Austr.: 109 (1836). —Harvey in F.C. **2**: 206 (1862).—Eyles in Trans. Roy. Soc. S. Africa **5**: 375 (1916) for name only, excluding specimen. —Forbes in Bothalia **4**: 963 (1948). —Mogg in

Macnae & Kalk, Nat. Hist. Inhaca Is., ed.2: 146 (1969) as "*polystrachys*". Type: South Africa, no locality, *Drège* s.n. (5462 in Herb. Meyer (B syntype)).

Tephrosia polystachya var. *hirta* Harv. in F.C. **2**: 206 (1862). —Forbes in Bothalia **4**: 964 (1948). Type: South Africa, KwaZulu-Natal coast, *Sutherland* s.n. (K).

Tephrosia polystachya var. *latifolia* Harv. in F.C. **2**: 206 (1862). —Burtt Davy, Fl. Pl. Ferns Transvaal, pt. 2: 378 (1932). —Forbes in Bothalia **4**: 965 (1948). Type: South Africa, Durban (Port Natal), *Gueinzius* 616 (K).

Probably usually short-lived perennial 60–150(200) cm high, erect, branched, sometimes becoming somewhat woody at the base. Young stems with appressed to spreading short brown hairs, particularly on the ridges of the stem. Leaves pinnate with (4)5–11 pairs of leaflets; petiole 0–2(2.7) cm long, petiole and rachis together (1.5)2.5–9 cm long; leaflets (0.8)1–2.5(3.5) × 0.2–1.0(1.3) cm, narrowly oblong or linear-elliptic to oblanceolate or obovate (see groups below), cuneate at the base, ± acute to rounded at the apex; upper surface glabrous or pubescent, lower surface appressed-pubescent; stipules 3–9(11) × 0.3–1 mm. Flowers in terminal and axillary inflorescences up to 15(25) cm long, the lateral ones often exceeding the leaves and ascending to form a pseudocorymb; bracts 2–5 × 0.5 mm, linear-triangular; pedicels 2–5 mm long. Calyx 2.5–7 mm long (see groups below), appressed-pubescent; lateral teeth half as long as to ± equalling the tube; upper teeth connate for about half their length. Petals 6–10 mm long, purplish to pink, or white and purple. Staminal tube joined above. Ovary pubescent; style glabrous, not twisted. Pods 33–52 × 3–5 mm, straight or slightly and gradually curved upwards, fairly thinly to densely pubescent with usually short greyish hairs. Seeds 7–8(10).

The species occurs from the Eastern Cape Province northwards to S Mozambique. It shows great variation and seems to intergrade into other species in South Africa. A taxonomic reassessment of the whole group is needed; meanwhile three groups have been recognized, two occurring in the Flora area (see Brummitt in Bol. Soc. Brot., sér.2, **41**: 233–234 (1968)).

Group 1.

Leaves with (4)5–8 pairs of leaflets; petiole (0.3)0.5–2.0(2.7) cm long; leaflets 1–2.5(3.5) × (0.3)0.4–1.0(1.3) cm, narrowly elliptic to oblanceolate; upper surface usually pubescent but occasionally glabrous; stipules (4)5–9(11) mm long. Lateral inflorescences not forming a pseudocorymb. Calyx (3)3.5–6(7) mm long. Petals (6)7–10 mm long. Pods (38)45–52 × 4–5 mm.

Mozambique. M: near freshwater lake, SE Maputo, fl.& fr. 4.iv.1947, *Hornby* 2621 (LMA, PRE).

Also in South Africa (Limpopo, Mpumalanga and KwaZulu-Natal Provinces) and Swaziland. Characteristically found at the margins of mountain forests; mainly 1220–1830 m, but recorded down to 215 m.

Group 2.

Leaves with (4)7–11 pairs of leaflets; petiole 0–1.5 cm long; leaflets 1–2.5 × 0.2–0.45 cm, narrowly oblong to linear-elliptic; upper surface glabrous or glabrescent; stipules 3–6 mm long. Lateral inflorescences overtopping the leaves and ascending to form a pseudocorymb. Calyx 2.5–3 mm long. Petals 6–8 mm long. Pods unknown.

Mozambique. M: Namaacha Falls, fl. 22.ii.1955, *Exell, Mendonça & Wild* 540 (LISC, SRGH).

Known only from the mountains in the Goba–Namaacha area of Mozambique, on basaltic rocks and savanna.

Plants referred to Group 2 show affinity with the third group (from the Eastern Cape Province of South Africa) but have a much shorter calyx, and also with *Tephrosia amoena* E. Mey. from KwaZulu-Natal.

8. **Tephrosia multijuga** R.G.N. Young in Ann. Transv. Mus. 14: 402 (6 July 1932). — Forbes in Bothalia 4: 981 (1948) in part. —Brummitt in Bol. Soc. Brot., sér.2, **41**: 262 (1968). Type: South Africa, Johannesburg, Houghton Ridge, *Moss* 7473 (PRE).

Tephrosia woodii Burtt Davy, Fl. Pl. Ferns Transvaal, pt.2: xxxii, 376, 378 (28 July 1932). Type: South Africa, Western Cape, Kei R., *Drège* s.n. (K holotype).

Annual or perhaps sometimes short-lived perennial, with an erect branching stem up to 60 cm high or sometimes with several procumbent trailing branching stems arising from a common base. Stems rather sparsely ± appressed-pubescent to subglabrous. Leaves pinnate with 3–9(10) pairs of leaflets; petiole 0.1–1.4(2–3) cm long, petiole and rachis together (2)3–6.5(8.5) cm long, pubescent like the stem; leaflets (1)1.5–2.5(3) × 0.15 × 0.4(0.55) cm, narrowly elliptic to linear-elliptic, ± cuneate at the base and acute or subacute at the apex; upper surface glabrous, lower surface thinly appressed-pubescent; stipules 3–5(10) × 0.2–0.5 mm, linear. Flowers in terminal (appearing leaf-opposed), rather slender, lax, ascending racemes up to 18 cm long; bracts 1.5–3 mm long, linear; pedicels 2–7 mm long. Calyx (1.5)2(3) mm long, ± funnel-shaped, shortly appressed-pubescent; teeth about equalling the tube. Petals 4–6 mm long, pink to purple or perhaps occasionally whitish. Staminal tube joined above. Ovary appressed-pubescent; style glabrous. Pods (27)30–40 × 3–4 mm, usually darkish brown, very shortly appressed-pubescent. Seeds (5)6–7, c.2.8 × 1.2 × 1.2, subcylindric-oblong, with a white aril c.0.6 mm long towards one end.

Mozambique. M: Namaacha Dist., Goba, Libombo Mts., bridge over R. Umbelúzi (Umbelusi), fl.& fr. 30.iv.1947, *Pedro & Pedrógão* 1040 (LMA, PRE).

Also in South Africa (Limpopo, Mpumalanga, Gauteng, KwaZulu-Natal, Free State and Eastern Cape Provinces). Recorded in the Flora area from a deep ravine in *Androstachys* forest, mixed woodland with *Afzelia*, *Combretum molle* and *Albizia adianthifolia* and secondary associations in rocky places; up to 250 m.

The species is usually erect with a single stem at the base, branching above. But two specimens from the Flora area, apparently referable here, are prostrate trailing plants with several stems from a common base. Both are from near Namaacha, *Barbosa & Lemos* 7535 (COI, LISC, LMA) and *Exell, Mendonça & Wild* 498 (BM, LISC, SRGH). Further collections might show that these could be separated from typical *T. multijuga*. Further investigation is also needed of the relationship of this species to *T. polystachya*, as some South African specimens seem to be intermediate between the two.

9. **Tephrosia elongata** E. Mey., Comment. Pl. Afr. Austr.: 111 (1836). —Harvey in F.C. **2**: 208 (1862). —Burtt Davy, Fl. Pl. Ferns Transvaal, pt.2: 276, 378 (1932). — Forbes in Bothalia 4: 977 (1948). —Brummitt in Bol. Soc. Brot., sér.2, **41**: 268 (1968). —Gillett in F.T.E.A., Legum., Pap.: 176 (1971). Type: South Africa, KwaZulu-Natal, between Umzinkulu and Umkomaas, *Drège* s.n. (B†).

Perennial suffrutex with a vertical stock and herbaceous annual shoots. Stems erect or decumbent, up to 50 cm high, or up to 90 cm in var. *tzaneenensis*, unbranched or sparingly branched, subglabrous or appressed to spreading-pubescent. Leaves with 1–2(3) pairs of leaflets; petiole 0.6–2.8 cm long, or as short as 0.1 cm in var. *tzaneenensis*, petiole and rachis together 0.8–4.5(7.0) cm long, pubescent like the stem; leaflets 3.5–9 (13 in var. *tzaneenensis*) × 0.3–1.5 cm, linear to linear-elliptic, linear-oblong or rarely linear-lanceolate, subcuneate at the base, acute to rounded at the apex; upper surface glabrous often with numerous minute gland dots between the veins, lower surface appressed to irregularly spreading pubescent; stipules (3)4–10(13) × 0.3–1 mm, linear. Flowers in terminal racemes up to 22(30) cm long; bracts 2–5 × c.0.5 mm; pedicels 2–5 (7 in fruit) mm long. Calyx (4)5–7 mm long, appressed to spreading-pubescent; lateral teeth equalling to exceeding the tube, upper teeth connate for $^1/_2$–$^2/_3$ their length to form an acute triangle. Petals 9–12(15) mm long, red, pink, salmon, orange or yellow; standard truncate to subcordate at the base. Staminal tube joined above. Ovary pubescent; style glabrous, with penicillate tip. Pods 55–70 × 3.5–4(5) mm, appressed-pubescent, tending to be glabrescent. Seeds 9–13(16), 2–3 × 2 × 1 mm, blackish, with a small rim aril at the middle of the longest side.

1. Hairs on stems, leaf rachides, petioles and lower leaf surface spreading ·······
   ·· **ii)** var. *lasiocaulos*
– Hairs appressed-pubescent ·· 2
2. Leaves with 1–2 pairs of lateral leaflets; petiole 6–28 mm long; leaflets up to 9 cm long ·· **i)** var. *elongata*
– Leaves with 1–3 pairs of lateral leaflets; petiole 1–7 mm long, or up to 10 mm in lower leaves; leaflets up to 13 cm long ················ **iii)** var. *tzaneenensis*

i) Var. **elongata** —Brummitt in Bol. Soc. Brot., sér.2, **41**: 270 (1968).

> *Tephrosia elongata* var. *pubescens* Sond. in Linnaea **23**: 30 (1850). —Harvey in F.C. **2**: 208 (1862), nom. illegit.
> *Tephrosia elongata* var. *glabra* Sond. in Linnaea **23**: 30 (1850). —Harvey in F.C. **2**: 208 (1862). —Baker f., Legum. Trop. Africa: 183 (1926). Type from South Africa (Gauteng).
> *Tephrosia sambesiaca* Taub. in Bot. Jahrb. Syst. **23**: 183 (1896). —Baker f., Legum. Trop. Africa: 182 (1926). Type: Malawi, top of Zomba, *Buchanan* 213 (K lectotype, chosen by Brummitt 1968).
> *Tephrosia dissitiflora* Baker in Bull. Misc. Inform., Kew **1897**: 257 (1897). Type: Malawi, Mt. Zomba, *Whyte* s.n. (K).

Erect or decumbent plants, up to 50 cm high. Stems, leaf rachides and petioles and leaflet lower surfaces appressed-pubescent. Leaves with 1–2 pairs of lateral leaflets; petiole 0.6–2.8 cm long; leaflets up to 9 cm long.

Malawi. S: Zomba Plateau, Malosa Mt., fl.& fr. 3.i.1967, *Hilliard & Burtt* 4153 (K).
Also in South Africa (Limpopo, Mpumalanga, Gauteng and KwaZulu-Natal Provinces), Swaziland and Tanzania. Within the Flora Zambesiaca area the variety is known only from the Zomba Plateau (1530–1830 m). Plants from here are generally more robust with broader leaves (mostly 10–15 mm) than in South Africa, but do not appear to be taxonomically separable.

ii) Var. **lasiocaulos** Brummitt in Bol. Soc. Brot., sér.2, **41**: 269 (1968). Type: Zimbabwe, Matobo Dist., Chesterfield Farm, *Miller* 5493 (K holotype, SRGH).

> *Tephrosia aurantiaca* R.G.N. Young in Ann. Transvaal Mus. **14**: 398 (1932) nom. illegit., non *T. aurantiaca* Harms (1900).
> *Tephrosia elongata* E. Mey. var. *pubescens* H.M.L. Forbes in Bothalia **4**: 978 (1948), in part, illegitimate name, non var. *pubescens* Sond. (1850).

Differs from var. *elongata* in having spreading pubescence on stems, leaf rachides and petioles and leaflet lower surfaces.

Zimbabwe. W: Matobo Dist., Quaringa Farm, fl. i.1961, *Miller* 7650 (K, SRGH). E: Nyanga Dist., opposite Montclair (Dannakay) turn off on way to Juliasdale, fl. 25.x.1946, *Rattray* 1025 (K, SRGH).
Also in South Africa (Limpopo, Mpumalanga, Gauteng and KwaZulu-Natal Provinces). Usually in grassland; 1450–1800 m.

iii) Var. **tzaneenensis** (H.M.L. Forbes) Brummitt in Bol. Soc. Brot., sér.2, **41**: 270 (1968). Type: South Africa, Limpopo Province, Tzaneen, *Pole-Evans* 4024 (PRE).

> *Tephrosia tzaneenensis* H.M.L. Forbes in Bothalia **4**: 977 (1948).

Erect plant, up to 90 cm high. Stems, leaf rachides and petioles, and leaflet lower surfaces appressed-pubescent. Leaves with 1–3 pairs of lateral leaflets; petiole 10–7 mm long, or up to 10 mm in lower leaves; leaflets up to 13 cm long.

Mozambique. M: Namaacha Dist., Goba, near rio Maiuáua, fl.& fr. 2.xi.1960, *Balsinhas* 159 (K, LMA, PRE).

Also in South Africa (Limpopo Province). Open woodland with *Pterocarpus* and *Peltophorum*.

Var. *tzaneenensis* is distinguished by its very shortly petioled leaves. Plants from around the type locality in Limpopo Province tend to have broader leaflets, rounded at the apex; a number of collections from the mountains in S Mozambique are referred to this variety on the basis of their very short petioles, but differ from the Limpopo Province plants in having very long (up to 13 cm), narrow leaflets acute at the apex.

T. elongata is often mistaken for *T. longipes* or *T. lurida*, from which it can be distinguished with certainty by its glabrous style with penicillate tip.

10. **Tephrosia dregeana** E. Mey. in Linnaea **7**: 169 (1832). —Harvey in F.C. **2**: 207 (1862). —Baker f., Legum. Trop. Africa: 189 (1926). —Forbes in Bothalia **4**: 966 (1948). —Torre in C.F.A. **3**: 151 (1962). —Schreiber in Merxmüller, Prodr. Fl. SW Afrika, fam. 60: 117 (1970). Type: South Africa, *Drège* s.n. (B† holotype).

    Tephrosia damarensis Engl., Bot. Jahrb. Syst. **10**: 29 (1888). —Harms in Engler, Pflanzenw. Afrikas **3**(1): 588 (1915). Type from Namibia.

    Tephrosia dinteri Schinz in Vierteljahrsschr. Naturf. Ges. Zürich **52**: 424 (1907). Type from Namibia.

Annual with a long slender taproot to perennial with a woody stock. Stems 10–60 cm high, often becoming woody at the base, with sparse to sometimes moderately dense, appressed or rarely ascending hairs. Leaves pinnate with (1)2–4(5) pairs of leaflets; petiole (1)1.3–3(4) cm long, petiole and rachis together (1.5)2.5–5.5(8.5) cm long, pubescent like the stem; leaflets (1.3)2–4.5(7) × 0.1–0.6(0.7) cm, linear to linear-elliptic or rarely narrowly elliptic, subcuneate at the base, acute to subobtuse at the apex; upper surface glabrous or appressed-pubescent, lower surface somewhat appressed-pubescent; stipules 2–7 × 0.3–0.8 mm, linear. Flowers in terminal racemes up to 35 cm long; bracts 1–2 × 0.3 mm, subulate; pedicels 2–7(10) mm long, very slender. Calyx 1.5–2.5(3) mm long, appressed-pubescent with greyish hairs; lateral teeth much shorter than to about equalling the tube, upper teeth joined for about half their length. Petals 5–6 mm long, purple to blue or pink, very rarely described as yellow. Stamen tube joined above. Ovary pubescent; style glabrous, with penicillate tip. Pods (15)20–25(30) × (3.5)4(4.5) mm, boat-shaped, closely appressed-pubescent, light creamy-brown to straw-coloured. Seeds (1)2–6, c.3.5 × 2.5 × 1 mm, subreniform-oblong, conspicuously flattened, light brown, the aril scarcely developed.

Botswana. N: Ngamiland Dist., 10 km N of Aha Hills, fl.& fr. 12.iii.1965, *Wild & Drummond* 6944 (K, SRGH). SW: 53 km NE of Ghanzi on Maun road, fl. 29.xii.1977, *Skarpe* 216 (K).

Also in South Africa, Namibia and Angola. *Combretum apiculatum, Terminalia prunioides, Sclerocarya* savanna on shallow sand overlying limestone; elsewhere from sandy and gravelly ground, often in river beds, on dolomite, schist, granite, etc.

Var. *capillipes* (Baker) Torre in C.F.A. **3**: 152 (1962), from the coastal provinces of Angola, differs in having even more slender, flexuous pedicels mostly (8)10–20 mm long and shorter petioles (5–10 mm). The collections from the Flora Zambesiaca area show some tendency towards this, having pedicels up to 9 mm long and petioles 9–12 mm. Similar plants occur in Namibia. The varieties are probably not worth formal recognition.

11. **Tephrosia limpopoensis** J.B. Gillett in Kew Bull. **13**: 418 (1958). Type from South
    Africa (Limpopo Province).

Perennial low shrub up to 1 m high, the older parts woody and defoliate. Young stems densely appressed-pubescent with grey hairs, older stems with rugose bark. Leaves with 1 or 2 pairs of lateral leaflets; petiole up to 1 cm long, petiole and rachis together up to 2 cm long; leaflets up to 2.6 × 0.6 cm, oblanceolate, rounded and ± recurved beneath at the apex; upper surface glabrous, lower surface sparsely appressed-pubescent; stipules 4–5 × c.0.5 mm, linear. Flowers in short few-flowered terminal racemes and in the axils of upper leaves; bracts c.2 mm long; pedicels 2–3 mm long. Calyx c.5 mm long, grey-pubescent; lateral teeth about equalling the tube, the two upper teeth joined for ± half their length. Petals 7–10 mm long. Upper stamen almost free. Ovary pubescent; style glabrous. Pods c.33 cm long, rather sparsely appressed-pubescent. Seeds 7–8.

Zimbabwe. S: Beitbridge Dist., 3 km from Beitbridge on West Nicholson road, 15.ii.1955, *Exell, Mendonça & Wild* 413 (BM).

Also in South Africa (Limpopo and Mpumalanga Provinces). Known in the Flora area from only the one collection cited, from *Colophospermum mopane* bush; 400 m.

This species is very poorly known, as is 40. *T. euchroa* from the same area. Both are low shrubs with whitish appressed indumentum on the stems. *T. limpopoensis* was described as having a glabrous style, and has only 2 pairs of leaflets in available material, whereas *T. euchroa* has a pubescent syle and 4–8 pairs of leaflets. Further collections are needed to be sure they are distinct species.

Some collections from Gwanda Dist. in S Zimbabwe (*Davies* 890 (SRGH), 2307 (K, SRGH) and 2318 (K)) seem to be intermediate between *T. limpopoensis* and *T. purpurea* var. *leptostachya* (see Brummitt in Bol. Soc. Brot., sér.2, **41**: 246 (1968)). More collections and field observations are required for satisfactory determination of its status.

12. **Tephrosia purpurea** (L.) Pers., Syn. Pl. **2**: 239 (1807). —Baker in F.T.A. **2**: 124
    (1871); —R.E. Fries, Wiss. Ergebn. Schwed. Rhod.-Kongo-Exped. **1**: 84 (1914).
    —Eyles in Trans. Roy. Soc. South Africa **5**: 376 (1916). —Baker f., Legum. Trop.
    Africa: 190 (1926). —Gomes e Sousa, Pl. Menyharth.: 72 (1936). —Forbes in
    Bothalia **4**: 974 (1948). —Suessenguth & Merxmüller in Proc. & Trans. Rhod.
    Sci. Assoc. **43**: 26 (1951). —Hepper in F.W.T.A., ed.2, **1**: 530 (1958). —Torre in
    C.F.A. **3**: 153 (1962). —Brummitt in Bol. Soc. Brot., sér.2, **41**: 235 (1968). —
    Gillett in F.T.E.A., Legum., Pap.: 186 (1971). Type: Sri Lanka, *Herb. Hermann* 1:
    t.37/301 (BM holotype).
    Cracca purpurea L., Sp. Pl.: 752 (1753).

Annual or perennial. Stems procumbent, ascending or erect, up to 70(100) cm high or sometimes (subsp. *altissima*) up to 150 cm, sometimes woody in the lower parts and forming a low bush, appressed or spreading-pubescent to densely white- or silvery-grey pubescent and sericeous. Leaves pinnate with (3)4–10(11) pairs of leaflets; petiole 0.1–3.5 cm long, petiole and rachis together (1.5)3–10(13) cm long; leaflets (0.5)0.8–4(4.5) × (0.2)0.3–1(1.3) cm, elliptic or obovate to oblong or oblanceolate, cuneate at the base, rounded to truncate at the apex; upper surface glabrous to silky, lower surface thinly appressed-pubescent to silky; stipules (0.5)1–8 × 0.2–1.5 mm, shortly triangular to linear-triangular. Flowers in terminal and leaf-opposed racemes 2–20(24) cm long; bracts 0.5–5 × 0.1–0.7 mm; pedicels 2–6(8) mm long. Calyx (1)2–4(6) mm long, appressed spreading-pubescent; lateral teeth approximately equalling the tube or rather longer. Petals (5)6–10 mm long, purple or pinkish. Staminal tube joined above. Ovary pubescent; style glabrous. Pods 26–60 × 4–6.5 mm, dark brown, shortly pubescent. Seeds (4)6–10(11), 3–4 × 2–2.5 × 1–1.2 mm, ± rhomboid or oblongoid, with a very small aril at the middle of one of the longer sides, not markedly closely packed, the distance between the centres of adjacent seeds being greater than the breadth of the pod.

1. Young stems and inflorescence axes densely appressed-pubescent with white or silvery-grey hairs (on sandy maritime or lake shores) · · · · · · · · · · · · · · · · · 2
 – Stems and inflorescence axes not white- or silvery-grey pubescent · · · · · · · · 3
2. Upper surface of leaflets subglabrous to thinly appressed-pubescent, not silky; pods c.4.5 mm broad · iv) subsp. *dunensis*
 – Upper surface of leaflets densely appressed-pubescent with silvery-grey hairs and ± silky; pods 4.5–6.5 mm broad · · · · · · · · · · · · · · · · · · · v) subsp. *canescens*
3. Plant 60–150 cm high, erect, slender, lax, not bushy; pods (40)48–60 mm long
· iii) subsp. *altissima*
 – Plant procumbent or ascending or ± bushy, up to 70(100) cm high; pods 26–50(56) mm long · 4
4. Plant forming an erect a low bush, woody in the lower parts; inflorescences (1)2–10(15) cm, short and compact; seeds usually 5 or 6, occasionally 7, very rarely 8 (native of Asia, recorded from towns in Zimbabwe and Malawi) · · · · · ·
· i) subsp. *purpurea*
 – Plant procumbent or ascending or sometimes (var. *delagoensis*) ± erect and bushy; inflorescences (2)5–18(25) cm long, usually lax; seeds 7–10 or rarely 6 (subsp. *leptostachya*) · 5
5. Plant ± erect and bushy, woody in the lower parts · · · · · · · · iic) var. *delagoensis*
 – Plant procumbent or ascending, the stems slender and herbaceous, sometimes the stock becoming woody · 6
6. Stems subglabrous to appressed-pubescent · · · · · · · · · · · · iia) var. *leptostachya*
 – Stems spreading-pubescent · iib) var. *pubescens*

i) Subsp. **purpurea** —Brummitt in Bol. Soc. Brot., sér.2, **41**: 242 (1968).

Annual or short-lived perennial. Stems stout, erect, woody in the lower parts, forming a low bush up to 70(100) cm high, thinly appressed or spreading-pubescent. Leaves with (3)5–10(11) pairs of leaflets; petiole (0.2)0.5–1.5(2) cm long, petiole and rachis together (1.5)2.5–9(10) cm long; leaflets 0.8–2.5(3.2) × 0.3–1(1.1) cm; upper surface glabrous to thinly pubescent; stipules 2–8(10) × 0.2–1 mm. Inflorescences (1)2–10(15) cm long, compact, the axis not white-pubescent; bracts 1–3 mm long. Petals (5)6–7(9) mm long. Pods 26–40(44) × 3.5–5 mm; seeds usually (4)6, occasionally 7, very rarely 8.

Zimbabwe. C: Harare, Agriculture Expt. Station, near laboratory, fr. 25.iv.1932, *Rattray* 510 (SRGH). **Malawi**. C: Lilongwe Nature Reserve, fr. 3.v.1984, *Patel, Seyani & Banda* 1492 (K, MAL). S: Zomba Town, fr. 5.iv.1984, *Banda & Salubeni* 2156 (K, MAL).
Native of Asia from W Pakistan to Indonesia; introduced in the Americas and in Africa, particularly West Africa where it is apparently extensively naturalized and spreading. Known in the Flora area only from Harare, Lilongwe and Zomba.

ii) Subsp. **leptostachya** (DC.) Brummitt in Bol. Soc. Brot., sér.2, **41**: 245 (1968). — Schreiber in Merxmüller, Prodr. Fl. SW Afrika, fam. 60: 119 (1970). —Gillett in F.T.E.A., Legum., Pap.: 186 (1971). Type: Senegal, *Perrottet* s.n. (G lectotype, selected by Brummitt 1968).
 Tephrosia leptostachya DC., Prodr. **2**: 251 (1825). —Baker f., Legum. Trop. Africa: 191 (1926). —Hutchinson & Dalziel, F.W.T.A., ed.1, **1**: 385 (1928).

Annual or short-lived perennial. Stems procumbent or ascending, weak and straggling, or sometimes (var. *delagoensis*) stouter, erect and ± bushy, thinly appressed to spreading-pubescent. Leaves with (3)4–10(11) pairs of leaflets; petiole (0.1)0.2–2.0(2.5) cm long, petiole and rachis together (1)2.5–9(10) cm long; leaflets (0.5)0.8–2.5(3.0) × 0.2–0.9(1.1) cm; upper surface glabrous to thinly pubescent; stipules 2–7(8) × 0.2–1 mm. Inflorescences (2)5–18(25) cm, lax, the axis not white-pubescent; bracts (1)2–5 mm long. Petals 5–8(10) mm long. Pods (20)32–50(56) × 3–4.5(5) mm; seeds (6)7–10(11).

Africa and Madagascar; probably introduced in the Americas.

iia) Var. **leptostachya** —Brummitt in Bol. Soc. Brot., sér.2, **41**: 245 (1968). —Gillett in F.T.E.A., Legum., Pap.: 186, fig.32/1–15 (1971).

Stems procumbent or ascending, usually ± straggling, subglabrous or with closely appressed hairs.

Botswana. SW: Kgalagadi Dist., 12 km NW of Tshabong along track to Tshane, fl.& fr. 29.iii.1980, *Skarpe* S-440 (K). SE: Central Dist., 60 km NW of Serowe, fr. 24.iii.1965, *Wild & Drummond* 7276 (SRGH). **Zimbabwe**. E: Chipinge Dist., lower Save (Sabi), E bank, Hippo Mine area, fl.& fr. 12.iii.1957, *Phipps* 594 (SRGH). S: near Beitbridge on Bulawayo road, fr. 20.iii.1967, *Corby* 1861 (K, SRGH). **Mozambique**. N: Mossuril Dist., Ilha de Moçambique and adjacent mainland (Mozambik), fr. 5.iv.1894, *Kuntze* (K). MS: Marromeu Dist., Chupanga (Shupanga), fl.& fr. 2.viii.1959, *Kirk* (K). M: town of Maputo (Lourenço Marques), fl.& fr. 25.ii.1945, *Sousa* 46 (LISC, PRE).

Widespread in tropical Africa; 0–1000 m.

Gonçalves (Garcia de Orta, sér.Bot **5**: 116, 1982) reported this variety also from Tete Province in Mozambique.

iib) Var. **pubescens** Baker in F.T.A. **2**: 125 (1871). —Baker f., Legum. Trop. Africa: 191 (1926). —Cronquist in F.C.B. **5**: 98 (1954). —Torre in C.F.A. **3**: 153 (1962). —Brummitt in Bol. Soc. Brot., sér.2, **41**: 246 (1968). —Gillett in F.T.E.A., Legum., Pap.: 188 (1971). Type: Tanzania/Mozambique, Rovuma R., *Meller* s.n. (K lectotype, selected by Brummitt 1968).

   Tephrosia laurentii De Wild., Miss. Laurent: 111 (1905); Pl. Bequaert. **3**: 330 (1925). — Baker f., Legum. Trop. Africa: 192 (1926). Type from Congo.

   Tephrosia transvaalensis Hutch. & Burtt Davy, Fl. Pl. Ferns Transv.: 376, 378 (1932). — Forbes in Bothalia **4**: 975 (1948) in part, excl. *Rogers* 6881. Type: South Africa, Mpumalanga, Komatipoort, fr. 16.xii.1897, *Schlechter* 11783 (K).

   Tephrosia burchellii, T. semiglabra, T. capensis, T. polystachya, etc. as used by many in South Africa.

Stems procumbent or ascending, usually ± straggling, covered with spreading or ascending hairs.

Caprivi Strip. Linyanti Area, c.80 km from Katima on road to Linyanti, fl.& fr. 27.vii.1958, *Killick & Leistner* 3134 (K, SRGH). **Botswana**. N: Chobe Dist., between Kasane Rapids and Nyungwe Hot Springs, fl. 2.viii.1950, *Robertson & Elffers* 101 (K, PRE). SE: Central Dist., Palapye, Moeng College, fr. 29.xi.1957, *de Beer* 519 (K, PRE, SRGH). **Zambia:** B: Sesheke Dist., fl. 1910, *Gairdner* 23 (K). N: Mbala Dist., Kalambo Falls, fl.& fr. 21.iv.1959, *Richards* 11297 (K, SRGH). W: Ndola, fl.& fr. 21.ii.1954, *Fanshawe* 860 (COI, K, LISC, SRGH). C: Lusaka Dist., Mt. Makulu Res. Station, near Chilanga, fl. 7.ii.1958, *Angus* 1843 (K). S: Choma Nat. Forest (Siamambo Forest Res.), fl.& fr. 22.ii.1960, *White* 7301 (K, SRGH). **Zimbabwe**. N: Mazowe Dist., Wengi R., Mtorashanga–Concession road, fl.& fr. 2.iii.1965, *Corby* 1254 (K). W: Matobo Dist., Matopos, fl.& fr. ii.1931, *Rattray* 269 (K, PRE, SRGH). E: Mutare Dist., Odzi, Mukuni Purchase Area, fl.& fr. iii.1961, *Davies* 2889 (K, SRGH). S: Mwenezi Dist., G'zan Native Purchase Area, fl.& fr. v.1955, *Davies* 1220 (K, PRE; SRGH). **Malawi**. C: Mchinji–Mkanda Road, fl., *Salubeni & Patel* 3601 (K, MAL). **Mozambique**. MS: Beira, fl.& fr. ii.1912, *Rogers* 5918 (K). M: Magude Dist., near Mapulanguene, fl.& fr. 17.ii.1953, *Myre & Balsinhas* 1540 (K, LISC, SRGH).

Widespread in tropical Africa from Ghana to Somalia and south to Namibia and South Africa. Disturbed grassy places; up to 1500 m.

iic) Var. **delagoensis** (H.M.L. Forbes) Brummitt in Bol. Soc Brot., sér.2, **41**: 248 (1968). Type: Mozambique, Maputo (Lourenço Marques), in sand, *Schlechter* 11521 (PRE holotype, COI, K).

> *Tephrosia delagoensis* H.M.L. Forbes in Bothalia **4**: 968 (1948).
> *Tephrosia indigofera* Bertol., Misc. Bot. **19**: 9, t.5 (1858). Type: Mozambique GI: Inhambane, *Fornasini* (BOLO holotype).

Stems ± stout, stiff, erect, forming a low bush up to 70(100) cm high, sometimes woody in the lower parts; young stems spreading to appressed-pubescent.

Botswana. N: Central Dist., NE corner of Makgadikgadi Pan, fl.& fr. 15.i.1959, *West* 3826 (LISC, PRE, SRGH). **Zambia.** B: banks of Zambezi, 16 km below Senanga, fl.& fr. 8.ii.1952, *White* 2029 (K). **Zimbabwe.** N: Kariba Dist., Zambezi Valley, Ume (Bumi) R. walk, fl.& fr. ix.1955, *Davies* 1486 (K, SRGH). S: Beitbridge Dist., Chikwarakwara, Limpopo R., fl.& fr. 23.ii.1961, *Wild* 5340 (K, LISC, PRE, SRGH). **Malawi.** S: Nsanje Dist., between Muona and Shire R., fl.& fr. 20.iii.1960, *Phipps* 2579 (K, PRE, SRGH). **Mozambique.** N: Nacala a Velha Dist., Fernão Veloso, fl.& fr. 17.v.1937, *Torre* 1425 (COI). Z: Marromeu Dist., Zambezi R., Lacerdónia pontoon, 14.xii.1971, *Pope & Müller* 609 (K, SRGH). MS: 5 km from Chemba on road to Tambara, fl.& fr. 23.iv.1960, *Lemos & Macuácua* 142 (COI, K, LISC, LMA, PRE, SRGH). GI: near Chibuto, road to Alto Changane, fl.& fr. 12.ii.1959, *Barbosa & Lemos* 8382 (COI, K, LISC, LMA, PRE, SRGH). M: km 7 on road Maputo (Lourenço Marques)–Ressano Garcia, fl.& fr. 17.ii.1945, *Sousa* 18 (PRE, LISC).

Confined to the Flora Zambesiaca area on the coastal plain of Mozambique, extending up the major river valleys. Lowland grassy and waste places and river banks; sea level to 600 m.

Var. *delagoensis* has the habit of subsp. *purpurea* from Asia but in other characters agrees with subsp. *leptostachya* from Africa. In its morphological and geographical position this variety is closely analogous with var. *daviesii* of *Tephrosia villosa*.

iii) Subsp. **altissima** Brummitt in Bol. Soc. Brot., sér.2, **41**: 250 (1968). Type: Mozambique, Nhaungue Mt. (Serra do Garuzo), fl.& fr. 5.iii.1948, *Barbosa* 1137 (LISC holotype).

Annual or perhaps sometimes short-lived perennial. Stems erect, slender, lax, up to 1.5 m high; young stems rather thinly pubescent with appressed or spreading hairs. Leaves with (5)6–8(9) pairs of leaflets; petiole 0.3–1.4 cm long, petiole and rachis together (4)5–10(12) cm long; leaflets (1.5)1.8–4(4.5) × 0.4–1(1.3) cm, elliptic-oblong, cuneate at the base, rounded to truncate at the apex; upper surface glabrous; stipules (3)4–8 × 0.2–1 mm. Inflorescences up to 20(24) cm long, lax, not white-pubescent; bracts 2–5 × 0.2–0.6 mm. Petals 7–10 mm long. Pods (40)48–60 × (4)4.5 cm; seeds 7–9.

Zimbabwe. E: Mutare town, Darlington, sand pits, 1130 m, fl.& fr. 10.ii.1955, *Chase* 5468 (BM, COI, K, LISC, SRGH). **Malawi.** S: Nsanje Dist., Malawe Hills, fl.& fr. 23.iii.1960, *Phipps* 2641 (K, PRE, SRGH). **Mozambique.** N: Nacala a Velha Dist., between Itoculo and Fernao Veloso, fl.& fr. 15.x.1948, *Barbosa* 2439 (LISC). MS: Manica Dist., Nhaungue Mts. (Serra do Garuzo), fl.& fr. 5.iii.1948, *Garcia* 539 (LISC).

Confined to the Flora Zambesiaca area, in the border mountains of E Zimbabwe border and adjacent Mozambique and apparently also in S Malawi and near the coast in N Mozambique; 1130 m in Zimbabwe.

iv) Subsp. **dunensis** Brummitt in Bol. Soc. Brot., sér.2, **41**: 251 (1968). —Gillett in F.T.E.A., Legum., Pap.: 188 (1971). Type: Tanzania, Uzaramo Dist., 16 miles N of Dar es Salaam, sand dunes, fl.& fr. 2.vii.1960, *Leach & Brunton* 10164 (K holotype, SRGH).

Perennial with procumbent or ascending to suberect stems, woody in the lower parts. Young stems densely white-pubescent with strongly appressed, or very rarely (one collection on L. Malawi) spreading, hairs. Leaves with (3)5–9(10) pairs of leaflets; petiole 0.7–3.5 cm long, petiole and rachis together 2–8.5 cm long; leaflets 1–2.7 × (0.3)0.4–0.9 mm; upper surface subglabrous to thinly appressed-pubescent, not silky; stipules (0.5)1–3(3.5) × 0.8–1.5 mm. Inflorescences (1)2–12(22) cm long, short and compact to long and lax, the axis densely white-pubescent; bracts 0.5–2(3) × 0.5(1) mm. Petals (5.5)6–8 mm long. Pods 30–45 × 4.5 mm; seeds 6–8.

Malawi. N: Rumphi Dist., foot of Chombe Mt., 490 m, fl. 8.vii.1973, *Pawek* 7184 (K). C: between Lake Malawi (Nyasa) Hotel and Senga Bay Hotel, sand hills beside L. Malawi, fl. 17.ii.1959, *Robson* 1637 (K, LISC). **Mozambique**. N: Mogincual Dist., Quinga beach, fr. 28.iii.1964, *Torre & Paiva* 11452 (LISC). Z: 32 km N of Quelimane, fl.& fr. 10.viii.1962, *Wild* 5872 (K, PRE, SRGH). MS: Beira, fl.& fr. iv.1921, *Dummer* 4669 (K).

Coast of Indian Ocean from Somalia to N Mozambique and Madagascar, and on Lake Malawi. Sand dunes on sea and lake shores, often a primary colonizer.

One specimen, *Robson & Steel* 1606 from Lake Malawi Hotel on the shores of L. Malawi, differs from all others seen in having spreading pubescence.

v) Subsp. **canescens** (E. Mey.) Brummitt in Bol. Soc. Brot., sér.2, **41**: 253 (1968). Type: South Africa, KwaZulu-Natal, Omsamculo, sandy places, *Drège* s.n. (B†, K).

Tephrosia canescens E. Mey., Comment. Pl. Afr. Austr.: 109 (1836). —Harvey in F.C. **2**: 204 (1862). —Forbes in Bothalia **4**: 962 (1948). —Mogg in Macnae & Kalk, Nat. Hist. Inhaca Is., ed.2: 146 (1969).

Perennial with procumbent branches, woody in the lower parts. Young stems densely appressed-pubescent with silvery-white hairs. Leaves with (3)4–7(8) pairs of leaflets; petiole (0.5)1–2.2(3) cm long, petiole and rachis together (2)3.5–7(8) cm long; leaflets 0.8–2(2.5) × 0.4–1 cm, the margins often undulate; upper and lower surfaces densely and closely appressed-pubescent with silvery hairs, silky; stipules 1–2(3) × 1–1.5 mm, ± broadly triangular. Inflorescences short and compact, 2–8(10) cm long; bracts 0.5(1) mm long, broadly triangular. Petals 6–7 mm long. Pods (30)35–47 × 4.5–6.5 mm; seeds 6–7(9).

Mozambique. N: Moma Dist., Nejovo Is., 16°33'S, 39°48'E, fl.& fr. 27.x.1965, *Gomes e Sousa* 4899 (K). MS: Beira beach, *Swynnerton* 1446 (K). GI: between Vilankulo (Vilanculos) and Mucoque, dunes on beach, fl.& fr. 26.viii.1955, *Myre & Carvalho* 2172 (PRE, SRGH). M: Maputo Dist., Inhaca Is., lighthouse, N shore dunes, fl.& fr. 7.x.1957, *Mogg* 27719 (K, SRGH).

Also in South Africa (KwaZulu-Natal). Coastal sand dunes, often a primary colonizer; 0–20 m.

Although the silky indumentum gives this plant a very distinctive appearance, it does not seem to differ significantly from *T. purpurea* sensu lato in other characters and is connected to it through a series of intermediates occupying the same habitat referred to subsp. *dunensis*. It seems most appropriate to regard it as a subspecies of *T. purpurea*. Field observations as to whether intermediates between subsp. *dunensis* and var. *delagoensis* occur in the vicinity of Maputo are required.

The record of *Tephrosia sparsiflora* H.M.L. Forbes from Botswana, cited in Lock (Leg. Trop. Afr. Check-list: 384, 1989), is presumably taken from the citation of *Mogg* 4902 by Forbes in Bothalia **4**: 971 (1948). That specimen is from Vryburg in the Northern Cape Province of South Africa, which was referred to as Bechuanaland at that time. The status of *T. sparsiflora* remains uncertain, but it belongs within the *T. purpurea* complex.

13. **Tephrosia malvina** Brummitt in Bol. Soc. Brot., sér.2, **41**: 256 (1968). —Gillett in F.T.E.A., Legum., Pap.: 189 (1971). Type: Zambia, Mbala Dist., Katula, top escarpment, fl.& fr. 12.v.1955, *Richards* 5674 (K holotype).

Perennial with a vertical rootstock and several slender, herbaceous, prostrate or ascending stems. Stems with short appressed to spreading hairs. Leaves pinnate with 3–7(8) pairs of leaflets; petiole 1–3.5(4.5) cm long, longer than the lowermost pair of leaflets except sometimes in the upper leaves, petiole and rachis together (3)5–8(12) cm long, pubescent like the stem; leaflets 1–2.6 × 0.4–0.8 cm, elliptic to elliptic oblong or sometimes lanceolate or oblanceolate, cuneate to rounded at the base, subacute to rounded at the apex; secondary and tertiary veins usually raised on the upper surface; upper surface glabrous or occasionally shortly pubescent, lower surface thinly appressed-pubescent; stipules 3–8 × 0.3–1 mm, linear-triangular. Flowers in terminal and leaf-opposed racemes (5)7–20 cm long, the axis stout and conspicuously ridged or narrowly winged; bracts 3–7 × 0.5–0.8 mm, linear-triangular, usually rather conspicuous; pedicels 3–6 mm long. Calyx (2.5)3–5 mm long, appressed or sometimes spreading-pubescent; lateral teeth slightly shorter than to 1¹/₂ times as long as the tube; upper teeth joined for up to half their length. Petals (6)7–10 mm long, mauve to pink. Staminal tube joined above. Ovary pubescent; style glabrous. Pods 33–47 × 4.5–5.5 mm, usually straw-coloured, shortly pubescent. Seeds 6(7).

Zambia. N: Mbala Dist., Kellet's Farm, fl.& fr. 25.ii.1955, *Richards* 4678 (K). W: Luano, fl.& fr. 21.ii.1966, *Fanshawe* 9524 (K). C: 78 km from Lusaka on road to Petauke, fl.& fr. 12.ii.1975, *Brummitt & Lewis* 14335 (K). **Malawi**. N: Mzimba Dist., c.5 km W of Mzimba cut-off, Mtangatanga Forest Res., fl.& fr. 9.iii.1978, *Pawek* 13939 (K, MAL, MO). C: 11 km NW of Dedza on road to Lilongwe, fl.& imm.fr. 18.ii.1970, *Brummitt* 8615 (K). S: Ntcheu Dist., lower Kirk Range, Chipusiri, fr. 17.iii.1955, *Exell, Mendonça & Wild* 960 (LISC, SRGH).

Also in Tanzania, Congo and Burundi. Usually in sandy and grassy places; 1100–1800 m in the Flora area, at 960 m in Tanzania.

This species closely resembles *T. purpurea* subsp. *leptostachya*, from which it differs in its broader, ± straw-coloured, usually 6-seeded pods, stout and ± winged inflorescence axis, stiffer leaves with prominent venation, and petioles usually exceeding the lower pair of leaflets in length.

14. **Tephrosia micrantha** J.B. Gillett in Kew Bull. **15**: 41 (1961). —Brummitt in Bol. Soc. Brot., sér.2, **41**: 254 (1968). —Gillett in F.T.E.A., Legum., Pap.: 188 (1971). Type: Tanzania, Songea Dist., by Nakawali R., 2.5 km SW of Kitai, *Milne-Redhead & Taylor* 9112 (K holotype).

Annual to biennial with a fairly stout taproot and usually tufted habit with several ascending, much branched stems arising from the base and 10–25 cm high, or sometimes mat-forming with more elongate ± procumbent branches up to 30 cm long. Stems pubescent with appressed to ascending or occasionally spreading hairs. Leaves pinnate with (2)3–6(9) pairs of leaflets; petiole 0.1–1.8(3) cm long, varying from much shorter to much longer than the lower pair of leaflets, petiole and rachis together (1)2–6(8) cm long, pubescent like the stem; leaflets (5)8–20(25) × (1)2–5(6) mm, oblanceolate or narrowly elliptic to linear-elliptic, subcuneate at the base, subacute to rounded at the apex; upper surface glabrous or appressed-pubescent; lower surface appressed-pubescent, sometimes densely so and conspicuously greyish, rarely ± silky; stipules 3–7(8) × 0.1–0.5 mm, subulate, usually persistent and often conspicuous near the base of the stems above the hypocotyl. Flowers in terminal and leaf-opposed or axillary racemes 1–6(11) cm long; bracts 2–5 × 0.2–0.5 mm, ± subulate; pedicels 2–3(4) mm long. Calyx (1.5)2–3 mm long, shortly appressed-pubescent; teeth about equalling the tube, the two upper ones joined for up to half their length. Petals 4–5(6) mm long, pink to purplish. Staminal tube connate above. Ovary pubescent; style glabrous. Pods 20–37 × 3.5–4.5 mm, brown to straw-coloured, very shortly pubescent with often geniculate hairs. Seeds (4)5–6(7), c.3 × 2 × 1.5 mm, subreniform-oblong.

Zambia. N: Mbala Dist., Chisungu Estate, fl.& fr. 20.iv.1959, *McCallum Webster* 840 (K). W: Kitwe Dist., Ichimpe Nat. Forest (Ichimpi), fl.& fr. 7.x.1964, *Mutimushi* 1087 (K). C: 10 km E of Lusaka, fl.& fr. 30.i.1956, *King* 291 (K). S: Mazabuka Dist., P. De Villiers Louw's farm, Choma to Pemba mile 10, fl.& fr. 15.ii.1960, *White* 7068 (K, SRGH). **Zimbabwe**. N: Gokwe South Dist., near source of Guye R., fl.& fr. 28.iii.1962, *Bingham* 192 (K, LISC, SRGH). C: Chegutu Dist., Poole Farm, fl.& fr. 27.ii.1952, *Hornby* 3281 (K, SRGH). E: Mutare Dist., near Nyamakari R., Burma Farm, Burma Valley, fl.& fr. 22.ii.1962, *Chase* 7632 (K, SRGH). **Malawi**. N: Mzimba Dist., Mbawa Expt. Station, fl.& fr. 5.iv.1955, *Jackson* 1580 (K, SRGH). C: Salima Dist., Chitala escarpment, fl. 12.ii.1959, *Robson* 1565 (K). S: Machinga Dist., lower Kasupe, fl.& fr. 13.iii.1955, *Exell, Mendonça & Wild* 829 (LISC, SRGH). **Mozambique**. N: Malema Dist., Mutuáli, fl.& fr. 12.v.1948, *Pedro & Pedrógão* 3348 (LMA). Z: Gurué Dist., Mt Currarre, next to R. Loussi, c.750 m, fl.& fr. 11.ii.1964, *Torre & Paiva* 10549 (LISC).

Also in Burundi and S Tanzania. In open woodland (*Brachystegia*), bush or grassy places, sometimes a weed in disturbed ground, roadsides, etc.; 500–1500 m.

15. **Tephrosia pumila** (Lam.) Pers., Syn. Pl. **2**: 330 (1807). —Torre in C.F.A. **3**: 154 (1962). —Brummitt in Bol. Soc. Brot., sér.2, **41**: 258 (1968). —Gillett in F.T.E.A., Legum., Pap.: 184 (1971). Type: Madagascar, *Sonnerat* s.n. (P holotype).

   Galega pumila Lam., Encycl. Méth. Bot. **2**: 599 (1786).
   Tephrosia purpurea var. *pumila* (Lam.) Baker in Hooker, Fl. Brit. India **2**: 113 (1876). — Baker f., Legum. Trop. Africa: 191 (1926). —Cronquist in F.C.B. **5**: 99 (1954).

Var. **pumila** —Brummitt in Bol. Soc. Brot., sér.2, **41**: 259 (1968). —Gillett in F.T.E.A., Legum., Pap.: 184, fig. 32/16 (1971).

Annual to short-lived perennial with procumbent or ascending stems up to 50(80) cm long or sometimes forming a low stiff bush up to 50 cm high. Stems with conspicuous, rather soft, brown or greyish, spreading hairs. Leaves pinnate with (2)3–5(6) pairs of leaflets; petiole 0.1–1(1.5) cm long, petiole and rachis together (1)1.5–4(6) cm long; leaflets 7–20(26) × (2)3–7(10) mm, obovate or obtriangular to oblanceolate, cuneate at the base, rounded to truncate at the apex; upper surface appressed-pubescent or rarely glabrous, lower surface usually densely appressed-pubescent with ± greyish shaggy hairs; stipules (1)2–7 × 0.3–1 mm, linear-triangular. Flowers in terminal and leaf-opposed, short racemes up to 7(10) cm long, the lowermost flowers usually subtended by foliage leaves, sometimes only one or two nodes subtended by bracts, and occasionally all flowers apparently subtended by foliage leaves; bracts 2–4 × 0.3–0.5 mm, linear-triangular; pedicels 2–6 mm long. Calyx 3.5–5(6) mm long, covered with longish, spreading, brown or grey hairs; teeth (1.5)2–4 times as long as the tube, ± linear, the two upper teeth joined for up to half their length. Petals 6–8 mm long, pink or sometimes mauve or rarely ± white. Staminal tube joined above. Ovary pubescent; style glabrous, abruptly deflexed at right-angles to the pod. Pods 32–42(45) × 3.5–4(4.5, or 6 in Madagascar) mm, shortly and fairly densely pubescent with ± spreading, often geniculate hairs. Seeds (9)11–13(15), 1.8–3 × 1.5–2 × c.1 mm, rhomboidal, with a very small aril in the middle of one of the edges, closely packed, the distance between the centres of adjacent seeds being less than the breadth of the pod.

Botswana. N: Ngamiland Dist., Thamalakane R., Okavango, fl.& fr. 13.iii.1961, *Richards* 14694 (K, SRGH). **Zambia**. E: Petauke Dist., Luangwa Valley, Lusengazi Camp, fl.& fr. 14.v.1963, *Verboom* 821 (SRGH). C: Mpika Dist., Luangwa Valley, Mfuwe, fl.& fr. 7.ii.1966, *Astle* 4501 (K). **Zimbabwe**. N: Mazowe Dist., top of Mupingi (Impinge) Pass, 16 km from Mvurwi (Umvukwes) towards Guruve (Sipolilo), fl.& fr. 10.ii.1982, *Brummitt & Drummond* 15843 (K, SRGH). W: Hwange Dist., Kazungula, fr. iv.1955, *Davies* 1118 (COI, K, SRGH). E: Chipinge Dist., Save Valley Expt. Station, fl.& fr. xi.1959, *Soane* 155 (K, LISC, PRE, SRGH). **Malawi**. N: Karonga Dist., Lupembe Farm, fl.& fr. 22.iv.1963, *Salubeni* 22 (K, SRGH). C: Salima–Balaka Road, 12 km S of

Chipoka, fr. 4.v.1980, *Blackmore & Banda* 1438 (K, MAL). **Mozambique**. MS: Gorongosa Nat. Park, Road 3–4 area, fr. iii.1969, *Tinley* 1766 (K, SRGH). M: Magude Dist., Chobela–Magude, fl.& fr. 10.iii.1942, *Viana* 25 (PRE).

Widespread in tropical Africa, from Ghana to Ethiopia and Somalia and south to Angola and the Flora Zambesiaca area, also in Madagascar and Comoro Islands. In grassland, waste places, river banks, etc., especially on disturbed soil; sea level–1500 m.

Drummond in Kirkia **8**: 228 (1972) has recorded this variety from S Zimbabwe.

Var. *ciliata* (Craib) Brummitt occurs from India to Indonesia and perhaps in Mauritius. It differs in having smaller flowers (calyx 2–3(4) mm long, petals 4–6(7) mm long) and fruit (25–34(40) mm long, seeds (8)9–11(12)). Var. *aldabrensis* (J.R. Drumm. & Hemsl.) Brummitt has appressed-pubescent to subglabrous stems; it is known at present only from coasts (sea cliffs, sandy foreshores) of Kenya, Tanzania, Zanzibar and Aldabra Island, but may possibly occur on the coast of Mozambique. The sharply deflexed style is very characteristic of the species.

16. **Tephrosia burchellii** Burtt Davy in Bull. Misc. Inform., Kew **1921**: 50 (1921). — Forbes in Bothalia **4**: 983 (1948) excluding *Moss & Rogers* 742. —Miller in J. S. Afr. Bot. **18**: 36 (1952) for name only. —Martineau, Rhod. Wild Fl.: 37 (1953) for name only. —Schreiber in Merxmüller, Prodr. Fl. SW Afrika, fam. 60: 117 (1970). Type: South Africa, Northern Cape, Griqualand West, *Burchell* 1932 (K).

   Tephrosia semiglabra sensu Forbes in Bothalia **4**: 982 (1948) in small part for Botswana (*Rogers* 6500).

Annual to short-lived perennial with decumbent or ascending stems 15–45 cm long. Stems clothed with soft whitish spreading hairs when young, glabrescent later. Leaves pinnate with (3)5–8(10) pairs of leaflets; petiole (0.2)0.5–1.5(2.0) mm long, petiole and rachis together 2–6.5(9) cm long, pubescent like the stem; leaflets (4)7–18 × (3)5–9 mm, obovate or obtriangular, cuneate at the base, broadly rounded to truncate or occasionally slightly emarginate at the apex; upper surface glabrous, lower surface rather softly appressed-pubescent; stipules 2–6 × 0.3–1.2(1.8) mm, linear-triangular. Flowers in ± short and few flowered, terminal (appearing leaf-opposed) racemes 1–8(12) cm long with up to 6(10) nodes; bracts 2–5 mm long, linear; pedicels 2–5 mm long. Calyx 4–6 mm long, with longish, grey, appressed to ascending hairs; teeth linear, 3–4 times as long as the very short tube, the two upper teeth slightly joined towards their base. Petals 5–7 mm long, pink or perhaps sometimes red or purplish. Staminal tube joined above. Ovary appressed-pubescent; style glabrous. Pods (25)32–40 × 3.5–4.2(4.5) mm, straw-coloured, thinly appressed-pubescent. Seeds (6)7–9(10), usually 8, c.3 × 2 × 1 mm, ± rhomboidal, brown, with the hilum and very small aril slightly excentric on one of the longer sides.

Botswana. SW: Kgalagadi Dist., Mahuditlhake (Mahudutlake) Pan, fl.& fr. 22–27.v.1967, *Cox* 312 (K). SE: Kgatleng Dist., Mochudi, fl.& fr. i.1914, *Harbor* in *Rogers* 6878 (K).

Also in South Africa (Limpopo, North West, Free State and Northern Cape Provinces) and Namibia. On sandy soil, in grassland and deciduous bushland.

Owing to taxonomic confusion in this group some previous records of this species in the Flora Zambesiaca area must be regarded as doubtful. It was said by Martineau (1953) to be very common in the Bulawayo area, but specimens from the Flora area have only been seen from SE Botswana. The description given by Miller (1952) for this (a shrublet 2 feet high with 3-seeded pod) seems also to be based on a misidentification. The species is best distinguished from *T. purpurea* by its strongly obovate or obtriangular leaflets and very long linear calyx teeth.

17. **Tephrosia coronilloides** Baker in F.T.A. **2**: 123 (1871). —Baker f., Legum. Trop. Africa: 195 (1926). —Gillett in Kew Bull. **13**: 120 (1958). —Torre in C.F.A. **3**: 155 (1962). —Brummitt in Bol. Soc. Brot., sér.2, **41**: 262 (1968). Type: Angola,

Pungo Andongo, *Welwitsch* 2080 (BM, K).

> *Tephrosia longana* Harms in Warburg, Kunene-Samb.-Exped. Baum: 259 (1903). —R.E. Fries, Wiss. Ergebn. Schwed. Rhod.-Kongo-Exped. **1**: 83 (1914) for name, excluding *Fries* 833a. Type: Angola, Napalanca, *Baum* 612 (K).
>
> *Tephrosia pallens* var. *angolensis* Baker f., Legum. Trop. Africa: 191 (1926). Type: Angola, rio Cassuango–Cuiriri, *Gossweiler* 3681 (BM holotype, K).

?Perennial, with ± erect branched stems 25–45 cm high. Stems with ascending to spreading brown hairs. Leaves pinnate with 5–10 pairs of leaflets; petiole 2–3(5) mm long, petiole and rachis together 1.5–7 cm long, pubescent like the stem; leaflets 1–1.7 × 0.3–0.7 cm, obovate to oblanceolate, subcuneate at the base, rounded to emarginate at the apex; upper surface thinly pubescent, lower surface densely pubescent; stipules 2.5–7 × 0.2–0.5 mm, linear. Flowers in terminal and axillary fairly lax racemes 3–10 cm long, and sometimes some in the axil of the uppermost leaves; bracts 2–3 × c.0.2 mm, linear; pedicels 3–5 mm long. Calyx 6–8 mm long, covered with ± villous brown hairs; lateral teeth 2–4 times as long as the tube, upper teeth free almost to the base. Petals 7–9 mm long, pink (possibly varying to red, orange or yellow?). Ovary pubescent; style glabrous with penicillate tip. Pods c.45 × 4.5–5 mm, brown, covered with brown hairs. Seeds 9–10, c.3.5 × 2 × 1.2 mm, brown with light markings, asymmetrically reniform, with a small aril c.0.2 mm long.

Zambia. B: Sesheke Dist., 16 km W of Katima Mulilo, 1952, *White* 2001 (FHO); Machili, fl.& fr. 24.ii.1961, *Fanshawe* 6319 (K, SRGH). **Zimbabwe**. C: Chikomba Dist., Wiltshire Native Purchase Area, fr. 30.iv.1965, *Corby* 1312 (K, SRGH).

Also in Angola. Within the Flora Zambesiaca area known only from the three collections cited. Recorded from sandy places.

The type collection of *T. pallens* var. *angolensis* Baker f. (BM, K) has the long calyx lobes characteristic of this species, but differs in having appressed greyish hairs on the stem, and obtusely pointed leaflets. It is referred with some doubt to *T. coronilloides*.

18. **Tephrosia lepida** Baker f. in Bull. Soc. Roy. Bot. Belg. **57**: 121 (1925); Legum. Trop. Africa: 195 (1926). —Cronquist in F.C.B. **5**: 101 (1954). —Brummitt in Bol. Soc. Brot., sér.2, **41**: 264 (1968). —Gillett in F.T.E.A., Legum., Pap.: 182 (1971). Type: Congo, Katanga, *Kassner* 2575 (BR syntype, BM, K), *Homblé* 358, *Ringoet* 417 & 501 (BR syntypes).

Annual or sometimes a short-lived perennial, often becoming somewhat woody towards the base. Stems erect or occasionally decumbent, branching, 30–100 cm high, with dense, appressed or very rarely somewhat ascending, pubescent, grey or brown hairs. Leaves pinnate with (2)4–7(8) pairs of leaflets; petiole (1)2–10(15) mm long, petiole and rachis together (0.5)2–5(7.8) cm long, pubescent like the stems; leaflets (0.7)1–2.5(3.5) × 0.2–0.8(0.9) cm, oblanceolate, or narrowly elliptic to sometimes linear-oblong, 3–5(6) times as long as broad, ± cuneate at the base, rounded to emarginate at the apex; upper surface glabrous, lower surface densely appressed-pubescent with silvery-grey hairs, sometimes subsericeous, usually contrasting markedly with the upper surface; stipules (1)2–4(5) × 0.2–0.5 mm, ± subulate. Flowers in terminal and axillary, usually fairly lax racemes (2.5)4–15(25) cm long and sometimes in the axils of the upper foliage leaves; bracts 1.5–3 × 0.2–0.5 mm; pedicels 1–5 mm long. Calyx 3–5 mm long, densely appressed-pubescent with dark brown to whitish hairs; lateral teeth $\frac{1}{2}$ as long as to almost equalling the tube, the upper teeth joined for $\frac{1}{2}$ to almost all their length. Petals (10)12–15(16) mm long, orange, red, pink or rarely somewhat purplish. Staminal tube firmly or loosely joined above. Ovary pubescent; style glabrous with penicillate tip. Pods 36–76 × (2.5)3–4 mm (see subspecies), markedly two-coloured, the marginal hairs brown or blackish and the hairs on the surfaces silver or grey. Seeds 8–12, 3–3.5 × 1.2–2 × c.1 mm, brown, oblong to almost fusiform, with a small aril excentrically placed on one of the longer sides.

Subsp. **lepida** —Brummitt in Bol. Soc. Brot., sér.2, **41**: 265 (1968). —Lock, Leg. Afr. Check-list: 376 (1989) in part.

Often perennial and woody towards the base, sometimes almost shrubby, but sometimes a slender annual. Leaflets (2)4–8(9) mm broad. Pods 36–48(51) × (2.5)3–3.5 mm, the margins usually brown and surfaces silvery-brown; seeds 8–10(11).

Zambia. N: Chinsali Dist., Shiwa Ngandu, fr. 4.vi.1956, *Robinson* 1583 (K, SRGH). W: Ndola, fr. 12.iv.1954, *Fanshawe* 1082 (K, LISC). C: Lusaka Dist., 2 km S of Shimabala, N of Kafue, fr. 13.iv.1958, *Angus* 1894 (K). S: Choma Dist., 16 km N of Choma, fr. 15.iv.1963, *van Rensburg* 1940 (K; SRGH). **Zimbabwe**. N: Gokwe South Dist., near Manyoni Vlei, Charama Plateau, fl.& fr. iii.1962, *Bingham* 171A (SRGH). C: Kadoma (Gatooma), fr. iii.1945, *McKinstry* 13 (K, SRGH).

Also in Congo (Katanga). Usually in woodland; 1220–1555 m.

Records from Burundi in Lock (1989), are based on misidentifications of *T. argyrotricha.*

Subsp. **nigrescens** Brummitt in Bol. Soc. Brot., sér.2, **41**: 265 (1968). —Gillett in F.T.E.A., Legum., Pap.: 182 (1971). Type: Zambia, Mbala Dist., Kumbula (Nmbulu) Is., Lake Tanganyika, steep side of island, fl.& fr. 11.iv.1955, *Richards* 5398 (K holotype).

Annual, slender and erect. Leaflets 2–5(7) mm broad. Pods (52)60–76 × 3.5–4 mm, the margins black or grey and surfaces dark grey or silvery, generally much darker than in subsp. *lepida,* and often with a less marked contrast between the margins and the surfaces; seeds (9)10–12.

Zambia. N: Mbala Dist., Kalambo Falls, fl.& fr. 29.iii.1955, *Exell, Mendonça & Wild* 1275 (BM, LISC, SRGH).

Apparently confined to Mbala and Mporokoso Districts in Zambia and S Tanzania (Ufipa and Iringa Dist.). In miombo woodland, bush, at roadsides, etc.; 730–1770 m.

19. **Tephrosia argyrotricha** Harms in Kuntze, Rev. Gen. Pl. **3** (2): 57 (1898). — Baker f., Legum. Trop. Africa: 191 (1926). —Gillett in Kew Bull. **13**: 120 (1958); in F.T.E.A., Legum., Pap.: 181 (1971). Type: Mozambique, "Mozambik" (Ilha de Moçambique and the adjacent mainland), 5.iv.1894, *Kuntze* s.n. (K, NY).

   Cracca argyrotricha (Harms) Kuntze, Rev. Gen. Pl. **3** (2): 57 (1898).
   Tephrosia burttii Baker f. in J. Bot. **71**: 341 (1933). Type: Tanzania, Dodoma Dist., between Kazikazi and Chaya stations, *B.D. Burtt* 4001 (BM holotype, K).
   Tephrosia argyrotricha var. *burttii* (Baker f.) J.B. Gillett in Kew Bull. **13**: 120 (1958).

Annual or perhaps sometimes a short-lived perennial, with erect, branched stems 45–150 cm high. Stems with dense, ± appressed, grey or silvery hairs. Leaves pinnate with (4)6–11 pairs of leaflets; petiole 0.1–0.3 cm long, petiole and rachis together 1.4–5 cm long, pubescent like the stem; leaflets (5)7–16 × (1)1.5–3.5(4.5) mm, oblanceolate to narrow-oblanceolate, 3–6 times as long as broad, subcuneate at the base, rounded to emarginate at the apex; upper surface glabrous or shortly pubescent, lower surface densely appressed-pubescent with silvery hairs; stipules 3–5(6) × c.0.2 mm, ± subulate. Flowers in terminal and axillary racemes 3–18 cm long and in the axils of up to 5 upper leaves; bracts 1–2.5 × c.0.2 mm, ± subulate; pedicels 1.5–3 mm long. Calyx 4–5 mm long, covered with longish, appressed to ascending brown or grey hairs; lateral teeth about equalling the tube, upper teeth connate for $^1/_2$–$^2/_3$ their length. Petals 7–9 mm long, orange, red or pink or sometimes somewhat purplish. Upper stamen free. Ovary pubescent; style glabrous with penicillate tip. Pods 32–47 × 3–3.5 mm, margins with dark brown hairs, surfaces with light brownish-grey to silvery appressed hairs. Seeds 8–9, 2–2.8 × 1–1.6 × 0.8–1 mm, oblongoid, with a very small aril excentrically placed on one of the longer sides.

Mozambique. N: Ribáuè, fl.& fr. 9.v.1948, *Pedro & Pedrógão* 3278 (LISC, LMA); Cabo Delgado, Montepuez, 23 km towards Nantulo, fl.& fr. 8.iv.1964, *Torre & Paiva* 11767 (K, LISC).

Also in Tanzania, as far north as Mwanza Dist., and in Burundi. Usually in miombo woodland.

The record for Zambia in Gillett (1958) and Lock (1989), is based on *Richards* 12591 from Kasama Dist., Chibutubutu, a specimen of *T. decora*.

20. **Tephrosia decora** Baker in F.T.A. **2**: 123 (1871). —Gillett in Kew Bull. **13**: 119 (1958). —Torre in C.F.A. **3**: 154 (1962). —Gillett in F.T.E.A., Legum., Pap.: 180 (1971). Type: Angola, Cuanza Norte, Pungo Andongo, *Welwitsch* 2098 (BM, COI, K).

   Tephrosia delicata Baker f., Legum. Trop. Africa: 192 (1926). —Torre in C.F.A. **3**: 154 (1962). Type: Zimbabwe, Odzani R. Valley, *Teague* 436 (K) in part, see Gillett in Kew Bull. **13**: 119–120 (1958).

   Tephrosia lateritia Merxm. in Proc. & Trans. Rhod. Sci. Assoc. **43**: 25 (1951). Type: Zimbabwe, Makoni Dist., Rusape, *Dehn* 743/52 (K, SRGH).

Annual with erect or decumbent-ascending branched stems, (15)30–80(100) cm high. Stems closely appressed-pubescent to occasionally ascending or spreading-pubescent, with usually light brown hairs. Leaves pinnate with (1)3–6(8) pairs of leaflets; petiole (0.2)0.3–1.2(2.2) cm long, petiole and rachis together (1)2–6(7) cm long, pubescent like the stem; leaflets (0.5)1–2.7(3.8) × 0.3–0.6(1.0) cm, narrowly elliptic to oblanceolate, 3–5(6) times as long as broad, ± cuneate at the base, obtuse to emarginate at the apex; upper surface glabrous or occasionally sparsely appressed-pubescent, lower surface densely appressed-pubescent with greyish or silvery hairs, sometimes ± silky; stipules 2–4(6) × 0.1–0.3(0.5) mm, ± subulate. Flowers in terminal and axillary, slender, lax racemes up to 15(19) cm long, sometimes also in the axils of the uppermost foliage leaves; bracts 1–2(4) × 0.3 mm, ± subulate; pedicels 2–3(5 in fruit) mm long. Calyx 3–4 mm long, densely pubescent with appressed or ascending brown hairs; lateral teeth equalling or somewhat longer than the tube, upper teeth joined for up to ¹/₄ their length or free to the base. Petals 6–8 mm long, orange, yellow, red or pink. Upper stamen free. Ovary pubescent; style glabrous with penicillate tip. Pods 32–52 × 3.5–4.5 mm, brown with appressed or ascending brown hairs. Seeds (9)10–11(12), 2.2–2.8 × 1.3–1.8 × 0.5–0.8 mm, brown or blackish, asymmetrically ovoid, with a conspicuous whitish aril 0.5–0.8 mm long on the longer side towards the narrower end.

Zambia. N: Kasama Dist., Misamfu, 6 km N of Kasama, fr. 5.iv.1961, *Angus* 2684 (K). W: Copperbelt, Mufulira Dist., fl. 19.ii.1949, *Cruse* 490 (K). C: Lusaka, Stuart Park, fl. 19.ii.1961, *Best* 282 (K, SRGH). E: Chipata Dist., fl. ii.1962, *Verboom* 472 (K, SRGH). S: Choma Dist., Tara Protected Forest Area, fl. 1.iii.1960, *White* 7513 (K, SRGH). **Zimbabwe**. N: Zvimba Dist., Trelawney, Tobacco Expt. Station, fl. 29.iii.1943, *Jack* 110 (PRE, SRGH). C: Harare Dist., Hatfield, fl. 17.ii.1957, *Whellan* 1176 (K, SRGH). E: Mutasa Dist., Odzani R. Valley, fl.& fr. 1915, *Teague* 436 in part (K). **Malawi**. N: Mzimba Dist., Mzuzu, Marymount, fl. 11.iii.1973, *Pawek* 6488 (K). C: Kasungu Dist., Chimaliro Forest Res., fr. 23.iv.1974, *Pawek* 8490 (K, MAL, MO, SRGH). S: Blantyre Dist., Shire Highlands, fl.& fr. 6.vii.1879, *Buchanan* (K). **Mozambique**. N: Malema Dist., Murralelo, Morgados farm, base of Inago Mt, fl. 19.iii.1964, *Torre & Paiva* 11254 (LISC). Z: between Régulo Ságura and Namarroi, 22.6 km from Régulo Ságura, fl.& fr. 17.ix.1949, *Barbosa & Carvalho* 4117 (K, LMA). T: between Vila Mouzinho and Zóbuè, 73.7 km from Vila Mouzinho, fl.& fr. 19.vii.1949, *Barbosa & Carvalho* 3699 (K, LISC, LMA).

Also in Congo, S Tanzania and Angola. Usually in woodland (*Brachystegia, Uapaca*), often on sandy soils; 720–1525 m.

Gillett in Kew Bull. **13**: 120 (1958) has suggested that plants with leaflets pubescent above occur only in the western parts of the distribution range of this species and might be the result of introgression with *T. coronilloides*. It now appears that such plants occur throughout the area of the species and represent merely a sporadic variant.

21. **Tephrosia linearis** (Willd.) Pers., Syn. Pl. **2**: 330 (1807). —Baker in F.T.A. **2**: 120 (1871). —Baker f., Legum. Trop. Africa: 189 (1926). —Robyns, Fl. Sperm. Parc Nat. Alb. **1**: 308, t.28 (1948). —Cronquist in F.C.B. **5**: 95 (1954). —Hepper in F.W.T.A., ed.2, **1**: 528, 531(1958). —Gillett in Kew Bull. **13**: 118 (1958). —Torre in C.F.A. **3**: 152 (1962). —Gillett in F.T.E.A., Legum., Pap.: 179 (1971). Type: Ghana, Ada, *Isert* s.n. (B holotype, C).

> *Galega linearis* Willd. in Sp. Pl., ed.4, **3**: 1248 (1803).
> *Cracca linearis* (Willd.) Kuntze, Rev. Gen. Pl. **1**: 175 (1891). —Hiern, Cat. Afr. Pl. Welw. **1**: 222 (1896).

Annual with a slender taproot, or perennial with a vertical stock, with erect, branching stems (15)40–100(160) cm high. Stems with appressed to spreading or occasionally deflexed hairs. Leaves pinnate with (1)3–6(8) pairs of leaflets; petiole 0.1–0.4(0.6) cm long, petiole and rachis together (1)1.5–5(6.5) cm long, pubescent like the stem; leaflets (10)15–40(50) × (0.8)1–4 mm, linear-oblong to linear, ± cuneate at the base, narrowly rounded at the apex; upper surface glabrous or rarely shortly pubescent, lower surface appressed-pubescent with silvery, usually ± silky hairs contrasting rather markedly with the upper surface; stipules (2)3–6(7) × 0.1–0.7(0.8) mm. Flowers in terminal and axillary, slender, lax racemes up to 12(20) cm long, and often also in the axils of the uppermost leaves; bracts 2–3 × 0.2 mm, linear; pedicels 2–4(5) mm long. Calyx 2–7 mm long (see varieties), appressed to ascending or rarely spreading-pubescent; lateral teeth equalling to 3(4) times longer than the tube, the two upper teeth joined for up to half their length or free to the base. Petals 6–11 mm long (see varieties), orange, yellow, red or pinkish. Upper filament free or lightly attached, at least near the base, to the stamen tube. Ovary pubescent; style glabrous with penicillate tip. Pods 40–55(60) × 3–4(4.5) mm, brown, with fairly dense ± loosely appressed hairs. Seeds 9–13(15), brown to black, 2.5–3 × 1.5–2 × 0.8–1 mm, oblongoid-ovoid, with a conspicuous U-shaped aril obliquely placed towards one end of the seed and markedly thickened and elongated at the bottom of the U.

Widespread in tropical Africa (see varieties).

A very variable species with two recognisable taxa that have previously been given either specific or subspecific rank. Although extremes appear markedly different there is a very broad intergradation and no clear geographical separation (though there is a tendency for different altitudinal preferences), and varietal rank seems more appropriate. In the Flora area, however, even this distinction is often highly arbitrary. Further investigation is required, including growth experiments under controlled conditions. Future collections should be annotated as to whether annual or perennial.

Var. **linearis** —Drummond in Kirkia **8**: 227 (1972).

> *Tephrosia pulchella* Hook.f., Niger Fl.: 299 (1849). —Baker in F.T.A. **2**: 120 (1871). — Baker f., Legum. Trop. Africa: 189 (1926). Type from Nigeria.

Annual with stems (15)40–100(130) cm high. Stems appressed- to ascending-pubescent. Leaflets (10)15–35(50) × (0.8)1–3 mm. Calyx 2–4(5) mm long, the lateral teeth 1–1.5(2) times as long as the tube. Petals 6–8(9) mm long. Upper stamen usually free.

Zambia. B: Mongu Dist., Barotse floodplain, Lizula Is., fl.& fr. 8.iv.1964, *Verboom* 1036 (K). N: Mbala Dist., Kasaba (Cassava) Sands, L. Tanganyika, fl. 17.ii.1959, *Richards* 10932 (K). C: Lusaka Dist., Chongwe R. near Constantia, fl.& fr. 30.xi.1972, *Kornaś* 2729 (K). E: Chipata Dist., Jumbe area, fl.& fr. 24.iii.1963, *Verboom* 813 (K, SRGH). S: Mazabuka, edge of Kafue floodplain, near van Wyk's farm, fl.& fr.

6.iii.1963, *van Rensburg* KBS 1606 (K, SRGH). **Zimbabwe**. E: Mutare Dist., municipal sand pits, Darlington suburbs, fl.& fr. 8.iii.1960, *Chase* 7281 (K, LISC, SRGH). **Malawi**. N: Karonga Dist., near Chitimba, fl.& fr. 24.iv.1969, *Pawek* 2318 (K). C: Nkhotakota (Kota Kota), fl.& fr. 2.vi.1963, *Verboom* 888 (K, SRGH). S?: without locality, 1891, *Buchanan* 596 (K). **Mozambique**. N: Mecubúri (Mucuburi), fl.& fr. 21.xi.1936, *Torre* 1069 (COI, LISC). Z: Lugela Dist., Namagoa, fl.& fr. i.ii.1943, *Faulkner* PRE 153 (K, PRE, SRGH). MS: Cheringoma coastal area, Zuni sector S of limit of Zambezi Delta, fl.& fr. v.1973, *Tinley* 2870 (K, SRGH). M: Marracuene Dist., Rikatla, fl.& fr. iii.1919, *Junod* 537 (LISC, PRE).

Also from W Africa (Senegal) to the Sudan and south to Angola and Mozambique; and in Madagascar. In grassland, woodland and disturbed ground; often at lower altitudes than var. *discolor*; below 1400 m.

Drummond in Kirkia **8**: 227 (1972) has recorded this variety from C Zimbabwe.

Var. **discolor** (E. Mey.) Brummitt in Bol. Soc. Brot., sér.2, **41**: 266 (1968). Type: South Africa, KwaZulu-Natal, Durban (Port Natal), *Drège* s.n. (K).

> *Tephrosia discolor* E. Mey., Comment. Pl. Afr. Austr.: 111 (1836). —Harvey in F.C. **2**: 207 (1862). —Burtt Davy, Fl. Pl. Ferns Transvaal, pt.2: 377, 378 (1932). —Forbes in Bothalia **4**: 969 (1948). —Cronquist in F.C.B. **5**: 96 (1954).
> *Tephrosia linearis* subsp. *discolor* (E. Mey.) J.B. Gillett in Kew Bull. **13**: 119 (1958). —Torre in C.F.A. **3**: 152 (1962).
> *Tephrosia longipes* sensu Baker in F.T.A. **2**: 120 (1871).

Perennial with stems (30)50–100(160) cm high. Stems with appressed, ascending, spreading or deflexed hairs. Leaflets (15)20–40(50) × (1)2–4 mm. Calyx 4–7 mm long, the lateral teeth (1.5)2–3(4) times as long as the tube. Petals (7)8–11 mm long. Upper stamen usually lightly attached, at least near the base, to the tube.

Zambia. N: Mbala Dist., Lake Chila, fl.& fr. 25.i.1952, *Richards* 566 (K). S: Choma Dist., Mapanza, fl.& imm.fr. 9.ii.1958, *E.A. Robinson* 2758 (K, SRGH). **Zimbabwe**. N: Zvimba Dist., Trelawney, Tobacco Res. Station, fl.& fr. 6.iv.1945, *Jack* 100 (SRGH). W: Matobo Dist., Farm Besna Kobila, fl.& fr. i.1954, *Miller* 2123 (K, LISC, PRE, SRGH). C: Marondera Dist., Delta Farm, fl.& fr. 5.iv.1950, *Wild* 3301 (K, SRGH). **Malawi**. N: Nyika Plateau, fl.& fr. ii.1903, *McClounie* (K). C: Lilongwe Dist., Namitete R. above bridge on Lilongwe–Chipata (Fort Jameson) road, fl.& fr. 5.ii.1959, *Robson* 1474 (K, LISC, SRGH). S: Mt. Mulanje, fl.& fr. 1891, *Whyte* (K). **Mozambique**. N: Ribáuè, 20 km from Ribáuè to Lalaua, on sr. Elio Silva's property, fl.& fr. 22.i.1964, *Torre & Paiva* 10108 (LISC). MS: Gondola Dist., Lion's Creek, between Beira and Mutare (Umtali), fl.& fr. 8.iv.1898, *Schlechter* 12208 (PRE). M: Matutuíne Dist., road from Ponta do Ouro beach to Salamanga, 3 km from Ponta do Ouro, fl.& fr. 20.ii.1952, *Barbosa & Balsinhas* 4792 (LISC, LMA); Polana road, fl.& imm.fr. 22.i.1941, *M.G. Hornby* 854b in part (K).

From Ethiopia through East Africa and Congo (mainly eastern) to Angola, the Flora Zambesiaca area and eastern parts of South Africa (Mpumalanga and KwaZulu-Natal). In grassland, woodland, disturbed or rocky ground; at the coast in KwaZulu-Natal but usually 1100–1600 m in the Flora area and to 2440 m in East Africa.

22. **Tephrosia richardsiae** J.B. Gillett in Kew Bull. **13**: 117 (1958). —Brummitt in Bol. Soc. Brot., sér.2, **41**: 267 (1968). —Gillett in F.T.E.A., Legum., Pap.: 178 (1971). Type: Zambia, Mpulungu, Lake Tanganyika, fl.& fr. 8.iii.1952, *Richards* 1079 (K holotype).

Herbaceous annual to woody perennial shrub, erect or prostrate, somewhat bushy, 30–100 cm high. Stems densely pubescent to tomentose with whitish to rusty-brown upwardly appressed to deflexed hairs. Leaves pinnate with (2)4–8 pairs of leaflets; petiole 0.3–1(1.5) cm long, petiole and rachis together (1)2–5(7) cm long, pubescent to tomentose like the stem; leaflets 1–3(3.7) × (0.3)0.4–0.7(0.9) cm, oblanceolate to narrowly obtriangular or narrowly elliptic, cuneate to rounded at the base, rounded to emarginate at the apex; upper surface glabrous or rarely the younger leaves appressed-pubescent and glabrescent, the lower surface densely appressed-pubescent with silvery hairs, silky; stipules 5–9(14) × 0.5–1 mm, linear, persistent. Flowers in dense heads 1–3(4) cm long, terminal on main or lateral branches and surrounded or exceeded by the upper leaves, and usually also some flowers in the axils of the 1–5 uppermost leaves; bracts 3–9 × 0.5–1 mm, linear; pedicels 2–3(5 in fruit) mm long. Calyx (5)6–9 mm long, densely pubescent with appressed to spreading hairs, grey, or brown on the teeth; lateral teeth equalling or somewhat exceeding the tube, upper teeth joined for less than $^1/_3$ their length or ± free to the base. Petals 12–18 mm long, yellow or orange to pinkish. Stamen tube connate above. Ovary pubescent; style glabrous, with penicillate tip. Pods 30–44 × 3.5–5 mm, lanate-tomentose with grey or brown hairs, those at the margin sometimes conspicuously darker than those on the surfaces. Seeds 6–9, c.2.8 × 2 × 1 mm, blackish-brown, ± ovoid, with a small rim aril towards the narrower end.

Subsp. **richardsiae** —Brummitt in Bol. Soc. Brot., sér.2, **41**: 267 (1968).

Stems with whitish to grey-brown appressed to ascending hairs. Calyx with appressed to ascending hairs. Pods (3.5)4–5 mm broad, with hairs uniformly grey or grey-brown, or those at the margin sometimes slightly darker.

Zambia. N: Kaputa Dist., Nsama, fl.& imm.fr. 4.iv.1957, *Richards* 9012 (K, SRGH). E: Lundazi Dist., Mica Mine Hill, fl. iii.1962, *Verboom* 610 (K). **Malawi**. N: Chitipa Dist., Kaseye Mission, 16 km E of Chitipa, fl. 5.iv.1969, *Pawek* 1944 (K).

Also in Tanzania (Iringa and Dodoma Districts). Usually on sandy or stony soil, often in disturbed ground, sometimes in plateau woodland; 550–1705 m.

Subsp. **erucifera** Brummitt in Bol. Soc. Brot., sér.2, **41**: 267 (1968). Type: Zambia N: Mpika Dist., low rocky hills by Serenje–Mpika road, 1200 m, fr. 6.iv.1961, *Richards* 14978 (K holotype, SRGH).

Stems with rusty-brown, deflexed or sometimes almost spreading hairs. Calyx with long irregularly spreading hairs. Pods 3.5–4 mm broad, the margins with dark brown hairs contrasting markedly with the grey hairs on the surfaces.

Zambia. N: Mpika Dist., Kaloswe, 62 km SW of Mpika, imm.fr. 24.vii.1930, *Hutchinson & Gillett* 4064 (K, LISC, SRGH). C: Serenje, fl. 18.ii.1955, *Fanshawe* 2091 (K).

Known only from Zambia, occurring further to the SW than does subsp. *richardsiae*. Low rocky hills and plateau woodland; 1200–1450 m.

23. **Tephrosia curvata** De Wild. in Ann. Mus. Congo, Bot., sér.4, **1**: 190, fig.46/1–8 (1903). —Baker f., Legum. Trop. Africa: 192 (1926). —Cronquist in F.C.B. **5**: 93 (1954). —Gillett in F.T.E.A., Legum., Pap.: 192 (1971). Type: Congo, Katanga, Lukafu, *Verdick* 403 (BR holotype).

    Tephrosia lessertioides Baker f. in Bull. Soc. Roy. Bot. Belg. **57**: 121 (1925); Legum. Trop. Africa: 204 (1926). Type from Congo.

    Tephrosia wittei Baker f. in Rev. Zool. Bot. Afric. **21**: 302 (1932). —De Wildeman & Staner, Contrib. Fl. Katanga, suppl. 4: 25 (1932). Type from Congo.

Perennial with a stout woody stock with several slender, erect or upcurved, branching stems 45–100 cm high. Stems glabrous, or very shortly and sparsely pubescent in the upper parts. Leaves pinnate with (3)5–8 pairs of leaflets; petiole 0.3–1.3 cm long, petiole and rachis together (1.5)2–7.5(9) cm long, glabrous to sparsely and shortly appressed-pubescent; leaflets 1–2.7 ×

0.15–4(5) cm, linear-elliptic or linear-oblong to almost linear, cuneate to rounded at the base, obtusely pointed to rounded at the apex; upper surface glabrous, lower surface very shortly appressed-pubescent; stipules 2–3.5 × c.0.2 mm, subulate. Flowers in terminal and lateral racemes, the lateral ones subtended either by foliage leaves or small bracts, the whole often appearing ± paniculate; bracts minute, 0.5–1 mm long; pedicels 2–3 mm long. Calyx 3–3.5 mm long, very shortly appressed-pubescent; teeth short and broad, the lateral ones $^1/_3$–$^1/_2$ as long as the tube, ± obtuse, the upper ones joined for most of their length and distinguishable only as two short apicula. Petals 9–11 mm long, usually white, occasionally pinkish or dull yellowish-brown; standard strongly reflexed. Staminal tube joined above. Ovary pubescent; style glabrous, with penicillate tip. Pods 20–40 × c.5 mm, asymmetrically clavate in outline, more rounded on the lower side near the apex, with 1–2 seeds in the upper part, from there tapered to the base, shortly appressed-pubescent.

Zambia. N: Mbala Dist., old road to Kasakalawe (Cascalawa) from Chemba village, L. Tanganyika, near Mpulungu, fl. 16.ii.1960, *Richards* 12493 (K, SRGH). W: Chingola, fl. 17.iii.1960, *Robinson* 3416 (K, SRGH); Kasempa, fl. 14.iii.1981, *Chisumpa* 701 (K).

Also in Congo (Katanga) and Tanzania (Ufipa Dist.). Usually in grassland and woodland on sandy or stony ground; 750–1300 m.

24. **Tephrosia aurantiaca** Harms in Bot. Jahrb. Syst. **28**: 402 (1900). —Baker f., Legum. Trop. Africa: 211 (1926). —Dewit in Bull. Soc. Roy. Bot. Belg. **84**: 73–81 (1951) in widest sense. —Gillett in Kew Bull. **13**: 115–117 (1958); in F.T.E.A., Legum., Pap.: 177 (1971). —non Young in Ann. Transv. Mus. **14**: 398 (1932). Type: Tanzania, Iringa Dist., Ufuagi, *Goetze* 746 (B†, BR).

Tephrosia aurantiaca subsp. *rufopilosa* Dewit in Bull. Soc. Roy. Bot. Belg. **84**: 79 (1951) nom. illegit. Type as for *Tephrosia aurantiaca*.

Perennial suffrutex with herbaceous, probably annual, erect stems 40–130 cm high, sparingly branched to more or less bushy. Stems tomentose with brown or sometimes greyish, crisped, ± matted hairs. Leaves pinnate with (2)3(5) pairs of leaflets; petiole (0.4)1–1.6 cm long, petiole and rachis together (1.6)4–6(8) cm long, pubescent like the stem; leaflets (2.5)4–6(7.5) × (0.5)1–2.3 cm, obovate to oblanceolate, ± cuneate at the base, obtuse to rounded or sometimes emarginate at the apex, the veins prominent and rather conspicuous on both surfaces; upper surface glabrous or very sparsely hairy, lower surface conspicuously grey pubescent especially on the veins; stipules (4)7–16 × 0.8–2(2.3) mm, linear to linear-triangular. Flowers in dense terminal and axillary racemes 2–10(15) cm long; terminal raceme usually exceeding the upper leaves; bracts 5–10 × 0.8–1.5 mm, usually persistent; pedicels 3–6 mm long; bracteoles present, linear, c.3 mm long. Calyx 9–13 mm long, densely pubescent with appressed or ascending, light to dark brown hairs; lateral teeth shorter than to about equalling the tube, the lower tooth usually distinctly longer than the tube, the two upper teeth connate for $^1/_2$–$^2/_3$ their length to form a broad triangle. Petals 16–21 mm long, orange. Stamen tube joined above. Ovary pubescent; style glabrous, reflexed above pod, with penicillate tip. Pods (known only immature) 50 × 6 mm, densely brown tomentose to subvillous, tending to have darker hairs along the sutures. Seeds c.6.

Zambia. N: south of Mbala (Abercorn), between Mbala and Lunzuwa, fl. 20.v.1958, *Angus* 1988 (K); Isoka Dist., Chisenga–Rumphi road, 112 km from Rumphi, fl. 17.ii.1961, *Richards* 14375 (K). **Malawi**. N: 1 km west of cross-roads from Karonga–Chitipa and Chisenga–Misuku, fl. & fr. 28.xii.1972, *Pawek* 6163 (K).

Also in S and SW Tanzania. Usually in miombo woodland; 1500–1800 m.

This species is similar – at least superficially – to 69. *T. tanganicensis* from the same area, but is a perennial with a more or less pedunculate inflorescence as distinct from an annual with subsessile inflorescence. The fact that *T. tanganicensis* has pubescent style and is placed in subgenus *Barbistyla* may suggest that the distinction between the two subgenera is not as clear as has been thought. Further investigation of species with large yellow flowers is desirable. See also the following species 25. *T. hockii*, which has two subspecies differing in style pubescence, and 30. *T. zoutpansbergensis*, where the pubescence is somewhat intermediate.

The record of Congo in Lock (1989) reflects the broad view of Dewit (1951) not adopted here; see also Gillett (1971). Good fruiting specimens are needed.

25. **Tephrosia hockii** De Wild. in Repert. Spec. Nov. Regni Veg. **11**: 545 (1913). — Baker f., Legum. Trop. Africa: 211 (1926). —Cronquist in F.C.B. **5**: 91 (1954). — Gillett in Kew Bull. **13**: 115–117 (1958). Type from Congo (Katanga).

  Tephrosia aurantiaca subsp. *hockii* (De Wild.) Dewit in Bull. Soc. Roy. Bot. Belg. **84**: 79 (1951).

Perennial suffrutex with herbaceous, probably annual, erect, unbranched or rarely 1-branched stems 45–75 cm high. Stems appressed-pubescent with grey or silvery hairs. Leaves pinnate with (1)2–4(5) pairs of leaflets; petiole (1.3)1.8–4.2 cm long, petiole and rachis together (2)4.5–11.2 cm long, pubescent like the stem; leaflets (3.5)4.5–8(9) × (0.6)1–2 cm, narrowly elliptic to oblanceolate, ± cuneate at the base, obtuse to rounded at the apex, veins prominent and conspicuous on both surfaces; upper surface glabrous, lower surface thinly appressed-pubescent; stipules (5)8–13 × (1)1.5–2 mm, linear to linear-triangular. Flowers in dense terminal racemes 2–10 cm long with usually also flowers in the axils of 1–3(4) upper leaves, occasionally with 1–2 short axillary racemes; terminal racemes usually exceeded by the upper leaves; bracts quickly falling; pedicels 4–10 mm long; bracteoles usually present, very short, c.0.7 mm long. Calyx 9–12 mm long, densely appressed-pubescent with grey hairs; lateral teeth about equalling the tube, lower tooth distinctly longer, the two upper teeth joined for most of their length. Petals (18)20–24 mm long, orange to yellow. Stamen tube joined above. Ovary pubescent; style glabrous or pubescent, with penicillate tip. Pods 70–100 × 8–10 mm, shortly appressed-pubescent to tomentulose. Seeds 6–9.

Subsp. **hockii** —Lock, Leg. Afr. Check-list: 374 (1989).

  Tephrosia lutea R.E. Fr., Wiss. Ergebn. Schwed. Rhod.-Kongo-Exped. 1: 81 (1914). Type: N Zambia, Mbala Dist., near Katwe, *Fries* 1205 (UPS holotype).

  Tephrosia rhizomatosa De Wild. in Bull. Soc. Roy. Bot. Belg. **57**: 125 (1925). Type from Congo (Katanga).

  Tephrosia aurantiaca subsp. *lutea* (R.E. Fr.) Dewit in Bull. Soc. Roy. Bot. Belg. **84**: 79 (1951).

  Tephrosia aurantiaca forma *cinerea* Dewit in Bull. Soc. Roy. Bot. Belg. **84**: 79 (1951). Type from Congo (Katanga).

  Tephrosia aurantiaca forma *fulvescens* Dewit in Bull. Soc. Roy. Bot. Belg. **84**: 79 (1951). Type as for *Tephrosia rhizomatosa*.

Style glabrous. Pods (seen in one specimen only) 100 × 8 mm.

Zambia. N: Kawambwa Dist., Kambia (Kambya), fl. 23.x.1963, *Mutimushi* 396 (K). Also in Congo (Katanga). In the Flora area known only from the type of *T. lutea* and the above specimen from a 2 acre colony on dambo margin in *Uapaca* woodland.

Subsp. **hirsutostylosa** (Dewit) J.B. Gillett in Kew Bull. **13**: 115 (1958). Type: Zambia, Mwinilunga Dist., 0.4 km (¹/₄ mile) S of Matonchi Farm, *Milne-Redhead* 2591 (fl.), 2591a (fr.) (K).

  Tephrosia aurantiaca subsp. *hirsutostylosa* Dewit in Bull. Soc. Roy. Bot. Belg. **84**: 79 (1951).

Style densely to rather thinly pubescent. Pods 70–80 × 9–10 mm.

Zambia. W: Kitwe, fl.& imm.fr. 7.x.1955, *Fanshawe* 2495 (K, SRGH).

Also in Congo (S Katanga). In woodland; 1200–1450 m.

26. **Tephrosia paniculata** Baker in F.T.A. **2**: 122 (1871). —De Wildeman in Bull. Soc.
Roy. Bot. Belgique **57**: 124 (1925). —Baker f., Legum. Trop. Africa: 207 (1926).
—Cronquist in F.C.B. **5**: 95 (1954). —Torre in C.F.A. **3**: 160 (1962). —Brummitt
in Bol. Soc. Brot., sér.2, **41**: 272 (1968). —Gillett in F.T.E.A., Legum., Pap.: 176
(1971). Type: Angola, Cuanza Norte, Pungo Andongo, *Welwitsch* 2075 (BM
holotype, K).

Subsp. **paniculata** Brummitt in Bol. Soc. Brot. sér.2, **41**: 273 (1968).

> *Tephrosia dimorphophylla* Baker in F.T.A. **2**: 116 (1871). —Harms in Engler, Pflanzenw.
> Afrikas **3**(1): 589 (1915). —Baker f., Legum. Trop. Africa: 207 (1926). Type: Angola, Huíla,
> Missão de Montino, *Welwitsch* 2073 (BM holotype, K).
> *Tephrosia eriosemoides* Oliv. in Trans. Linn. Soc., London **29**: 57 (1872). —R.E. Fries, Wiss.
> Ergebn. Schwed. Rhod.-Kongo-Exped. 1: 83 (1914). —Harms in Engler, Pflanzenw. Afrikas
> **3**(1): 589 (1915). —Robyns, Fl. Sperm. Parc Nat. Alb. **1**: 310 (1948). —Brenan, Check-list
> For. Trees Shrubs Tang. Terr.: 446 (1949). Type: Tanzania, Bukoba Dist., Karagwe, *Grant*
> 414 (K holotype).
> *Tephrosia preussii* Taub. in Bot. Jahrb. Syst. **23**: 182 (1896). —Harms in Engler, Pflanzenw.
> Afrikas **3**(1): 589 (1915). —Hepper in F.W.T.A., ed.2, **1**: 529, 531 (1958). Type: Cameroon,
> *Preuss* 629 (BM holotype, K).
> *Tephrosia schizocalyx* Taub. in Bot. Jahrb. Syst. **23**: 183 (1896). —Harms in Engler,
> Pflanzenw. Afrikas **3**(1): 589 (1915). Type: Malawi, Shire Highlands, 1885, *Buchanan* 494
> (K lectotype, chosen by Brummitt 1968).
> *Tephrosia melanocalyx* Baker in Bull. Misc. Inform., Kew **1897**: 258 (1897), non Welw. ex
> Baker in F.T.A. **2**: 106 (1871). Type: Malawi, near Chitipa (Fort Hill), *Whyte* s.n. (K).
> *Tephrosia lelyi* Baker f., Legum. Trop. Africa: 207 (1926). Type: Nigeria, Naraguta, *Lely*
> 572 (K holotype).
> *Tephrosia nigrocalyx* Baker f., Legum. Trop. Africa: 208 (1926). Type as for *T. melanocalyx*.
> *Tephrosia paniculata* var. *schizocalyx* (Taub.) J.B. Gillett in Kew Bull. **13**: 115 (1958). —
> Torre in C.F.A. **3**: 161 (1962).

Annual or short-lived perennial, with erect much-branched stems 30–130(240) cm high, or
sometimes described as decumbent or as a weak scrambler to 3.3 m high, sometimes becoming
woody towards the base. Stems with spreading or sometimes appressed, brown or sometimes
grey pubescence. Leaves unifoliate or trifoliate, or some leaves with 2 pairs (rarely 3 pairs) of
lateral leaflets (see note below); petiole 0.3–1(1.5) cm long, petiole and rachis together
(0.2)0.8–2(4.5) cm long, pubescent like the stem; lateral leaflets (1.8)2.5–4(5) ×
(0.5)0.8–1.6(2.3) cm, elliptic to linear-elliptic or narrowly oblanceolate, subcuneate to almost
rounded at the base, subacute to rounded at the apex, the terminal leaflets (3)4–8(9) ×
(0.7)1–2(3) cm, 1.2–2 times as long as the lateral ones; usually greyish-green, the veins
prominent beneath, upper surface usually fairly pubescent, rarely glabrous, lower surface
usually rather densely greyish pubescent; stipules (3)4–8(10) × (0.5)0.8–1.5(2) mm, ± linear.
Flowers in terminal and axillary fairly dense racemes up to 16(22) cm long, often with the
lowermost flowers in the axils of the uppermost leaves; bracts 5–14 × 0.8–2 mm, linear-
triangular to linear-lanceolate; pedicels 1.5–4(5) mm long. Calyx (7)9–13(14) mm long, fairly
densely pubescent with ascending to spreading hairs often grey or white towards the base of the
calyx, brown or blackish on the teeth, or sometimes rather conspicuously blackish throughout;
teeth 2–3 times as long as the tube, long-acute, the two upper ones joined for about half their
length or sometimes free nearly to their base, the lowermost one slightly longer than the others.
Petals (10)12–16 mm long, orange, red, scarlet, salmon or rarely mauve-purple. Staminal tube
joined above. Ovary pubescent; style glabrous, with penicillate tip. Pods (30)35–50 × 4–6 mm,
villous with light to very dark brown, spreading or loosely appressed hairs. Seeds (7)9–12(13),
± transverse, c.3 × 1.5 × 1 mm, ovoid, reddish- or light brown, with a rather conspicuous lateral
aril c.1 mm long towards the narrower end.

Zambia. B: Kalabo, fl.& imm.fr. 11.ii.1963, *Angus* 3574 (FHO, K). N: Mbala Dist., Lumi R., fl. 31.v.1957, *Richards* 9938 (K, SRGH). W: Mwinilunga Dist., Kalene Hill, fl. iii–iv.1929, *J.M. Marks* 7 (K). C: Walamba, fr. 22.v.1954, *Fanshawe* 1227 (K, SRGH). E: Chipata, fl.& fr. 1.vi.1958, *Fanshawe* 4499 (K). **Zimbabwe**. N: Mazowe Dist., Glendale, fr. 23.iii.1944, *McCall* in *GHS* 9784 (K, SRGH). C: Goromonzi, fl. 27.iii.1947, fr. 17.iv.1947, *Jack* in *GHS* 16234, 16356 (K, SRGH). E: Mutare Dist., roadside below Charleswood, Vumba Mts, fl. 3.iv.1960, *Chase* 7318 (K, LISC, SRGH). **Malawi**. N: Rumphi Dist., between Khondowe (Kondowe) and Karonga, fl.& fr. vii.1896, *Whyte* 319 (K). C: Ntchisi Dist., foot of Nswanswa Hill, fl. 2.v.1980, *Brummitt, Banda & Patel* 15598 (K, MAL). S: Zomba Dist., Mpita Estate, Thondwe, near Namadzi R., fr. 11.vii.1985, *Kwatha & Tawakali* 252 (K, MAL). **Mozambique**. N: Lago Dist., Maniamba, Serra de Jéci (Géci), fl. 29.v.1948, *Pedro & Pedrógão* 4046 (LMA).

Also from W Africa (Sierra Leone) to Uganda and Kenya and south to Angola and the Flora Zambesiaca area. Usually on river banks, swamp margins, dambos, etc.; 1155–1740 m.

Gonçalves (1982) has reported this species also from Tete Province of Mozambique.

Plants from SW Tanzania and from S Malawi (Shire Highlands) and adjacent Mozambique, (Tete, near Ulónguè (Vila Coutinho), *Torre & Paiva* 11124) tend to have predominantly unifoliolate leaves and have been distinguished on this basis as a separate variety, *T. paniculata* var. *schizocalyx* (Taub.) J.B. Gillett (see synonymy above). It is, however, difficult to draw a clear distinction, some plants having the leaves on the main stems 3–5 foliolate and on the lateral branches unifoliolate.

Subsp. *holstii* (Taub.) Brummitt has a tendency to subcapitate inflorescences and smaller flowers, and occurs east of the Rift Valley in Ethiopia, Kenya and Tanzania.

27. **Tephrosia acaciifolia** Baker in F.T.A. **2**: 106 (1871). —Harms in Engler, Pflanzenw. Afrikas **3**(1): 587 (1915). —Baker f., Legum. Trop. Africa: 180 (1926). —Forbes in Bothalia **4**: 954 (1948). —Torre in C.F.A. **3**: 146 (1962). — Schreiber in Merxmüller, Prodr. Fl. SW Afrika, fam. 60: 117 (1970). —Gillett in F.T.E.A., Legum., Pap.: 173 (1971). —Lock, Leg. Afr. Check-list: 367 (1989) as "*acaciaefolia*". Type: Angola, Pungo Andongo, forest margins between Bumba and Condo, *Welwitsch* 2071 (K holotype). FIGURE 3.3.**26**.

Tephrosia melanocalyx Baker in F.T.A. **2**: 106 (1871). —Baker f., Legum. Trop. Africa: 180 (1926). —Torre in C.F.A. **3**: 146 (1962). Type from Angola.

Tephrosia salicifolia Schinz in Vierteljahrsschr. Naturf. Ges. Zürich **52**: 425 (1907). Type: South Africa, Gauteng, *Schlechter* 4193 (?lectotype).

Annual or short-lived perennial herb up to 60(130) cm high, with simple to much branched stems. Stems closely appressed- to ascending-pubescent. Leaves unifoliolate; petiole 0–2 mm long; lamina 5–17 × 0.4–1.7 cm (or those on shorter branches somewhat smaller), linear-lanceolate to linear, (5)7–14(18) times as long as broad, the upper surface glabrous, the lower surface appressed- or rarely spreading-pubescent; stipules up to 5 mm long, triangular-linear. Flowers in elongate lax racemes up to 15 cm long, or commonly the lower flowers in the axils of the upper 1–2 foliage leaves, rarely with foliage leaves in the middle of the inflorescence; pedicels 1–3 mm long. Calyx (2.5)3–5(6) mm long, brown pubescent, the teeth acute, equal or up to twice as long as the tube, the two upper ones joined for $^1/_2$–$^2/_3$ their length. Petals orange to red or pink, 2–3 times as long as the calyx; standard conspicuously brown pubescent on outside, 8–11(12) mm long, the limb suborbicular, subtruncate at the base; wings and keel slightly shorter than the standard. Stamen tube rather loosely fused dorsally. Ovary pubescent; style glabrous, with penicillate tip. Pods (25)40–60 × 2.5–3.5 mm, closely to loosely appressed-pubescent, the hairs on the sutures dark brown to blackish and conspicuously darker than the grey or whitish hairs covering the surfaces.

Caprivi Strip. Caprivi side of Zambezi R. near Andara Mission Station, fr. 23.ii.1956,

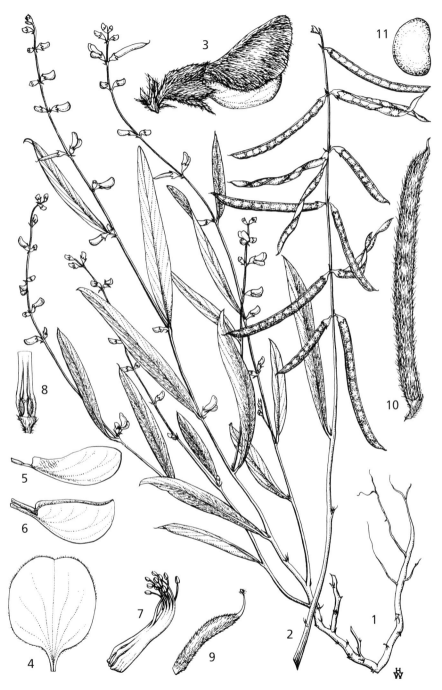

Fig. 3.3.**26**. TEPHROSIA ACACIIFOLIA. 1, habit ($\times$ $^1/_2$), from *Noak* 120 and *Best* 281; 2, fruiting branchlet ($\times$ $^1/_2$), from *Hornby* 2869 and *Fanshawe* 3247; 3, flower ($\times$ 3); 4, standard ($\times$ 3); 5, wing ($\times$ 3); 6, keel ($\times$ 3); 7, stamens ($\times$ 3); 8, upper side of base of staminal tube ($\times$ 3); 9, gynoecium ($\times$ 3), 3–9 from *Noak* 120; 10, pod ($\times$ 1 1/3), from *Hornby* 2869; 11, seed ($\times$ 4), from *Fanshawe* 3247. Drawn by Heather Wood.

de Winter & Marais 4811 (K). **Zambia**. C: Lusaka town, Stuart Park, fl.& imm.fr. 19.ii.1961, *Best* 281 (K, SRGH). E: Chipata Dist., Ngoni Reserve (area), fl. ii.1962, *Verboom* 463 (K, PRE, SRGH). S: Choma Dist., between Choma and Masuku Mission, fl. 21.ii.1963, *van Rensburg* 1398 (K, SRGH). **Zimbabwe**. N: Gokwe South Dist., c.5 km from the Swiswi river crossing, Gokwe–Binga road, fr. 31.iii.1962, *Bingham* 216 (K, SRGH). W: Nkayi Dist., Gwampa Forest Land (Reserve), fl. 2.ii.1955, *Goldsmith* 47/55 (PRE, SRGH). C: Chegutu Dist., Poole Farm, fl.& fr. 20.viii.1948, *Hornby* 2869 (K, SRGH). E: Mutasa Dist., Honde Valley, Mpanga R., fl. 3.iii.1949, *Chase* 1377 (K, SRGH). **Malawi**. N: Mzimba Dist., c.1.6 km S of M14 at Lunyangwa R., fl. 7.iii.1976, *Pawek* 10909 (K). C: Lilongwe, near capital site, imm.fr. 29.iii.1970, *Brummitt & Little* 9516 (K). S: Machinga Dist., lower Kasupe, fl.& fr. 13.iii.1955, *Exell, Mendonça & Wild* 833 (SRGH). **Mozambique**. MS: Barué Dist., c.16 km N of Catandica (Vila Gouveia) on road to Tete, fr. 1.vi.1938, *Pole Evans & Erens* 491A (K, PRE).

Also in Tanzania, Angola, Namibia and South Africa (Limpopo, Mpumalanga and Gauteng Provinces). Woodland and open ground, often at roadsides, etc. 1050–1400 m.

Gonçalves (Garcia de Orta **5**: 114, 1982) has recorded the species from Tete Province of Mozambique.

Variation in this species requires further investigation. Most plants known from the Flora area appear to be annual, much branched in lower and upper parts, with the calyx 3–4 mm long and leaves 5–11(12.5) cm long. Plants from the SW of the distribution area, from Angola to South Africa and including the specimen cited from the Caprivi Strip, tend to be perennial with a number of stems arising from a more or less woody stock, the stems being simple or sparingly branched in the upper part only, with the calyx 4–5(6) mm and the leaves (7)10–17 cm long. When further collections are available it may prove possible to recognize the two forms as subspecies.

28. **Tephrosia kindu** De Wild. in Ann. Mus. Congo, Bot., sér.4, **1**: 191 (1903). — Cronquist in F.C.B. **5**: 89 (1954). —Torre in C.F.A. **3**: 147 (1962). Type from Congo (Katanga).

Perennial herb with usually several stems arising from a short erect stock (perhaps also rhizomatous, Cronquist 1954). Stems erect, usually branched in the upper part, closely appressed-pubescent. Leaves unifoliolate; petiole 0–2 mm long; lamina up to 8 × 0.6 cm, linear-oblong or elliptic to linear, 9–18 times as long as broad, the upper surface glabrous, the lower surface silvery-sericeous; stipules up to 2 mm long, subulate. Flowers in lax terminal racemes with the lower ones in the axils of up to 5 of the uppermost leaves, mostly solitary; pedicels 2–4 mm long; bracteoles usually clearly visible, c.0.5 mm long. Calyx 5–6 mm to the tip of the lower tooth, brown appressed-pubescent; teeth unequal, the two upper ones together forming a broad triangle, connate for $^1/_2$–$^2/_3$ their length, almost as long as the tube and $^1/_2$–$^2/_3$ as long as the lower tooth, the lateral teeth intermediate in size. Petals orange or pinkish-orange, approximately 3–5 times as long as the calyx; standard 16–20 mm long, suborbicular-cuneate, somewhat curved upwards, light brown pubescent on the back; wings and keel slightly shorter than the standard. Stamen tube open dorsally. Ovary pubescent; style glabrous; stigma pencil-like. Pods 40–50 × 4 mm, appressed-pubescent with the dark brown sutures contrasting with the light straw-coloured surfaces.

Zambia. N: Mbala Dist., Chilongowelo, fl. 4.iii.1952, *Richards* 870 (K). W: Mwinilunga Dist., near Musangila R., fl. 25.i.1938, *Milne-Redhead* 4327 (K).

Also in Angola, Congo and SW Tanzania (Ufipa Dist.). In miombo and *Uapaca-Monotes* woodland and secondary associations; 1300–1500 m.

The specific name is derived from a Congolese vernacular name for the species.

29. **Tephrosia forbesii** Baker in F.T.A. **2**: 116 (1871). —Harms in Engler, Pflanzenw. Afrikas **3**(1): 589 (1915). —Baker f., Legum. Trop. Africa: 205 (1926). —Forbes in Bothalia **4**: 956 (1948). —Brummitt in Bol. Soc. Brot., sér.2, **41**: 275 (1968). Type: Mozambique, Maputo, Delagoa Bay, *Forbes* s.n. (K holotype).

A slender, probably short-lived perennial or perhaps biennial. Stems decumbent to suberect, up to 0.4(1) m high, branching in the lower part; young stems appressed- or spreading-pubescent. Leaves unifoliolate; petiole c.1 mm long; lamina up to 8 × 1.5 cm, linear-oblong or linear-lanceolate, 6–10 times as long as broad, the upper surface glabrous, the lower surface appressed or spreading-pubescent; stipules triangular-linear. Flowers in groups of 1–3 in axils of foliage leaves; pedicels 1–15(17) mm long. Calyx 4–10 mm long, pubescent, the teeth long-acute, subequal, 1.5–4 times as long as the tube, the two upper ones only slightly joined at the base. Petals pink or somewhat violet, or perhaps yellow outside, equalling or exceeding the calyx; standard 7–13 mm long, broadly spathulate to obovate-cuneate; wings and keel ± equal in length, slightly shorter to about equalling the standard, with the claw about half as long as the limb. Stamen tube open dorsally. Ovary pubescent on sutures, glabrous or pubescent on surfaces; style glabrous. Pod 20(27)–40 × 3.5–5 mm, sutures pubescent, surfaces glabrous or pubescent.

1. Ovary and pod pubescent on surfaces ················ i) subsp. *forbesii*
– Ovary and pod glabrous on surfaces ····························· 2
2. Calyx 4–8 mm long; petals 8–10 mm long; pods c.3.5 mm broad; larger leaves on each plant 35–55 × 4–6 mm ······················· ii) subsp. *inhacensis*
– Calyx (6)7–14 mm long; petals 7–13 mm long; pods c.4.5 mm broad; larger leaves on each plant 55–80 × (6)7–15 mm ············· iii) subsp. *interior*

i) Subsp. **forbesii** —Brummitt in Bol. Soc. Brot., sér.2, **41**: 276 (1968).

> *Tephrosia junodii* De Wild. in Ann. Mus. Congo, Bot., sér.5, **1**: 261 (1906). Type: Mozambique, Maputo (Lourenço Marques), *Junod* 168 (BR holotype).

Leaves ± linear-oblong to linear-elliptic, the larger ones 35–60(70) × 4–7(9) mm. Pedicels 2–5(6) mm long. Calyx (4)5–7(8) mm long, the teeth 2–3 times as long as the tube. Petals 8–11 mm long, slightly to distinctly exceeding the calyx. Pods 25–40 × 3.5–4(4.5) mm, pubescent on surfaces.

Mozambique. GI: near Xai Xai (Vila João Belo), fl.& fr. 10.xii.1940, *Torre* 2307 (LISC). M: Maputo (Lourenço Marques), commonage, fl.& fr. 18.i.1948, *Faulkner* Kew 183 (K, SRGH).

Apparently restricted to coastal districts of S Mozambique. Sandy open ground; 0–50 m.

ii) Subsp. **inhacensis** Brummitt in Bol. Soc. Brot., sér.2, **41**: 277 (1968). Type: Mozambique, Inhaca Is., W coast ridge, fl.& fr. 30.i.1962, *Mogg* 29869 (K holotype, SRGH).

> *Tephrosia forbesii* and *Tephrosia acaciifolia* sensu Mogg in Macnae & Kalk, Nat. Hist. Inhaca Is., ed.2: 146 (1969).

Leaves linear-oblong, the larger ones 35–55 × 4–6 mm. Pedicels 1–3(5 in fruit) mm long. Calyx 4–8 mm long, the teeth $1^1/_2$–$2^1/_2$ times as long as the tube. Petals 8–10 mm long, clearly exceeding the calyx. Pods c.3.5 mm broad, glabrous on surfaces.

Mozambique. M: Maputo Dist., Inhaca Is. S of Research Station, fl.& fr. 30.i.1958, *Noel* 98 (K, PRE).

Known only from Inhaca Island. Grassland, forest remnant and fresh-water swamp; 0–50 m.

iii) Subsp. **interior** Brummitt in Bol. Soc. Brot., sér.2, **41**: 277 (1968). Type: Zimbabwe, Chiredzi Dist., Mozambique border, Sango (Vila Salazar), 26.iv.1961, *Drummond & Rutherford-Smith* 7543 (SRGH holotype, K).

Leaves ± linear-oblong, or slightly broader in the basal third, the larger ones on each plant 55–80 × (6)7–10 mm. Pedicels 1–15 mm long. Calyx (6)7–14 mm long, the teeth 3–5 times as long as the tube. Petals 7–13 mm long, clearly exceeding or about equalling the calyx. Pods c.4.5 mm broad, glabrous on surfaces.

Zimbabwe. S: Chiredzi Dist., Mozambique border, Sango (Vila Salazar), fl.& fr. 26.iv.1961, *Drummond & Rutherford-Smith* 7534 (K, SRGH). **Mozambique**. GI: Gaza Dist., Zimbabwe border, Vila Eduardo Mondlane (Malvérnia), 26.iv.1961, *Drummond & Rutherford-Smith* 7533 (K, SRGH).

Also in South Africa (Limpopo, Mpumalanga and KwaZulu-Natal Provinces). Mixed tree savanna on sandstone, sandy ground by roadsides; 200–300 m.

Plants referred to subsp. *interior* show a wide range of variation in flower size and particularly in pedicel length. The two collections cited have short pedicels 1–5 mm long, but *Drummond* 7730 (K, SRGH) from Kapateni, Chiredzi Dist. and specimens from South Africa have pedicels 5–15 mm long. Only three collections are known from the Flora area.

30. **Tephrosia zoutpansbergensis** Bremek. in Ann. Transvaal Mus. **15**: 242 (1933). — Obermeyer, Schweickerdt & Verdoorn in Bothalia **3**: 239 (1937). —Forbes in Bothalia **4**: 973 (1948). —Lock, Leg. Afr. Check-list: 386 (1989) as "*zouspansbergensis*". Type: South Africa, Soutspansberg, Zoutpan, *Bremekamp & Schweickerdt* 279 (TM, P).

Perennial herb or sometimes a shrub 60–100 cm high, woody in the lower parts. Young stems densely appressed-pubescent with silvery silky hairs. Leaves pinnate with 2–4 pairs of leaflets; petiole 0.8–2(3.2) cm long, petiole and rachis together 1.8–7.2 cm long; leaflets (1.5)2–3.8 × (0.5)0.8–1.2(1.4) cm, ± oblanceolate, cuneate at the base, obtuse to retuse at the apex; upper surface glabrous, lower surface appressed-pubescent; stipules 3–5 × 0.5–1 mm, linear, silky like the stem. Flowers in fairly dense terminal racemes up to 15 cm long in fruit; bracts 3–4 × c.0.5 mm long, linear; pedicels 2–3 mm long. Calyx 6–8 mm long, covered with short, greyish, spreading hairs; teeth equalling to twice as long as the tube, the two upper connate for about ³/₄ their length. Petals (10)12–15 mm long, yellow or perhaps sometimes cream coloured. Upper stamen loosely attached to the staminal sheath at least along one side. Ovary densely pubescent, prolonged into a densely pubescent beak terminated by the glabrous and upwardly reflexed style, with penicillate tip. Pods 8–16 × 5–6 mm, shortly tomentose, brown or greyish. Seeds 1 or 2.

Botswana. SE: Palapye, Moremi, Tswapong Hills, fl.& fr. 6.iii.1957, *de Beer* T20 (K, SRGH). **Zimbabwe**. S: Beitbridge Dist., Tschiturapadsi (Chiturupazi), fl.& fr. 25.ii.1961, *Wild* 5392 (K, PRE, SRGH); Tshiturapadsi (Chiturupadzi) Store, fl.& fr. 18.iii.1967, *Corby* 1843 (K, SRGH).

In the Flora Zambesiaca area only known from the collections cited; also in South Africa (Limpopo Province). Sandy places and rocky outcrops; 400–1000 m.

31. **Tephrosia miranda** Brummitt in Bol. Soc. Brot., sér.2, **41**: 387 (1968). Type: Mozambique, Serra da Mesa, 6 km from Nampula, granite rocks, 500 m, fl.& fr. 3.iv.1964, *Torre & Paiva* 11599 (LISC holotype, K).

Shrub 2–3 m high. Stems grey ascending-tomentose. Leaves mostly with 5–7 pairs of leaflets but the upper ones becoming successively reduced to a single leaflet; petiole 3–7 mm long, petiole and rachis together up to 8 cm long, grey tomentose; leaflets up to 33 × 13 mm, obovate to oblanceolate, cuneate to rounded at the base, rounded at the apex; upper surface thinly pubescent with fine delicate hairs, lower surface loosely appressed-pubescent with grey hairs; stipules 7–9 × 1.5–2 mm, linear-triangular, tomentose, falling early. Flowers in terminal and axillary racemes up to 10 cm long, with the nodes subtended by reduced (mostly unifoliolate) foliage leaves which are successively smaller in size towards the apex where they appear as grey tomentose linear-elliptic bracts c.5 mm long; pedicels 8–11 mm long. Calyx 4.5–5.5 mm long, shortly appressed grey pubescent; upper teeth fused for most of their length to form an acute triangle free for c.0.7 mm at the apex. Petals dark reddish; standard 14–16 mm long, cuneate at the base, wings and keel petals slightly shorter. Stamen tube joined above. Ovary appressed-tomentose; style pubescent, upwardly reflexed. Pods c.35 × 7.5 mm, slightly curved and somewhat constricted laterally between the seeds, shortly and densely pubescent, apparently indehiscent but breaking up irregularly, the epicarp flaking off from the endocarp. Seeds 3–5, c.5 × 3 × 1.5 mm, brownish; aril small and inconspicuous.

Mozambique. N: Nampula Dist., Mt Nassapo, 23 km from Nampula towards Meconta, fl. 13.i.1964, *Torre & Paiva* 9929 (LISC).

Known only from the two cited collections in Nampula District. Granite rock outcrops; 400–500 m.

The pods on the one known fruiting specimen appear to be most unusual for this genus, as described above. More fruiting material is required.

The affinities of the species have been rather obscure, but similarity of general appearance and the strongly upwardly refexed style suggest relationship with *T. zoutpansbergensis* and *T. villosa*, despite its apparently pubescent style.

32. **Tephrosia punctata** J.B. Gillett in Kew Bull. **13**: 126 (1958); in F.T.E.A., Legum., Pap.: 196 (1971). —Brummitt in Bol. Soc. Brot., sér.2, **41**: 349 (1968). Type: Zambia, Mbala (Abercorn), *Bullock* 3792 (K holotype).

　　Tephrosia kasikiensis sensu Cronquist in F.C.B. **5**: 116 (1954) in part for specimen *Ritschard* s.n.

Annual, or perhaps sometimes biennial herb, with erect or perhaps sometimes decumbent-ascending stems 0.5–1.5 m high, sparingly to profusely branched; branches (inflorescences) exceeding the subtending leaves. Stems shortly appressed-pubescent. Leaves pinnate with 4–6 pairs of leaflets, or the upper ones reduced to 1–3 pairs; petiole (0.3)0.8–2.2(3) cm long, the petiole and rachis together 3–7(9) cm long, appressed-pubescent; leaflets (1.3)1.7–3.6 × (0.2)0.5–1(1.3) cm, elliptic to oblong, rounded to slightly emarginate at the apex, cuneate to rounded at the base, the upper surface glabrous, the lower surface shortly appressed-pubescent, both surfaces with minute blackish gland dots between the veins; stipules 6–10 × 0.5–1 mm, linear-subulate. Flowers in terminal and axillary elongate racemes (6)12–32 cm long; bracts (1)2–4(5) mm long, linear-subulate, falling early; pedicels 2–4(5) mm long, shortly appressed-pubescent. Calyx 2–3 mm long (to tip of lower tooth), the campanulate part 1–1.5 mm, shortly and closely appressed-pubescent and with golden pellucid gland dots, the lateral teeth about equalling the campanulate part, the two upper teeth very short and distinguishable only as a pair of more or less subulate projections c.0.3 mm long. Petals 8–11 mm long, purplish or pinkish or the keel whitish, the standard brown pubescent outside; all with golden pellucid gland dots; standard ± reniform to suborbicular, ± cordate at the base, rather strongly upwardly curved. Stamen tube fused above. Ovary pubescent; style usually pubescent on both surfaces. Pods 40–65 × 3–4.5(5) mm, with whitish hairs on the surfaces and darker hairs on the sutures when very immature, at maturity covered with light, closely appressed hairs. Seeds 8–10, mature ones known only in subsp. *redheadii*, almost oblong, slightly narrower at the radicular end, rounded at both ends, with a conspicuous elongate aril covering the hilum and extended about half the length of the seed.

Subsp. **punctata** —Brummitt in Bol. Soc. Brot., sér.2, **41**: 349 (1968).

Plant usually with copious ascending axillary inflorescences forming a pseudopanicle, producing abundant flowers. Leaf petiole and rachis together 4–6.5(11) cm long; leaflets (3)4–7(10) mm broad.

Zambia. N: Mbala Dist., old road to Uningi Pans, fl. 1.iii.1955, *Richards* 4731 (K).

Also in Congo and S and SW Tanzania. In miombo woodland and grassland, usually on sandy soils; 1500–1830 m.

Richards 1365 from Mbala Dist., Mombashi Plantation on Kasama road, 1830 m (K), is indistinguishable from subsp. *punctata* except in its remarkable standard which is hooded, curving downwards instead of upwards at the apex, and lacks the characteristic brown hairs on the upper surface except on the central part where they are sparse. The flowers are described as white.

Subsp. **redheadii** Brummitt in Bol. Soc. Brot., sér.2, **41**: 349 (1968). Type: Zambia, Mwinilunga Dist., just S of Matonchi Farm, *Milne-Redhead* 4636 (K holotype).

Plant rather sparingly branched, with lax axillary racemes much less floriferous than subsp. *punctata*. Leaf petiole and rachis together 5–9 cm long; leaflets (5)8–12(14) mm broad.

Zambia. N: Kasama Dist., Mungwi, fl.& fr. 14.iii.1962, *Robinson* 5017 (K; SRGH). W: Mwinilunga Dist., just S of Matonchi Farm, fl. 19.ii.1938, *Milne-Redhead* 4636 (K).

Also in Congo (Katanga). Usually in miombo woodland; 1300–1400 m.

33. **Tephrosia kasikiensis** Baker f. in Rev. Zool. Bot. Afric. **21**: 301 (1932). —De Wildeman & Staner, Contrib. Fl. Katanga, Suppl. **4**: 24 (1932). —Cronquist in F.C.B. **5**: 116 (1954) excluding *Ritschard* specimens. —Brummitt in Bol. Soc. Brot., sér.2, **41**: 347 (1968). —Gillett in F.T.E.A., Legum., Pap.: 196 (1971). Type: Congo, Katanga, Kasiki, *De Witte* 382 (BM holotype).

Annual or ?short-lived perennial with erect branched stems 1–2.3 m high. Stems appressed to spreading-pubescent, hairs brown or grey. Leaves pinnate with 4–6 pairs of leaflets, or the upper leaves with 1–3 pairs; petiole 0.8–5 cm long, the petiole and rachis together 5–12.5 cm long, their pubescence resembling that of the stem; leaflets (1.5)2–5(7) × 0.4–1.5(2) cm, narrowly elliptic to obovate or oblanceolate, subacute to slightly emarginate at the apex, cuneate to rounded at the base; upper surface glabrous, lower surface rather thinly appressed-pubescent or sometimes also with scattered ascending hairs; stipules 4–9 × 0.5–1 mm, linear. Flowers in terminal and axillary elongate racemes 8–16(24) cm long, or in dense heads (see subspecies); bracts 2–5 × 0.4–0.7 mm, ± linear, falling early; pedicels (3)4–7 mm long, grey pubescent to villous. Calyx 3.5–5(6) mm long, shortly ascending-pubescent to grey villous; the lateral teeth about or almost equalling the campanulate part, the two upper teeth very short and distinguishable only as a pair of ± subulate projections up to 0.5 mm long. Petals 8–12 mm long, purplish or the keel white, the standard brown pubescent outside, all, or at least the keel, with usually golden pellucid gland dots; standard ± reniform or laterally oblong, broader than long, rather strongly upwardly curved. Stamen tube fused above. Ovary pubescent; style pubescent. Pods 40–55 × 4–5 mm, when immature with pale brown tomentum on the surfaces and darker on the sutures, at maturity with spreading darkish hairs over surfaces. Seeds (6)7–8, c.4.5 × 2 × 1.5 mm, oblong and rounded at the ends.

Zambia. N: Mbala Dist., Ndundu Hill, fl.& fr. 6.v.1959, *Richards* 11370 (K). W: Chingola, fl.& fr. 25.viii.1954, *Fanshawe* 1483 (K); 78 km along Solwezi–Mwinilunga road, 5.5 km from main road going NW, fl. 14.ii.1995, *Nawa, Harder, Zimba & Luwiika* 35 (K).

Also in eastern Congo (Katanga) and SW Tanzania. From margins of mountain and riverine forests; up to 2100 m.

Plants referred to subsp. *kasikiensis* show considerable variation and further collections may show that subspecies can be recognized (see Brummitt 1968). On the other hand, the characters distinguishing *T. senegaensis* Baker f. (Leg. Trop. Afr. **1**: 192, 1926) may seem insufficient to separate two species, and it may be preferable to sink *T. kasikiensis* into that species. The one collection described as subsp. *chinsaliana* Brummitt (type: Zambia N: Chinsali Dist., Shiwa Ngandu, fl. 5.vi.1956, *E.A. Robinson* 1609 (K holotype, SRGH) seems to be intermediate between them. More collections are needed, especially of fruits.

34. **Tephrosia elata** Deflers, Voy. Yemen: 130 (1889). —Gillett in F.T.E.A., Legum., Pap.: 202 (1971). Type: Yemen, El Mekhader, 1800 m, *Deflers* 653 (P holotype).

Annual or perhaps sometimes biennial, with a taproot usually branching horizontally. Stems erect, sometimes slightly woody towards the base, ± branched above, 0.5–1.5(2) m high, minutely closely appressed-pubescent to (subsp. *elata*) spreading-pubescent. Leaves pinnate with (5)6–10(14) pairs of leaflets; petiole (0.5)1.0–2.6(4.5) cm long, petiole and rachis together (5)8–15(24) cm long, minutely appressed-pubescent; leaflets (2.5)3–6(8.5) × (0.6)0.8–1.7(2.5) cm, oblanceolate or elliptic to linear-elliptic, cuneate to rounded at the base, rounded to emarginate at the apex; upper surface glabrous, lower surface shortly and closely appressed-pubescent; stipules (6)9–16 × (0.5)0.8–1.4(2.5) mm, linear, often falcate. Flowers in terminal and axillary racemes (7)14–30(35) cm long; bracts (4)6–11 × 0.2–1(2) mm, linear to linear-triangular, or narrowly lanceolate, falling early; pedicels 3–8(10) mm long. Calyx (3.5)4–6(8) mm long, densely and shortly appressed-pubescent, the lateral teeth shorter than to scarcely longer than the tube, the lower tooth equalling or slightly exceeding the tube, the two upper teeth joined, shorter than the lateral and joined for most of their length to form a broad triangle with only 0.2–1 mm free at the apex. Petals (10)12–15(18) mm long, purple or reddish; standard truncate to cordate at the base. Stamen tube joined above. Ovary pubescent; style pubescent on the upper surface. Pods (35)55–70(80) × 4–5.5 mm, pale straw-coloured or (subsp. *elata*) with brownish hairs, glabrescent to densely spreading-pubescent. Seeds (9)10–12(13).

Subsp. **elata** —Brummitt in Bol. Soc. Brot., sér.2, **41**: 339 (1968).

Tephrosia rigida Baker in F.T.A. **2**: 114 (1871), non Spanoghe (1841). Type: Uganda, Madi, *Grant* s.n. (K).

Tephrosia bequaertii De Wild. in Bull. Soc. Roy. Bot. Belg. **57**: 117 (1925). —Baker f., Legum. Trop. Africa: 199 (1926). —Robyns, Fl. Sperm. Parc Nat. Alb. **1**: 311 (1948). Type: Congo, Kivu, Kengele (Beni), foot of Ruwenzori Mtns, *Bequaert* 3949 (BR).

Tephrosia mbogaensis De Wild. in Bull. Soc. Roy. Bot. Belg. **57**: 123 (1925). —Baker f., Legum. Trop. Africa: 199 (1926). Type: Congo, Lake Albert, Mboga, *Bequaert* 3084 (BR).

Tephrosia heckmanniana sensu lato. —Cronquist in F.C.B. **5**: 115 (1954) in part.

Stems and inflorescence axes often ridged, appressed-pubescent to (var. *tomentella* Brummitt) tomentose. Pods (35)40–60(65) × 4.5–5.5 mm, straw-coloured or slightly darker and usually clothed with spreading brown hairs.

Zimbabwe. N: cited in Drummond in Kirkia **8**: 227 (1972). C: Goromonzi Dist., Chinamora C.L. (Chindamora Reserve), fl.& fr. 15.iv.1922, *Eyles* 3384 (K, SRGH); Chinamora C.L., Domboshawa Rock, fr. 6.iii.1974, *Mavi* 1546 (K, SRGH).

Also in the Arabian Peninsula (Yemen), Ethiopia to eastern Congo and Tanzania, mainly to the west of the Great Rift Valley. Collections known from the Flora area lie well outside the main distribution area of this subspecies are presumed to have been introduced, perhaps cultivated as a fish poison or for making rope.

Subsp. **heckmanniana** (Harms) Brummitt in Bol. Soc. Brot., sér.2, **41**: 343 (1968). Type: Tanzania, Njombe Dist., Mt. Ukangu, *Goetze* 889 (K).

Tephrosia heckmanniana Harms in Bot. Jahrb. Syst. **30**: 326 (1901). —Baker f., Legum. Trop. Africa: 196 (1926). —Brenan, Check-list For. Trees Shrubs Tang. Terr.: 445 (1949).

—Cronquist in F.C.B. **5**: 115 (1954) in part. —Torre in C.F.A. **3**: 156 (1962). —Chapman, Veg. Mlanje Mt.: 54 (1962). —Gillett in F.T.E.A., Legum., Pap.: 202 (1971). —Gonçalves in Garcia de Orta, sér.Bot. **5**: 115 (1982). —Lock, Leg. Afr. Check-list: 374 (1989).

Tephrosia emarginato-foliolata De Wild., Pl. Bequaert. **3**: 328 (1925). Type: Congo, Katanga, near Lubumbashi (Elizabethville), *de Giorgi* s.n. (BR holotype).

Tephrosia multinervis Baker f., Legum. Trop. Africa: 201 (1926). Type: Congo, Katanga, Kundelungu Mt, *Kassner* 2736 (K).

Stems and inflorescence axes smooth, minutely and closely appressed-pubescent. Pods (40)55–70(80) × 4–4.5(5) mm, usually conspicuously straw-coloured, glabrescent or with inconspicuous short irregular hairs.

Var. **heckmanniana** —Brummitt in Bol. Soc. Brot., sér.2, **41**: 343 (1968).

Petiole and rachis together (5)8–15(21) cm long; leaflets (2.5)3–6(7.5) cm long; stipules (0.5)0.8–1.4 mm broad. Bracts (4)6–11 × 0.2–1(1.5) mm, linear to linear-triangular, falling early. Calyx (3.5)4–6(7) mm long.

Zambia. N: Mbala Dist., Kalambo Falls, fl.& fr. 21.iv.1959, *Richards* 11319 (K). W: Mwinilunga Dist., between R. Kalalima (Kamwezhi) and R. Isongela (Isongailu), fl. 17.ii.1938, *Milne-Redhead* 4624 (K). C: c.13 km E of Lusaka, fl. 2.iii.1956, *King* 335 (K). E: Chipata Dist., Chipangali area, fr. ii.1962, *Verboom* 482 (K, SRGH). S: Gwembe Dist., Zambezi escarpment c.80 km from Mochipapa, fl.& fr. 13.iii.1962, *Astle* 1509 (K, SRGH). **Zimbabwe**. N: Mutoko Dist., Mudzi Dam, fl. 16.ii.1962, *Wild* 5675 (K, PRE, SRGH). C: Chegutu Dist., Serni Drift, fl.& fr. 9.iii.1948, *Fynn* in GHS 19178 (K, SRGH). E: Chimanimani Dist., Junction Tea Room, Cashel road, fl.& fr. 31.i.1963, *Lady Drewe* 86 (SRGH). S: Buhera, fl. 7.ii.1954, *Masterson* 37 (SRGH). **Malawi**. N: Nkhata Bay Dist., Chikale beach, fr. 19.vii.1970, *Pawek* 3604 (K). S: Mulanje Mt., Likabula House, fl. 8.iii.1937, *Lawrence* 280 (K). **Mozambique**. N: Lago Dist., Metangula, fr. 22.v.1948, *Pedro & Pedrógão* 3839 (LMA). Z: 16 km SW of Gurué, fl. 7.vii.1942, *Hornby* 1162 (PRE).

Also in Congo (Katanga), Burundi, S Tanzania and Angola. In miombo woodland, mushitu fringes, etc.; 450–1750 m.

See also note on specimen from Tete Province mentioned at the end of this subspecies.

Var. **abercornensis** Brummitt in Bol. Soc. Brot., sér.2, **41**: 345 (1968). Type: Zambia, Mbala Dist., steep path from Chisungu homestead to firebreak, fl. 14.iv.1952, *Richards* 1470 (K holotype).

Petiole and rachis together up to 24 cm long; leaflets up to 8.5 cm long; stipules 2–2.5 mm broad. Bracts 7–9 × 2 mm, narrowly lanceolate. Calyx 6–8 mm long, the lateral teeth slightly, and the lower one distinctly, exceeding the tube.

Zambia. N: Mbala Dist., Chilongowelo, in bush just above the Nindi Still, fl.& imm.fr. 10.iv.1952, *Richards* 1375 (K).

Known only from Mbala Dist. Miombo woodland and fringing bushland with long grass; 1440–1500 m.

In several characters var. *abercornensis* appears to be intermediate between typical subsp. *heckmanniana* and *T. nyikensis*, and might represent the result of past introgression between these two. *T. nyikensis* is not recorded from this district.

A specimen from Mozambique (T: 65 km from Ulónguè (Vila Continho) towards Zóbuè, fl.& fr. 10.iii.1964, *Torre & Paiva* 11123 (LISC)) is probably also referable to subsp. *heckmanniana* but has conspicuously longer peduncles (up to 24 cm), broader stipules (2.5 mm) and broader pods (70 × 6 mm). A specimen from W Zambia (Luano Forest Reserve, Chingola, fl.& fr. 13.iii.1961, *Mutimushi* 114 (SRGH)) is apparently referable to var. *heckmanniana* but differs from all others of this subspecies in having spreading pubescence:

35. **Tephrosia nyikensis** Baker in Bull. Misc. Inform., Kew **1897**: 257 (1897) in part. —Baker f., Legum. Trop. Africa: 212 (1926). —Brenan in Mem. New York Bot. Gard. **8**: 252 (1953). —Cronquist in F.C.B. **5**: 115 (1954). —Brummitt in Bol. Soc. Brot., sér.2, **41**: 335 (1968). —Gillett in F.T.E.A., Legum., Pap.: 203 (1971). Type: Malawi, Nyika Plateau, *Whyte* s.n. (K holotype, fruiting specimen chosen here as lectotype).

> *Tephrosia congestiflora* Harms in Mildbraed, Wiss. Ergebn. Deutsch. Zentr.-Afrika Exped., Bot. **3**: 254 (1911). —Baker f., Legum. Trop. Africa: 213 (1926). Type: Rwanda, Lake Mohasi, *Mildbraed* 445 (B† lectotype, chosen by Baker f. 1926).

Subsp. **nyikensis** —Brummitt in Bol. Soc. Brot., sér.2, **41**: 335 (1968). —Gillett in F.T.E.A., Legum., Pap.: 203 (1971). —Moriarty, Wild Fl. Malawi: 132, fig.66/5 (1975).

Annual or perhaps sometimes a short-lived perennial. Stems virgate, erect, branched, up to 2 m high, sometimes somewhat woody towards the base, with short closely appressed hairs interspersed with longer ascending or spreading brown hairs. Leaves pinnate with (3)6–10 pairs of leaflets; rachis and petiole together (3)7–15(21) cm long, pubescence appressed or like that of the stem (but see note below); leaflets 3.5–6(7.5) × (0.5)0.7–1.4(1.8) cm, narrowly elliptic or oblong to oblanceolate, subacute to broadly rounded or rarely emarginate at the apex, cuneate at the base, the upper surface glabrous, the lower surface shortly and closely appressed-pubescent; stipules 8–13(16) × (1.5)2–4.5 mm, ± falcate. Flowers in terminal and axillary racemes, varying from a compact ovoid head to up to 25 cm long; bracts 6–13(16) × 1–2(2.5) mm, linear-lanceolate, falling at flowering; pedicels (2)3–8(11) mm long; bracteoles rarely present (seen only in 2 Tanzanian specimens). Calyx (to tip of lower tooth) 7–10 mm long, with dark brown or rarely grey longish appressed to spreading hairs, the lateral teeth narrowly triangular and 1–2 times as long as the tube, the two upper teeth narrowly triangular and connate for about $^1/_4$ of their length. Petals 11–15(17) mm long, purple or reddish or the keel white; standard truncate to cordate at the base, upwardly curved towards the apex, brown pubescent outside. Stamen tube fused above. Ovary pubescent; style shortly pubescent on the upper surface, sometimes sparingly on the lower surface also. Pods 65–90 × 5–7.5 mm, linear, light brown covered with shortly spreading dark brown hairs, with 10–13 seeds.

Malawi. N: Nkhata Bay Dist., Viphya, 37 km SW of Mzuzu, fl. 15.iii.1969, *Pawek* 1860 (K). C: Dedza Dist., Chongoni Forestry School, fl.& fr. 5.iv.1967, *Salubeni* 635 (K, MAL, SRGH). **Mozambique**. N: Lago Dist., Maniamba, Serra Jéci (Geci), fl.& fr. 29.v.1948, *Pedro & Pedrógão* 4110 (LMA).

Also in S Tanzania. Woodland, forest fringes, grassland and open ground in mountain areas; (600)1100–2150 m.

Subsp. *victoriensis* Brummitt & J.B. Gillett, found in Congo, Rwanda, Burundi, Uganda, Kenya and possibly NW Tanzania, has only appressed pubescence on the stems, broader inflorescence bracts and a shorter calyx.

The variation of this species and its relationship to *T. interrupta* and *T. elata* require further investigation. Subsp. *nyikensis*, as here defined, is very variable and the few collections known from the Flora area appear very heterogeneous. Plants from the Nyika, Masuku and Rungwe (Tanzania) plateaux appear to have a much more compact inflorescence than the others, and a tendency to shorter leaves. Three collections from near Mzuzu in Malawi N: (*Pawek* 3515, 11483 and *Brummitt*

11037), and one collection from Ntchisi Dist. in Malawi C: (*Verboom* 869), are exceptional in having an elongate inflorescence and short dense brown tomentum. The specimen cited above from Mozambique has pods 7.5 mm broad, the broadest known in this species.

36. **Tephrosia interrupta** Engl., Hochgebirgsfl. Trop. Afrikas: 260 (1892). —Baker f., Legum. Trop. Africa: 195 (1926). —Gillett in Kew Bull. **13**: 127–129 (1958); in F.T.E.A., Legum., Pap.: 204, fig.34 (1971). Type: N Ethiopia, N side of Mt Sholoda, 1837, *Schimper* 344 (B† holotype, K).

Virgate shrub 2–4(6) m high. Stems subglabrous and smooth to densely appressed- or spreading-pubescent. Leaves pinnate with 3–13(15) pairs of leaflets; petiole 0.3–2 cm long, petiole and rachis together (1)2–11(17) cm long, resembling the stem in pubescence; leaflets (1)1.5–4 × (0.2)0.4–0.8(1.2) cm, ± elliptic or oblanceolate, subcuneate at the base, obtuse or rounded at the apex; upper surface glabrous, lower surface appressed-pubescent; stipules 4–17 × 1–2 mm, linear-triangular, often falling early. Inflorescences various – see subspecies. Calyx (4)5–10(13) mm long, pubescent to tomentose or villous with brown or grey, appressed to usually spreading hairs; upper teeth connate for ½–¾ length. Petals 17–24(28) mm long, usually purple to deep violet, sometimes blue, greenish, red, pinkish or white tinged with purple. Stamen tube joined above. Ovary pubescent; style pubescent on one or both sides. Pods (35)45–100(110) × 4.5–8 mm, brown loosely appressed-pubescent with straight or geniculate hairs. Seeds (6)8–14, 4–5 × 2–3 × 1.5 mm, oblongoid-ellipsoidal or rarely rhomboidal, with a white aril 0.8–1 mm long in the middle of one of the longer sides.

Subsp. **elongatiflora** J.B. Gillett in Kew Bull. **13**: 127 (1958); in F.T.E.A., Legum., Pap.: 206 (1971). Type: Tanzania, Rungwe Dist., Kyimbila, *Stolz* 240 (K holotype).
    Tephrosia interrupta sensu Brenan in Mem. New York Bot. Gard. **8**: 252 (1953).

Stems for the most part subglabrous and smooth, often becoming more pubescent in the upper parts towards the inflorescence. Leaves with (5)6–13 pairs of leaflets; rachis and petiole together (2)4–11(17) cm long; stipules 4–9 × c.1 mm, linear-triangular, quickly falling. Flowers in terminal and axillary racemes, each raceme usually elongate (up to 30 cm long) but sometimes much condensed (1–2 cm), always pedunculate and subtended by a foliage leaf; bracts subtending fascicles 4–5 × 0.8 mm, linear-triangular; pedicels (8)10–14 mm long; bracteoles apparently absent. Calyx 5–8 mm long, pubescent or tomentose with appressed to spreading, grey or brown hairs. Petals 18–22 mm long. Style pubescent on both sides. Pods (60)68–100(110) × 4.5–5.5(6) mm. Seeds (8)10–14.

Zambia. E: Nyika Plateau, 2–3 km SW of Rest House, 2150 m, fr. 30.x.1958, *Robson & Angus* 479 (K, LISC, SRGH). **Malawi**. N: Mzimba Dist., Mzuzu, Marymount, 1372 m, fl. 30.vii.1969, *Pawek* 2578 (K). C: Ntchisi Dist., Ntchisi (Nchisi) Mt., 1400 m, fl.& fr. 24.vii.1946, *Brass* 16899 (K, SRGH). **Mozambique**. N: near Lichinga (Vila Cabral), fl. 6.vii.1934, *Torre* 188 (COI, K, LISC).
    Also in S Tanzania. Recorded from miombo woodland, damp places and evergreen forest fringe; 1300–2150 m.

Subsp. **mildbraedii** (Harms) J.B. Gillett in Kew Bull. **13**: 128 (1958); in F.T.E.A., Legum., Pap.: 206 (1971). Type: Rwanda, Rugege, *Mildbraed* 737 (B† holotype).
    Tephrosia nyikensis sensu Baker in Bull. Misc. Inform., Kew **1897**: 257 (1897) in part, excluding lectotype.
    Tephrosia mildbraedii Harms in Mildbraed, Wiss. Ergebn. Deutsch. Zentr.-Afrika Exped., Bot. 3: 255 (1911). —Baker f., Legum. Trop. Africa: 213 (1926). —Brenan in Mem. New York Bot. Gard. **8**: 252 (1953). —Cronquist in F.C.B. **5**: 114 (1954).

Tephrosia atroviolacea Baker f. in Bull. Soc. Roy. Bot. Belg. **57**: 115 (1925); Legum. Trop. Africa: 214 (1926). Type: Congo, Katanga, *Bequaert* 6169 (BR holotype).

Tephrosia doggettii Baker f., Legum. Trop. Africa: 213 (1926). Type: Uganda, Ruwenzori, *Doggett* s.n. (K holotype).

Stems tomentose to villous with brown spreading hairs, or (sometimes in Flora area) subglabrous and smooth for the most part. Leaves with 6–10 pairs of leaflets; rachis and petiole together 2–8 cm long; stipules (7)11–17 × 1–2 mm, linear-lanceolate, usually not quickly falling. Flowers in dense, terminal and sometimes also ± axillary heads, which are each probably aggregations of several sessile condensed lateral racemes; bracts within each head up to 18 × 3 mm, narrowly lanceolate; pedicels 6–12 mm long; bracteoles apparently absent. Calyx 7–10(12) mm long, tomentose to villous with grey or brown hairs. Petals 17–28 mm long. Style pubescent on both sides. Pods (40)50–70(7.5) × 5.5–8 mm. Seeds (6)8–10.

Malawi. N: Rumphi Dist., Nchenachena Spur, Nyika Plateau, fl. 20.viii.1946, *Brass* 17352 (K); Mzuzu Dist., Lunyangwa Forest Res., fl. 19.viii.1984, *Balaka & Kaunda* 584 (K).

Also in the mountains of W Uganda, eastern Congo, Burundi, Rwanda, Tanzania (Mahali Mts. and highlands N of Lake Malawi), reaching its southernmost point in the Flora Zambesiaca area on the Nyika Plateau where it may be sympatric with subsp. *elongatiflora*. Open grassy places at high altitudes; 1800–2590 m.

Subsp. *interrupta* is intermediate between subsp. *elongatiflora* and subsp. *mildbraedii* in its inflorescence characters, having the lateral racemes of the former contracted into 3–13 sessile or subsessile dense heads forming a secondary compound raceme, the bracts within each head being elliptic to suborbicular-acuminate and up to 7 mm broad. It occurs from central Tanzania to Ethiopia.

37. **Tephrosia bracteolata** Guill. & Perr. in Guillemin, Perrottet & Richard, Fl. Seneg. Tent.: 194 (1832). —Baker in F.T.A. **2**: 116 (1871). —Baker f., Legum. Trop. Africa: 201 (1926). —Cronquist in F.C.B. **5**: 112, t.8 (1954). —Hepper in F.W.T.A., ed.2, **1**: 529, 530 (1958). —Torre in C.F.A. **3**: 158 (1962) excluding *Gossweiler* 4260. —Brummitt in Bol. Soc. Brot., sér.2, **41**: 331 (1968). —Gillett in F.T.E.A., Legum., Pap.: 199 (1971). Types: Senegal, *Perrottet* 208 (P syntype, BM) & Lamsar, *Leprieur* s.n. (P syntype).

Tephrosia concinna Baker in F.T.A. **2**: 112 (1871). —Baker f., Legum. Trop. Africa: 190 (1926). Type: Nigeria, on R. Niger, *Baikie* (K holotype).

Cracca bracteolata (Guill. & Perr.) Kuntze, Rev. Gen. Pl. **1**: 174 (1891). —Hiern, Cat. Afr. Pl. Welw. **1**: 221 (1896). —Pires de Lima in Bol. Soc. Brot., sér.2, **2**: 137 (1924) for synonym, excluding specimen (e.g. *T. reptans* var. *microfoliata*).

Var. **strigulosa** Brummitt in Bol. Soc. Brot., sér.2, **41**: 333 (1968). —Gillett in F.T.E.A., Legum., Pap.: 200 (1971). Type: Zambia, Mpulungu, Lake Tanganyika, pebbly beach, fl.& fr. 8.iii.1952, *Richards* 1049 (K holotype).

Annual with a short taproot to perennial. Stems erect, 1–3 m high, with ascending branches in the upper part, densely covered with rather stiff hairs ascending at an angle of about 45°. Leaves pinnate with (8)10–15(17) pairs of leaflets; petiole 0.6–1.6 cm long, petiole and rachis together up to 30 cm long, with ascending hairs like the stem; leaflets (2)3.5–8.0(11.5) × 0.3–0.7(1.1) cm, linear-oblong to linear, ± cuneate at the base, acute to rounded or rarely truncate at the apex; upper surface glabrous, lower surface somewhat silvery-grey appressed-pubescent; stipules 5–10(14) × (1)1.5–4(5) mm, linear-triangular, pubescent. Flowers in terminal and axillary long-peduncled racemes up to 30 cm long; bracts 4–7 × 1–3 mm, ovate-acuminate, somewhat purplish but covered with appressed grey hairs, conspicuous and somewhat imbricate at bud stage but falling at flowering to expose similar ovate bracts subtending the fascicle of flowers at each node; pedicels 3–7 mm long. Calyx 3–4 mm long, densely covered with appressed or ascending grey or brown (often both) hairs; the two upper teeth short, joined for $^1/_2$–$^3/_4$ length, the lateral teeth shorter than the tube, the lower tooth about 2–3 times as long as the lateral and exceeding the tube. Petals 12–16 mm long, red or pink to purplish. Stamen tube joined above. Ovary pubescent; style pubescent. Pods 58–88 × (4)4.5–5.5 mm, brown, shortly pubescent with loosely appressed or irregular hairs. Seeds 12–19, obliquely transverse, each one rhomboid, light brown with dark brown markings.

Zambia. N: Mbala Dist., Kumbula (Mbulu) Is., L. Tanganyika, fl.& fr. 17.ii.1955, *Richards* 4514 (K, SRGH). **Malawi**. N: 14.5 km W of Karonga, fl.& fr. 22.iv.1975, *Pawek* 9498 (K, MAL, MO). C: Dedza Dist., Mtakataka (Ntaka-taka), fr. 11.iv.1969, *Salubeni* 1326 (K, MAL, SRGH).

Also in Congo, central and SW Tanzania and E Angola. In the Flora area known principally from the shores of Lakes Tanganyika and Malawi on wet ground in their immediate vicinity (although two specimens cited from Malawi are more inland); 590–850 m. Recorded from banks of ditches and floodplains in Tanzania and from savanna in Congo.

Var. *bracteolata* differs in having the stem hairs closely appressed, and occurs from Senegal and Ethiopia to Bas Congo, Uganda and N Tanzania; the species is also found in the Cape Verde Islands. The two varieties are readily distinguishable and largely geographically separated, and might be considered to merit subspecific rank.

38. **Tephrosia nana** Kotschy in Schweinfurth, Reliq. Kotschy.: 20 (1868). —Baker in F.T.A. **2**: 109 (1871). —Baker f., Legum. Trop. Africa: 188 (1926). —Gillett in Kew Bull. **13**: 130 (1958). —Torre in C.F.A. **3**: 155 (1962). —Brummitt in Bol. Soc. Brot., sér.2, **41**: 329 (1968). —Gillett in F.T.E.A., Legum., Pap.: 208 (1971). Types: Sudan/Ethiopia border, Matamma, Gallabat area, *Schweinfurth* 1871 (K syntype, BM); Sudan, Fesoglu, *Boriani* 109 (W syntype).

Tephrosia barbigera Baker in F.T.A. **2**: 113 (1871). —Harms in Mildbraed, Wiss. Ergebn. Deutsch. Zentr.-Afrika Exped., Bot. **3**: 253 (1911). —Fries, Wiss. Ergebn. Schwed. Rhod.-Kongo-Exped. **1**: 82 (1914). —De Wildeman in Bull. Soc. Roy. Bot. Belg. **57**: 115 (1925). —Baker f., Legum. Trop. Africa: 196 (1926); in J. Bot. **66**, Polypet. Suppl.: 106 (1928). — Cronquist in F.C.B. **5**: 111 (1954). —Hepper in F.W.T.A., ed.2, **1**: 528, 530 (1958). Type: Angola, *Welwitsch* 2096 & 2097 (K syntypes).

Tephrosia nana var. *angolensis* Baker in F.T.A. **2**: 110 (1871). Type from Angola.

Tephrosia polysperma Baker in Trans. Linn. Soc., London **29**: 55 (1872). Type: Uganda, Mengo Dist., vii.1862, *Grant* s.n. (K holotype).

Cracca barbigera (Baker) Kuntze, Rev. Gen. Pl. **1**: 174 (1891). —Hiern, Cat. Afr. Pl. Welw. **1**: 220 (1896).

Cracca barbigera var. *bakeriana* Hiern, Cat. Afr. Pl. Welw. **1**: 220 (1896) nom. illegit. Type as for *Tephrosia nana* var. *angolensis*.

Annual with a short taproot and decumbent-ascending to erect branched stems 60–120(150) cm high (or much shorter outside the Flora area, see note below). Stems fairly densely covered with soft, brown, ascending or spreading hairs. Leaves pinnate with 4–9 pairs of leaflets; petiole 0.5–2.5(3) cm long, petiole and rachis together 4–14(18) cm long, pubescent like the stem; leaflets 2.5–5.5(6.5) × (0.5)0.7–1.8(2) cm, elliptic or obovate to oblanceolate (in other areas often oblanceolate to linear-oblong), cuneate to rounded at the base, rounded to emarginate at the apex; upper surface glabrous or appressed-pubescent (see note below), lower surface appressed-pubescent; stipules 6–12(17) × (0.4)1–3 mm, linear-triangular. Flowers in mostly terminal, sometimes also axillary, usually rather short and compact racemes, mostly 3–15 cm long, or often with the lowermost flowers in the axils of the upper leaves; bracts 4–9(12) ×

(0.2)0.4–1(2) mm, pubescent, soon falling; pedicels 3–7 mm long. Calyx (2.5)3–5 mm long, appressed-pubescent to conspicuously villous with strongly deflexed hairs especially on the lower side, the teeth with much shorter and darker hairs towards their margins and so apparently with a dark outline; upper teeth connate for about half their length, lateral teeth short (c.1 mm) and acute to subobtuse. Petals (10)13–19 mm long, purple to brownish or pink. Stamen tube joined above. Ovary pubescent; style pubescent. Pods (40)50–70 × 5–6 mm (or shorter and narrower in W Africa), brown, shortly pubescent with ± irregular hairs. Seeds (13)15–18(21), obliquely transverse.

Zambia. N: Samfya Dist., L. Bangweulu, Kamindas, fl. 5.x.1911, *R.M. Fries* 889 (UPS). W: Copperbelt, Mufulira Dist., fl.& fr. 17.iii.1956, *Fanshawe* 2848 (K, SRGH). **Mozambique**. Z: near Ile, fl. 1.iv.1943, *Torre* 5037 (K, LISC).

Widely distributed from Senegal and Ethiopia to Angola and the Flora Zambesiaca area. In woodland, grassland, on river banks, in disturbed habitats, etc.; 400–1250 m.

A very common species in Congo but apparently only just extending into the Flora Zambesiaca area. The specimen cited from Mozambique is the only one known from the Flora area outside the parts of Zambia bordering on Katanga, and seems to be well removed from the species main distribution area.

This species has conspicuous brown spreading hairs on the stems and leaves, and may be confused with *T. caerulea* subsp. *caerulea*. It is distinguished by its annual habit, usually broader leaflets, shorter calyx, and rather conspicuously diagonal seeds. The latter character is shared with the related *T. bracteolata* which has narrower leaflets and broadly ovate inflorescence bracts.

This species has long been known as *T. barbigera* but as Gillett (1958) has pointed out this does not seem to be specifically separable from the dwarf (10–15 cm) plants in the Sudan and Ethiopia which were earlier described as *T. nana*. These dwarf forms are probably merely phenotypic modifications induced by arid conditions, but further investigation would be desirable.

Variation in the species is discussed further by Brummitt (1968). Specimens from Zambia, together with those from adjacent Katanga, resemble plants from West Africa in having leaflets glabrous above, while in the rest of Congo and East Africa the leaflets in the great majority of plants are pubescent above. One collection from Mozambique has leaflets pubescent above, unlike those from Zambia. Specimens from W Zambia have generally much broader leaflets (13–18 mm) than are found throughout the rest of the range of the species.

39. **Tephrosia longipes** Meisn., in Hooker, London J. Bot. **2**: 87 (1843). —Harvey in F.C. **2**: 208 (1862). —Baker in F.T.A. **2**: 120 (1871) for name only. —Young in Ann. Transvaal Mus. **14**: 398 (1932). —Forbes in Bothalia **4**: 980 (1948). — Gillett in Kew Bull. **13**: 125 (1958) in part. —Brummitt in Bol. Soc. Brot., sér.2, **41**: 308 (1968). Type: South Africa, Durban (Port Natal), side of Tafelberge mountain, *Krauss* 20 (K holotype).

    Cracca longipes (Meisn.) Kuntze, Rev. Gen. Pl. **1**: 175 (1891). —Hiern, Cat. Afr. Pl. Welw. **1**: 222 (1896) for name only, excluding specimen.

Annual or short-lived perennial, erect to 1.6 m from a taproot, or suffrutex with many ascending stems to 0.5 m from a woody rootstock. Stems shortly but densely appressed-pubescent with grey hairs, or glabrescent in lower parts, or (var. *swynnertonii*, var. *drummondii*) with ascending to spreading pale brownish hairs. Leaves with (0)1–7(8) pairs of leaflets; petiole 0.3–7.5 cm, much shorter to longer than the rachis, the petiole and rachis together (1.5)2.5–13(18) cm, pubescent like the stem; leaflets (1.5)2.5–8(11) × 0.15–7(14) mm, linear or linear-elliptic to oblong, terminated by a usually blackish and often somewhat recurved mucro; upper surface glabrous or rarely pubescent, lower surface appressed-pubescent;

stipules (3)4–14(17) × (0.1)0.3–1 mm, linear, blackish. Flowers in terminal or leaf-opposed long-peduncled racemes (2)8–25(30) cm long, the flowers sometimes (var. *swynnertonii*) conspicuously clustered into a compact inflorescence at the tip of the peduncle; pedicels 3–8 mm. Calyx 3–6(8) mm long, shortly appressed to ascending grey or brown-pubescent or villous; upper teeth joined for $^{1}/_{3}$ to $^{1}/_{2}$ length; lateral teeth about equalling or slightly exceeding the tube. Petals pink to purple; standard 12–18(22) mm. Stamen tube joined above. Ovary pubescent to tomentose; style pubescent. Pods (28)50–88 × 4–5 mm, pubescent with irregular geniculate hairs or rarely (var. *lissocarpa*) with regular straight appressed hairs; seeds (5)10–22.

The species as interpreted here includes a wide range of morphological forms. The extremes (var. *longipes*, var. *drummondii*) look so different from each other that it seems scarcely credible that they should be treated as conspecific. Drawing satisfactory boundaries, however, is problematic. Most specimens included fall into two groups with different growth habits – those which appear to be annuals to short-lived perennials with stems up to 1.4 m high from a tap root, and those which are suffrutices with numerous stems up to 50 cm from a woody rootstock. In preliminary notes in 1968 I maintained these as separate species, *T. longipes* and *T. lurida* respectively. There are associated leaf characters (see key below) and they have different geographical ranges, albeit with a broad overlap in Zimbabwe and South Africa. It has to be admitted, however, that the characters are not as clear-cut as one might like in recognizing species, and a number of specimens are difficult to place one way or another. It is not clear whether the tap-rooted perennials may mature into suffrutices, or whether the different habits may be due to environmental conditions. Young in Ann. Transvaal Mus. **14**: 402 (1932) has noted that the two forms hybridize in South Africa.

Gillett in Kew Bull. **13**: 125 (1958) reduced *T. lurida* to a variety of *T. longipes*, but in F.T.E.A. in 1971 he treated the two taxa as separate species. Schrire, however, in his synopsis of South African species in Bothalia **17**: 11 (1987), has combined the two without any taxonomic recognition, noting that the characters used to separate them are based on ecological variables and cannot be upheld taxonomically. The compromise solution of treating them as varieties of one species is adopted here, bearing in mind particularly their geographical differences.

It appears that within each of these varieties some local differentiation of populations has taken place. The most distinct of these is the variant of the annual to short-lived perennial growth form found in the Eastern Highlands of Zimbabwe and adjacent Mozambique, where plants have a very characteristic compact inflorescence, ascending indumentum and broader leaflets. This is so distinctive that in 1968 I gave it subspecific rank under *T. longipes*, but in view of its restricted geographical range it is here regarded as only a variety, var. *swynnertonii*. This decision is supported by the occurrence in the same geographical area of a variant of the suffruticose growth form which also is very distinctive in its spreading indumentum (and reduced leaflet number), var. *drummondii*. Another conspicuous indumentum variant with the suffruticose habit, occurring at the extreme southwest of the distribution of *T. longipes*, is here also maintained at varietal level, var. *lissocarpa*. One species with five varieties in the Flora area seems to be the optimal taxonomy in our present state of knowledge. It is possible that a sixth variety may be recognizable in South Africa (Gauteng, Northern Cape and KwaZulu-Natal), var. *uncinata* Harv., differing from var. *lurida* in having broader oblong-oblanceolate leaflets 4–7 mm broad and truncate to retuse at the apex, with a shorter and more sparse indumentum on all parts. Further field observations and experimental work is needed.

1. Stems and leaf rachis with ascending to spreading usually pale brownish hairs; if plant over 50 cm high, flowers ± crowded into a compact inflorescence (highlands of E Zimbabwe and adjacent Mozambique) · · · · · · · · · · · · · · · · 2
   - Stems and leaf rachis with appressed grey hairs; flowers not crowded into a compact inflorescence · 3
2. Bushy herb to subshrub usually 1–1.6 m high; leaves with 3–6 pairs of leaflets; flowers crowded into a ± compact inflorescence · · · · · · · · · **ii**) var. *swynnertonii*
   - Decumbent ascending herb to 50 cm; leaves unifoliolate or trifoliolate; flowers not crowded · **v**) var. *drummondii*
3. Branching herb to subshrub (0.5)0.7–1.4 m high; stems usually 2–5 mm diameter; leaves with 3–8 pairs of leaflets; petiole 0.5–4.2(5.5) cm, usually shorter than the rachis · **i**) var. *longipes*
   - Suffrutex with woody rootstock and numerous ascending stems to 0.5 m high and 1–1.5(2) mm diameter; leaves with 1–3(4) pairs of leaflets; petiole (1.5)2.5–7.5 cm, usually much longer than the rachis · · · · · · · · · · · · · · · · 4
4. Fruits irregularly pubescent to villous with spreading often geniculate hairs · **iii**) var. *lurida*
   - Fruits with closely appressed parallel hairs · · · · · · · · · · · · · · **iv**) var. *lissocarpa*

i) Var. **longipes**

Tephrosia pseudolongipes Baker f., Legum. Trop. Africa **1**: 199 (1926). Type: Zimbabwe, Bromley, *Walters* 2207 (K holotype, SRGH).

Tephrosia longipes var. *icosisperma* Brummitt in Bol. Soc. Brot., sér.2, **41**: 313 (1968). Type: Mozambique, Gondola Dist., Chimoio, Floresta de Nhamissanguere, near Gondola road, fl.& fr. 17.ii.1948, *Garcia* 262 (LISC holotype).

Annual or short-lived perennial herb to subshrub (0.5)0.7–1.2(1.4) m high from a taproot or slender rootstock. Stems and leaf rachis densely appressed-pubescent with grey hairs. Leaves with 3–7(8) pairs of leaflets, each (1)2–5(6) mm broad and ± linear; petiole (0.5)0.9–4.2(5.5) cm, usually shorter than the rachis. Inflorescence elongate, lax.

Zambia. C: between Kafue and Lusaka, fl.& fr. 17.viii.1946, *Gouveia & Pedro* 1683 (LMA). S: Mazabuka Dist., Yates Jones Farm, near Choma, fl.& fr. 27.i.1960, *White* 6506 (FHO, K). **Zimbabwe**. N: Gokwe Dist., Charama escarpment, near turn-off on Gokwe–Charama road, fl.& fr. 18.iii.1962, *Bingham* 177 (K, LISC, SRGH). W: Matobo Dist., Farm Besna Kobila, fl.& fr. i.1959, *Miller* 5736 (K, SRGH). C: Marondera Dist., Injina, fl.& fr. 27.ii.1946, *Collins* 14 (K, SRGH). E: Mutare Dist., Odzani R. valley, fl.& fr. 1914, *Teague* 17 (K). S: Masvingo Dist., Makaholi Expt. Farm, fl.& fr. 23.iii.1948, *D.A. Robinson* 309 (K, SRGH). **Mozambique**. MS: Savana de Maronga, fl.& fr. 2.viii.1945, *Simão* 427 (LISC). GI: Chibuto Dist., near Chibuto, road towards Alto Changane, fl.& fr. 12.ii.1959, *Barbosa & Lemos* 8381 (COI, K, LISC, LMA). M: near Maputo (Lourenço Marques), fl.& fr. ii.1946, *Pimento* 17304 (LISC, SRGH).

Also in South Africa. In *Brachystegia* woodland, grassland and sandveld, occasionally in mopane woodland, and in sandy places near the coast; 20–1200 m.

Plants from Mozambique tend to have longer pods, (62)70–88 mm, with more seeds (16) 17–22, than those elsewhere (pods (40)50–70(77) mm with 10–16(17) seeds). They have been separated as a variety, see synonymy above, but formal recognition now seems inappropriate.

ii) Var. **swynnertonii** (Baker f.) Brummitt, stat. nov.

Tephrosia swynnertonii Baker f., Legum. Trop. Africa **1**: 202 (1926)). Type: Zimbabwe, near Chirinda, 1100 m, *Swynnerton* 369 (BM holotype, K).

Tephrosia longipes subsp. *swynnertonii* (Baker f.) Brummitt in Bol. Soc. Brot., sér.2, **41**: 315 (1968).

Tephrosia grandiflora sensu Baker f. in J. Linn. Soc., Bot. **40**: 54 (1911). —sensu Eyles in Trans. Roy. Soc. S. Africa **5**: 375 (1916).

Short-lived perennial herb (rarely annual?) to subshrub 1–1.6 m high, from a taproot or slender rootstock. Stems (at least on ridges) and leaf rachis with dense ascending to spreading usually light brown hairs. Leaves with (3)4–6 pairs of leaflets, these 4–8(10) mm broad and linear to oblong; petiole 0.3–2 cm, much shorter than the rachis. Inflorescence compact, crowded at end of peduncle or occasionally the lowermost flowers ± remote from others.

Zimbabwe. E: Chipinge Dist., Ngungunyana Forest Area (Gungunyana Forest Res.), fl.& fr. ii.1962, *Goldsmith* 42/62 & 51/62 (K, LISC, SRGH). **Mozambique**. MS: Chimanimani Mts., base of SE slopes, fl. 25.iv.1974, *Pope & Müller* 1297 (K, SRGH).

Known only from E Zimbabwe and adjacent Mozambique. Open grassland; 900–1160 m. Numerous intermediates between this and var. *longipes* occur in the same area.

iii) Var. **lurida** (Sond.) J.B. Gillett in Kew Bull. **13**: 125 (1958) in part. —Torre in C.F.A. **3**: 158 (1962) for name only. Type: South Africa, Gauteng Prov., Mooi R., Magaliesberg & Crocodile R., *Zeyher* 456 (K syntypes).

Tephrosia lurida Sond. in Linnaea **23**: 30 (1850). —Harvey in F.C. **2**: 208 (1862). —Eyles in Trans. Roy. Soc. S. Africa **5**: 375 (1916) for name only. —Baker f., Legum. Trop. Africa: 200 (1926) for name only. —Young in Ann. Transvaal Mus. **14**: 402 (1932). —Forbes in Bothalia **4**: 980 (1948). —Suessenguth & Merxmüller in Proc. & Trans. Rhod. Sci. Assoc. **43**: 26 (1951) for name only. —Martineau, Rhodesia Wild Fl.: 37 (1953). —Brummitt in Bol. Soc. Brot., sér.2, **41**: 317 (1968). —Gillett in F.T.E.A., Legum.-Pap.: 197 (1971), excl. *T. laxiflora.*

Cracca lurida (Sond.) Kuntze, Rev. Gen. Pl. **1**: 175 (1891).

Tephrosia angustissima Engl., Bot. Jahrb. Syst. **10**: 29 (1888). —Forbes in Bothalia **4**: 979 (1948). —Torre, C.F.A. **3**: 158 (1962) for name only? Type: South Africa, Northern Cape (British Bechuanaland), near Kuruman, summit of Mhana peak, fr. ii.1886, *Marloth* 1086 (PRE isotype).

Tephrosia dowsonii Baker f., Legum. Trop. Africa 1: 190 (1926). Type: Kenya, Nairobi, 1675 m, 1916, *Dowson* 519 (K).

Suffrutex with numerous decumbent-ascending stems to 50 cm high from a woody rootstock. Stems and leaf rachis densely appressed-pubescent with grey hairs. Leaves with 1–3(4) pairs of leaflets, these 1–4(5) mm broad and ± linear; petiole (1.5)2.5–7.5 cm, usually much longer than the rachis. Inflorescence elongate, lax.

Zimbabwe. N: Mvurwi Range (Umvukwe Mts), fl.& fr. 5.iii.1961, *Richards* 14566 (K, SRGH). W: Bulawayo, fl.& imm.fr. ii.1906, *Eyles* 1206 (PRE, SRGH). C: Marondera (Marandellas), fr. 5.vi.1947, *Newton* 60 (SRGH). E: Nyanga Dist., Pungwe Falls Farm, fl.& imm.fr. xi.1957, *Miller* 4691 (K, SRGH).

Also in Kenya and South Africa (Northern Cape and North West Provinces). Grassland and rocky places; 1270–1830 m.

Specimens from Kenya referred to this variety are widely separated geographically from the main area of distribution in the N South Africa and Zimbabwe. They often have more diffuse, branching habit and shorter leaflets and may be taxonomically separable, perhaps as a further variety. One specimen from the Flora area, Zimbabwe E: Chimanimani Mts., below hut, fl.& fr. 29.v.1959, *Noel* 2158 (SRGH), closely resembles these Kenyan plants.

Gillett in F.T.E.A. (1971) included *T. laxiflora* in *T. lurida*, but this is distinct in having terminal and axillary inflorescences, unlike the *T. longipes* complex which has terminal and leaf-opposed inflorescences.

iv) Var. **lissocarpa** (Brummitt) Brummitt, comb. nov.

> *Tephrosia lurida* var. *lissocarpa* Brummitt in Bol. Soc. Brot., sér.2, **41**: 320 (1968). Type:
> South Africa, Limpopo Prov., Soutspansberg Dist., c.15 km E of Louis Trichardt,
> Rustfontein farm, fr. 9.x.1955, *Schlieben* 7339 (K).

Differs from var. *lurida* in its pods which are covered with longish parallel closely appressed
hairs and are often shorter (down to 28 mm) with fewer seeds (as few as 5 per pod).

Zimbabwe. W: Matobo Dist., Farm Besna Kobila, fl.& fr. iii.1954, *Miller* 2284 (K,
SRGH).

Within the Flora Zambesiaca area known only from the collection cited; c.1300 m.

v) Var. **drummondii** (Brummitt) Brummitt, comb. nov.

> *Tephrosia lurida* var. *drummondii* Brummitt in Bol. Soc. Brot., sér.2, **41**: 321 (1968). Type:
> Zimbabwe, Chimanimani Dist., Glencoe Forest Res., fl.& fr. 23.xi.1955, *Drummond* 4976 (K
> holotype, PRE, SRGH).

Differs from var. *lurida* in its stem, leaf petiole and rachis, inflorescence axis and calyx being
densely spreading-pubescent to tomentose, and leaflet under-surface with irregular geniculate
hairs. Leaves unifoliolate or ± trifoliolate (always long-petioled); leaflets up to 12 mm broad.
Pods as in var. *lurida*.

Zimbabwe. E: Chimanimani Dist., Tarka Forest Reserve, fl.& fr. x.1968, *Goldsmith*
155/68 (K, SRGH).

Known from the Flora Zambesiaca area only from the two collections cited, from
steep grassy slopes; c.1300 m. Two similar collections are known from South Africa
(Mpumalanga and Gauteng Provinces).

The variety looks very distinct and might perhaps be better regarded as a separate
species. More collections are required. In its spreading indumentum and broad
leaflets it differs from var. *lurida* in the same way that var. *swynnertonii* differs from var.
longipes, and it occurs in the same region as var. *swynnertonii*. It differs markedly from
var. *swynnertonii* in its habit, mostly unifoliate leaves, and lax inflorescence.

Tephrosia longipes × reptans

A specimen from C Zambia (Chakwenga headwaters, 100–120 km E of Lusaka, fl.&
fr. 27.iii.1965, *E.A. Robinson* 6515 (K)) appears to be intermediate between *T. longipes*
var. *longipes* and *T. reptans* var. *reptans*, both of which have been collected by Robinson
at this locality. It is regarded as a probable hybrid between them. It has the erect habit
of *T. longipes*, but is probably annual although woody towards the base, and it differs
markedly from *T. longipes* in its indumentum of spreading hairs as in *T. reptans*.
Hybrids are relatively rare in Papilionoid legumes, but this case is quite convincing
in view of the occurrence of both putative parents with it.

40. **Tephrosia euchroa** Verdoorn in Bothalia **3**: 239 (1937). —Forbes in Bothalia **4**:
    969 (1948). —Gillett in Kew Bull. **13**: 417 (1958). —Schrire in Bothalia **17**: 11
    (1987). Type: South Africa, Limpopo Province, Soutpansberg, rocky NW slopes,
    Obermeyer, Schweickerdt & Verdoorn 73 (PRE holotype, K).

Probably a short-lived perennial with a stout taproot and branching stems 25–100 cm high
forming a low bush, somewhat woody towards the base. Stems usually fairly densely appressed-
pubescent with grey hairs, rarely spreading-pubescent or subglabrous. Leaves pinnate with
(3)4–8 pairs of leaflets; petiole 0.5–2.5(3.3) cm long, petiole and rachis together 3–10 cm long,
pubescent like the stem; leaflets 1–3.5(4.5) × 0.3–0.8(1) cm, narrowly elliptic to linear-elliptic,
rounded to subcuneate at the base, acute to rounded at the apex; upper surface glabrous or

appressed-pubescent, lower surface appressed-pubescent; stipules 5–10 × (0.4)1–2 mm, linear-triangular. Flowers in terminal racemes, sometimes with the lowermost flowers in the axils of the uppermost leaves; bracts 3–5 × 0.5–1 mm, linear-triangular, falling early, somewhat purplish with grey pubescence; pedicels 2–4 mm long. Calyx (3)4–6(7) mm long, usually somewhat purplish with appressed to spreading grey hairs; upper teeth connate for about half their length; lateral teeth about equalling the tube, triangular-acuminate. Petals (13)14–18(20) mm long, pink or red or sometimes purplish-red. Stamen tube connate above. Ovary pubescent; style pubescent. Pods (35)60–65 × 4.5–5 mm, shortly and rather sparsely appressed-pubescent. Seeds 11–13, c.3.5 × 2 × 1.3 mm, ± oblong, with a small white aril in the middle of one of the long sides, the testa with a conspicuous irregular pattern of dark brown markings on a light brown background.

Zimbabwe. S: Beitbridge Dist., Tshiturapadsi (Chiturupazi), fl.& fr. 22.ii.1961, *Wild* 5335 (K, PRE, SRGH); Tshiturapadsi (Chiturupadzi) Store, fl.& fr. 18.iii.1967, *Corby* 1864 (K, SRGH); Umzingwane R., Gem Farm, fl.& fr. 7.ii.2000, *Timberlake & Cunliffe* 4434 (K, SRGH).

Also in South Africa (Limpopo Province). In South Africa recorded from rocky places, low bushveld and Mopane veld on black turf soils; 300–450 m.

Apart from the collections cited above only one other record of this species is known from the Flora area: Zimbabwe, Chiredzi Dist., Runde (Lundi) R., Chipinda Pools, fl.& fr. 22.i.1961, *Goodier* 67 (K, SRGH). However, this collection differs markedly from all others in having subglabrous stems, longer petioles on the lower leaves (2.5–3.3 cm), smaller calyx (3 mm), and smaller pods (35–40 mm) with minutely pubescent or subglabrous surfaces. Further collections may show that this represents a separate taxon.

41. **Tephrosia faulknerae** Brummitt in Bol. Soc. Brot., sér.2, **41**: 282 (1968). Type: Mozambique, Lugela Dist., 'Namagoa, Mocuba and Moebede road, Lugela', fl. 23–24.iii.1949, *Faulkner* Kew 404 (K holotype, COI, K, SRGH)*.

A moderately large bush up to 2 m or more high. Stems densely greyish-white or brown appressed- or ascending-pubescent, at least when young. Leaves with (6)8–11 pairs of leaflets; petiole 1–4(7) mm long, petiole and rachis together 5–12 cm long, appressed-pubescent or the lower side glabrescent; leaflets 1–3.2 × 0.5–1 cm, oblong-oblanceolate, rounded to subcuneate at the base, ± truncate and strongly mucronate at the apex; upper surface glabrous, lower surface rather thinly and shortly appressed or irregularly pubescent; stipules 6–10 × 1.5–2.5 mm, narrowly triangular, pubescent. Flowers in mostly terminal racemes 7–10 cm long, very shortly pedunculate or with the lowermost flower in the axil of the upper leaf; bracts 5–8 × 0.7–1 mm, linear-triangular; pedicels c.3 mm long. Calyx 4–5 mm long, shortly and rather irregularly pubescent; two upper teeth joined for ³/₄ length to form an acute triangle. Petals mauve; standard 15–17 mm long, about equalled by the keel. Stamen tube joined above. Ovary appressed-pubescent; style pubescent. Pods 50–64 × 4–5 mm, puberulent to very shortly irregularly pubescent, brown. Seeds 12–14.

Mozambique. N: Nampula, fl.& fr. 17.ii.1937, *Torre* 1192 (COI). Z: Lugela Dist., Namagoa and Muobede road, fr. 3.iv.1949, *Faulkner* Kew 404 (COI, K, SRGH).

Not known outside the Flora Zambesiaca area. Recorded from woodland and sisal plantations; 150–400 m.

---

* Three sheets are labelled Kew 404. The data on the label of sheet 1 of Faulkner Kew 404 suggests that there were three collections: the first made on 23 March at Namagoa (which was at that time considered to be in Mocuba Dist.), the second on 24 March made on Muobede road in Lugela Dist. and the third, in fruit, on 3 April, said to be from the 'same plants'. Two different habitats are given on the label: 'woodland' and 'plantation'. The holotype is sheet 1, but it is not clear from which locality or habitat this was collected. It is not certain that sheets 2 and 3, or the duplicates of this number at COI and SRGH, are from exactly the same locality. Strictly speaking they may not be true isotypes, despite having the same number as the holotype.

42. **Tephrosia reptans** Baker in F.T.A. **2**: 121 (1871). —Oliver in Trans. Linn. Soc., London **29**: 56, t.27 (1872). —Baker f., Legum. Trop. Africa: 197 (1926). — Suessenguth & Merxmüller in Proc. & Trans. Rhod. Sci. Assoc. **43**: 26 (1951). — Brummitt in Bol. Soc. Brot., sér.2, **41**: 283 (1968). —Gillett in F.T.E.A., Legum., Pap.: 200 (1971). —Schrire in Bothalia **17**: 11 (1987). Types: Tanzania, Tabora Dist., near Tabora, *Grant* s.n. (K lectotype), and Mozambique, Chupanga (Shupanga), *Kirk* s.n. (K).

    Tephrosia kirkii Baker in F.T.A. **2**: 115 (1871). —Baker f., Legum. Trop. Africa: 200 (1926). Type: Mozambique, Mungari (Luawe) R. mouth, *Kirk* s.n. (K holotype).

    Cracca reptans (Baker) Kuntze, Rev. Gen. Pl. **1**: 175 (1891).

Annual or short-lived perennial with the rootstock often becoming somewhat woody, with several prostrate or decumbent-ascending, straggling, branched stems up to 1 m or more long or (var. *microfoliata*) with the main stem ± erect with usually decumbent branches from near the base. Stems with conspicuous, brown, spreading hairs up to c.1 mm long, or (var. *arenicola*, var. *microfoliata*) appressed-pubescent to glabrous. Leaves with 4–9 pairs of leaflets; petiole 0.2–2.4 (4.4 on occasional leaves) mm long, petiole and rachis together 4–11(13) cm long, pubescent as on the stem; leaflets (1)1.5–4(5.5) × (0.2)0.5–1(1.3) cm, oblanceolate to elliptic-oblong or rarely linear-elliptic or linear-oblong, rounded to subcuneate at the base, rounded to truncate or emarginate at the apex, mucronate; upper surface shortly pubescent, lower surface rather shortly appressed or irregularly pubescent; stipules 4–13(16) mm long, linear to linear-triangular, ± persistent. Flowers in terminal and axillary, long-peduncled racemes (4)7–22(30) cm long; bracts 3–7 × 0.2–0.6 mm, linear-triangular; pedicels 2–4(6) mm long. Calyx (3)4–7(8) mm long, shortly spreading (or appressed in var. *arenicola* and var. *microfoliata*) brown or grey pubescent; two upper teeth connate for $^1/_3$–$^1/_2$ length, all teeth long-acute and ± filiform towards the apex. Petals pink to purple or blue, the keel pale; standard 12–17 mm long, the other petals somewhat shorter. Stamen tube joined above. Ovary shortly pubescent; style pubescent. Pods (38)50–65(76) × 4–5 mm, puberulent to very shortly pubescent with irregular to ± parallel hairs, straw-coloured to brown. Seeds (10)14–16(18).

1. Stems and leaf rachides spreading-pubescent · · · · · · · · · · · · · · · i) var. *reptans*
– Stems and leaf rachides glabrous or appressed-pubescent · · · · · · · · · · · · · · 2
2. Perennial with robust stems usually woody and ± prostrate towards the base; seeds 11–13(16) per pod · ii) var. *arenicola*
– Annual with rather slender stems, not woody, the main stem ± erect but often with decumbent branches from the base; seeds (12)15–18 per pod · · · · · · · · · ·
· iii) var. *microfoliata*

i) Var. **reptans** —Brummitt in Bol. Soc. Brot., sér.2, **41**: 284 (1968). —Gillett in F.T.E.A., Legum., Pap.: 200 (1971). —Schrire in Bothalia **17**: 11 (1987).

    Tephrosia carvalhoi Taub. in Bot. Jahrb. Syst. **23**: 183 (1896) as "*carvalhi*". Type: Mozambique, Gorongosa, 1885, *Carvalho* s.n. (B holotype, COI).

    Tephrosia godmaniae Baker f., Legum. Trop. Africa: 194 (1926) as "*godmanae*". Type: Zimbabwe, Harare (Salisbury), *Godman* 144 (BM holotype).

Annual to perennial, the branches prostrate or decumbent-ascending. Stems and leaf rachides with conspicuous spreading hairs. Leaves with 4–7(9) pairs of leaflets of varying shape; stipules (4)8–13(16) mm long. Calyx with ± spreading hairs. Pods with (10)14–16 seeds.

Zambia. C: Lusaka Dist., Chakwenga headwaters, 100–129 km E of Lusaka, fl.& fr. 27.iii.1965, *E.A. Robinson* 6515 (K). S: Choma Dist., Mochipapa, near Choma, fl. 10.iii.1962, *Astle* 1497 (K, SRGH). **Zimbabwe**. N: Zvimba Dist., Trelawney, Tobacco Res. Station, fl.& fr. 29.iii.1944, *Jack* 67 (K, PRE, SRGH). W: Bulawayo, Beacon Hill, fl.& fr. 11.ii.1967, *Best* 578 (K, SRGH). C: Chegutu Dist., Poole Farm, fl.& fr. 4.iii.1948, *Hornby* 2868 (K, LISC, SRGH). E: Mutare Dist., Marange Communal Land (Maranke

Reserve), fl.& fr. 10.ii.1953, *Chase* 4764 (BM, COI, LISC, SRGH). S: Masvingo Dist., Makaholi Expt. Station, fr. 15.iii.1978, *Senderayi* 260 (K, SRGH). **Malawi**. N: Nkhata Bay, by Rest House, fl.& fr. 12.v.1970, *Brummitt* 10625 (K). C: Ntcheu (Ncheu), Masasa (Msase) escarpment, Dedza–Golomoti road, fl.& fr. 19.iii.1955, *Exell, Mendonça & Wild* 1037 (BM, LISC, SRGH). S: Shire valley, 16 km W of Zomba, Lulanya Hill overlooking Chingali, fl.& fr. 12.iii.1977, *Brummitt & Patel* 14826 (K). **Mozambique**. N: Ngauma Dist., Massangulo, fl.& fr. iv.1933, *Gomes e Sousa* 1350 (COI, K). Z?: Zambezi, fl. ix.1866, *Kirk* (K). T: cited from Tete Province by Gonçalves (in Garcia de Orta **5**: 116, 1982). MS: Gorongosa, fl. 1884–5, *Carvalho* s.n. (COI).

Also in Ethiopia, Burundi, Uganda, Kenya, Tanzania, NE South Africa and Madagascar. In grassland and veld, usually on sandy soils; 180–1525 m.

ii) Var. **arenicola** Brummitt in Bol. Soc. Brot., sér.2, **41**: 288 (1968). —Gillett in F.T.E.A., Legum., Pap.: 200 (1971). Type: Tanzania, Mpanda Dist., Kibwesa Point, bare sand dunes, fl.& fr. 17.vii.1958, *Juniper & Jefford* 48 (K holotype).

Perennial, with robust stems usually woody and ± prostrate towards the base. Stems and leaf rachides glabrous to shortly and rather sparsely appressed-pubescent. Leaves with 5–9 pairs of leaflets; leaflets mostly 3–5 times as long as broad, narrowly elliptic to oblanceolate; stipules 4–8 mm long. Calyx with short appressed hairs. Pods with 11–13(16) seeds.

Zambia. N: Mbala Dist., Nkamba (Kamba) Bay, shore of L. Tanganyika, fl.& fr. 15.iv.1957, *Richards* 9196 (K). **Malawi**. N: Karonga, sandy foreshore, fl.& fr. 4.vii.1952, *Williamson* 21 (BM). C: Salima Dist., Senga Bay, near Grand Beach Hotel, fl.& imm.fr. 20.iii.1977, *Grosvenor & Renz* 1289 (K, SRGH). **Mozambique**. N: Lago Dist., Metangula, fr. 24.v.1948, *Pedro & Pedrógão* 3864 (LMA). T: cited from Tete Province by Gonçalves (in Garcia de Orta **5**: 116, 1982).

Also in Tanzania, Kenya and Burundi. This appears to be the usual variety found on sandy lake shores.

The specimen cited by Brummitt (1968) from Mwenembwe on the Nyika Plateau in Malawi at 2440 m (fl. ii–iii.1903, *McClounie* 113 (K)), is now certainly believed to have been mislabelled and to have been collected on the lakeshore, probably near Karonga.

iii) Var. **microfoliata** (Pires de Lima) Brummitt in Bol. Soc. Brot., sér.2, **41**: 287 (1968). Type: Mozambique, near Palma, *Pires de Lima* 34 (PO).
    Cracca bracteolata sensu Pires de Lima in Broteria, sér.Bot. **19**: 123 (1921).
    Cracca bracteolata var. *microfoliata* Pires de Lima in Bol. Soc. Brot., sér.2, **2**: 137 (1924).
    Tephrosia iringae sensu Cronquist in F.C.B. **5**: 105 (1954) for Mozambique specimens (*Faulkner* 151, 226).

Annual with rather slender stems, not woody, the main stem usually ± erect but often with decumbent branches from near the base. Stems and leaf rachides shortly appressed-pubescent. Leaves with 4–9 pairs of leaflets; leaflets 1–4 × 0.2–0.6(0.8) cm, mostly 5–9 times as long as broad, linear-oblong; stipules 5–9(12) mm long. Calyx shortly appressed-pubescent. Pods with (12)15–18 seeds.

Mozambique. N: Nampula, fl.& fr. 2.ii.1937, *Torre* 1223 (COI). Z: Lugela Dist., Namagoa, fl.& fr. 26.iii.1948, *Faulkner* 226 (COI, K, PRE, SRGH).

Known only from N Mozambique. Grassland; 150–450 m.

43. **Tephrosia caerulea** Baker f., Legum. Trop. Africa: 197 (1926). —Brummitt in Bol. Soc. Brot., Sér. 2, **41**: 290 (1968). —Schrire in Bothalia **17**: 11 (1987). Type: Zambia, Mazabuka, *Woods* 51 (BM holotype).

Perennial with a vertical woody basal stock and herbaceous, robust, rather sparingly branched, erect stems up to 1.4 m high, or an annual up to 1 m with a taproot. Stems conspicuously covered with long, spreading, often geniculate brown hairs 1–2.5 mm long or with greyish-white appressed hairs. Leaves with 5–9 pairs of leaflets; petiole (1.2)2.5–5 cm long, petiole and rachis together (7)12–18(23) cm long, with hairs similar to those on the stem; leaflets (3)4–7.4 × (0.3)0.5–1.2(1.4) cm, linear-oblong or linear-elliptic to linear, rounded to subcuneate at the base, truncate to emarginate at the apex; upper surface glabrous, lower surface covered with appressed hairs; stipules (9)12–15(19) × (0.8)1–1.2(1.5) mm, linear or somewhat falcate, ± persistent. Flowers in mostly terminal racemes (8)12–26(32) cm long; bracts 3–6 × 0.2–0.6 mm, ± linear; pedicels 3–4 (6 in fruit) mm long. Calyx 5–6(7) mm long, with long, brown, ascending to spreading villous hairs, the tube 2–2.5 mm long, equalling or exceeding the lateral teeth; upper teeth joined for $^1/_3$–$^2/_3$ the length. Petals pink to pale purplish or bluish, the keel very pale, the standard brown pubescent outside; standard 13–18 mm long, the other petals somewhat shorter. Stamen tube joined above. Ovary appressed-pubescent; style pubescent. Pods 55–70(90) × 4.5–5.5 mm, rather thinly covered with long appressed hairs overlying much shorter appressed hairs, ± straw-coloured. Seeds (13)14–17(19).

Subsp. **caerulea** —Brummitt in Bol. Soc. Brot., sér.2, **41**: 290 (1968). —Schrire in Bothalia **17**: 11 (1987).

Usually perennial with a woody rootstock. Stems and leaf rachis with long spreading, often geniculate, brown hairs 1–2.5 mm long.

Botswana. N: Ngamiland Dist., c.3 km SE of Shakawe, fl. 24.i.1956, *de Winter* 4406 (K, M). **Zambia**. B: Sesheke Dist., Sichinga Forest, fl. 28.xii.1952, *Angus* 1057 (FHO, K, PRE). C: South Luangwa Nat. Park, Mfuwe, fl. 13.iii.1969, *Astle* 5601 (K). S: Choma Dist., Mochipapa to Sinazongwe, mile 39.4, fl.& fr. 2.iii.1960, *White* 7565 (FHO, K). **Zimbabwe**. N: Hurungwe Dist., km 349.5 Harare to Chirundu, Urungwe Safari Area, S of Chirudzu, fl.& imm.fr. 14.ii.1981, *Philcox, Leppard & Dini* 8534 (K). W: Bulawayo, fl.& fr. v.1915, *Rogers* 13414 (K). **Malawi**. C: Lilongwe, new capital site, fl.& fr. 23.iii.1970, *Brummitt & Little* 9336 (K). S: Mangochi Dist., Chipoka, Mangochi (Fort Johnston), fr. 22.iii.1956, *Banda* 235 (BM, K, SRGH). **Mozambique**. N: Mandimba Dist., Ungane (Ngami), N of Mandimba, fl. 10.i.1942, *Hornby* 1118 (LISC, PRE).

Also in SW Tanzania north to Mpanda. Usually on sandy soils or rocky ground, in woodland or on lake or river banks; 300–1200 m.

Subsp. **otaviensis** (Dinter) Schreiber & Brummitt in Bol. Soc. Brot., sér.2, **41**: 292 (1968). —Schreiber in Merxmüller, Prodr. Fl. SW Afrika, fam. 60: 117 (1970). — Schrire in Bothalia **17**: 11 (1987). Type: Namibia, Hereroland, Klein Otavi, *Dinter* 5747 (NH, K photo).

Tephrosia otaviensis Dinter in Repert. Spec. Nov. Regni Veg. **30**: 204 (1932).
Tephrosia longipes sensu Schreiber in Mitt. Bot. Staatssamml. München **16**: 298 (1957).

?Annual or biennial with a taproot. Stem and leaf rachis with appressed or ascending greyish-white hairs.

Botswana. N: Ngamiland Dist., Tsodilo Hills, fl.& fr. 2.v.1975, *Müller & Biegel* 2305 (K, SRGH).
Also in Namibia. Mixed woodland, edge of dry stream; 1100 m.

44. **Tephrosia euprepes** Brummitt in Bol. Soc. Brot., sér.2, **41**: 292 (1968). —Schrire in Bothalia **17**: 11 (1987). Type: Zimbabwe, Umguza Dist., Nyamandhlovu Pasture Res. Station (Bongolo Farm), fl.& fr. 12.ii.1948, *West* 2694 (SRGH holotype, K).

Tephrosia reptans sensu Baker f., Legum. Trop. Africa: 197 (1926) as regards *Lugard* 175, 152.

A slender annual with a long, thin taproot; stems simple or sparingly branched in the lower part only, 20–50(90) cm high. Stems sparsely to fairly densely pubescent, with at least some hairs towards the base of the stem long and conspicuously spreading, usually brown, very rarely with all hairs appressed. Leaves often few in number and restricted to the lower part of the plant, with (1)2–4(5) pairs of leaflets; petiole (1.5)3–8 cm long, petiole and rachis together (2)3–15 cm long, with rather sparse long hairs and usually at least some of these conspicuously spreading; leaflets (1)2–5(7) × (0.2)0.4–0.9(1.3) cm, narrowly lanceolate or elliptic-oblong to linear, rounded to subcuneate at the base, rounded or often conspicuously truncate or emarginate at the apex; upper surface glabrous, lower surface appressed-pubescent with longish hairs; stipules (3)5–11(15) × 0.3–0.8(1) mm, linear, usually conspicuously persistent towards the base of the plant after the fall of the leaves. Flowers in simple, elongate, lax racemes up to 25 cm long; bracts up to 6 × 0.3 mm, linear, inconspicuous; pedicels 3–6 mm long. Calyx (3.5)4.5–7 mm long, with long brown loosely appressed to spreading hairs; upper teeth about equalling the tube, connate for ¹/₂–²/₃ length. Petals pink or pale purplish; standard 15–20 mm long. Stamen tube joined above. Ovary appressed-pubescent; style pubescent. Pods (42)58–72 × 4–5 mm, appressed-pubescent with usually long hairs overlying shorter ones, straw-coloured. Seeds 8–12(13).

Botswana. N: Ngamiland Dist., Khwebe (Kwebe) Hills, fl.& fr. 14.ii.1898, *Mrs. Lugard* 175 (K). **Zambia**. B: Sesheke Dist., Masese, fl.& imm.fr. 14.iii.1961, *Fanshawe* 6424 (SRGH). E: Chipata Dist., Luangwa Valley, Chinzombo Lagoon (Chizombo), imm.fr. 20.ii.1969, *Astle* 5499 (K). S: Monze Dist., Lochinvar Nat. Park, fr. 14.iv.1972, *van Lavieren, Sayer & Rees* 850 (K, SRGH). **Zimbabwe**. W: Hwange Dist., near Matetsi R., fl.& fr. 28.ii.1963, *Wild* 6052 (BM, K, SRGH). E: Mudzi Dist., Nyanga, Lawleys Concession, fl.& fr. 19.ii.1954, *West* 3359 (K, PRE, SRGH). **Mozambique**. T: Tete, fl. i.1932, *Sofia Pomba Guerra* 10 (COI).

Not known outside the Flora Zambesiaca area. The species appears to have a somewhat disjunct distribution, being found on Kalahari sands and on basalt (in grassland, mopane scrub, etc.) in W Zimbabwe, N Botswana and SW Zambia, and also in lowveld areas in NE Zimbabwe and adjacent Mozambique to the Luangwa; 600–1100 m. The two areas are somewhat similar climatically and edaphically.

In its typical form on Kalahari sands, this species is an annual with a much more delicate habit than *T. caerulea*, from which it also differs in its fewer leaflets, petiole usually equalling or exceeding the leaf rachis, and its shorter and fewer-seeded pods. Some specimens, however, seem to be rather intermediate between these two species. It is possible that some more robust specimens originally included in *T. euprepes* may be nearer to *T. caerulea*.

45. **Tephrosia ringoetii** Baker f. in Bull. Soc. Roy. Bot. Belg. **57**: 126 (1925); Legum. Trop. Africa: 197 (1926). —Brummitt in Bol. Soc. Brot., sér.2, **41**: 294 (1968). Type: Congo, Katanga, Shinsenda, fl.& fr. 29.iii.1912, *Ringoet* 5 (BR holotype).

   Tephrosia jelfiae Baker f., Legum. Trop. Africa: 202 (1926). Type: Zambia, Luwingu, fl. iv.1922, *Jelf* 23 (BM holotype).

   Tephrosia longipes var. *ringoetii* (Baker f.) J.B. Gillett in Kew Bull. **13**: 125 (1958).

   Tephrosia stormsii sensu Cronquist in F.C.B. **5**: 106 (1954) in part.

Annual with a short taproot and tufterd habit with branching stem 30–100 cm high. Stems, at least the lower parts, conspicuously clothed with spreading, irregular, usually whitish, geniculate hairs up to 2 mm long. Leaves with (1)2–4(5) pairs of leaflets; petiole 5–9(13) cm long or on the upper leaves sometimes as short as 2 cm, petiole and rachis together (7)9–20(23) cm long, or on the upper leaves sometimes as short as 3.5 cm, with spreading whitish hairs like those on the stem; leaflets (2)4–9(10.5) × 0.2–0.8(1) cm, ± linear, rounded to subcuneate at the base, acute to rounded at the apex; upper surface glabrous, lower surface appressed- or irregularly pubescent;

stipules (4)8–16(18) × (0.4)1–2.4 mm, linear-triangular, sometimes ± falcate, persistent. Each inflorescence axis terminated by a dense capitulum of flowers surrounded by conspicuous bracts which usually equal or exceed the calyces of the flowers, also often with flowers at 1–3(5) nodes below the terminal capitulum; bracts (4)6–8(9) × 0.5–1 mm, linear-triangular, purplish with villous whitish hairs, conspicuous and persistent; pedicels 3–5(7) mm long. Calyx 3.5–6 mm long, usually purplish with villous greyish spreading hairs; upper teeth joined for not more than half their length; upper and lateral teeth about equalling the tube in length, the lower tooth up to 1.5 mm longer. Petals pale to deep pink or occasionally purplish; standard (10)11–15(16) mm long, ± cuneate at the base. Staminal tube joined above. Ovary pubescent, particularly on the sutures; style pubescent. Pods 48–65 × (3.8)4–5 mm, usually minutely puberulent as well as having short, irregular or loosely appressed, usually geniculate hairs. Seeds 11–15(16).

Zambia. N: Mpika Dist., by Serenje–Mpika road, fl.& fr. 5.iv.1961, *Richards* 14983 (K). W: Copperbelt, Mufulira Dist., fl.& fr. 4.v.1934, *Eyles* 8254 (K, SRGH).

Also in Congo (Katanga). In miombo woodland, grassland, mushitu, etc.; 1150–1350 m.

This species can usually be recognised by its tufted annual habit, linear leaves, and particularly the compact inflorescence at the summit of a long peduncle (though there are frequently also 1 or 2 flowers at one node below this) with spreading grey hairs on the calyx.

46. **Tephrosia stormsii** De Wild. in Ann. Mus. Congo, Bot., sér.4 **1**: 189 (1903). — Baker f., Legum. Trop. Africa: 203 (1926). —Cronquist in F.C.B. **5**: 106 (1954) in part excl. *T. ringoetii.* —Brummitt in Bol. Soc. Brot., sér.2, **41**: 297 (1968). — Gillett in F.T.E.A., Legum., Pap.: 198 (1971). Type: Tanzania, Mpanda Dist., Karema, *Storms* s.n. (BR holotype).

> *Tephrosia longipes* sensu Gillett in Kew Bull. **13**: 125 (1958) in part.

Annual with a taproot and rather slender, branched stems (40)60–140(180) cm high. Stems glabrous or with sparse appressed hairs in parts, or (var. *pilosa*) with rather conspicuous grey spreading hairs. Leaves with (1)2–4(6) pairs of leaflets; petiole (0.3)2–7(10) cm long, petiole and rachis together 5–18 cm long, pubescent like the stem; leaflets (2.5)4–11(14) × (0.1)0.2–0.9(1.1) mm, linear to linear-oblong, or the lower ones sometimes ± elliptic, the apex acute to rounded or occasionally emarginate; upper surface glabrous, lower surface appressed-pubescent; stipules (3)4–12 × 0.3–0.8 mm (up to 17 mm long in var. *pilosa*), linear, sometimes falcate. Flowers in lax terminal and axillary racemes up to 18(25) cm long, the upper branch racemes often subtended by only a bract instead of a foliage leaf and so forming a compound inflorescence, the branches usually forming an angle of about 45° or more, their axis slender (0.2–1 mm diameter) and usually rather flexuous and curving upwards, glabrous to shortly appressed-pubescent; bracts 2–4(5) × 0.2–0.5 mm, linear-triangular, rather inconspicuous and usually falling early; pedicels 3–6 mm long. Calyx 3–5.5 mm long, usually shortly appressed-pubescent with greyish hairs, or (var. *pilosa*) with longer spreading hairs; upper teeth usually joined for most of their length to form an obtuse or acute triangle, or (particularly in var. *pilosa*) free for more than half their length, the lateral teeth usually about equalling the tube. Petals pink to sometimes purplish, or the keel much paler and often white; standard 12–16(18) mm long. Stamen tube joined above. Ovary pubescent; style pubescent. Pods 48–70(82) × 4–5 mm, shortly pubescent with usually irregular geniculate hairs, sometimes with rather regular short appressed hairs. Seeds 10–16(17).

Var. **stormsii** —Brummitt in Bol. Soc. Brot., sér.2, **41**: 298 (1968). —Gillett in F.T.E.A., Legum., Pap.: 198 (1971).

> *Tephrosia lurida* sensu Eyles in Trans. Roy. Soc. S. Africa **5**: 375 (1916) as regards *Eyles* 265.
> *Tephrosia eylesii* Baker f., Legum. Trop. Africa: 200 (1926). Type: Zimbabwe, Mazowe Dist., Bernheim Hill, *Eyles* 265 (BM holotype, SRGH).

Stems glabrous to sparsely appressed-pubescent. Leaflets (2.5)4–11(14) × (1)2–6(9) mm, linear to linear-oblong; lower surface shortly and closely appressed-pubescent. Calyx appressed-pubescent with usually short greyish hairs; upper teeth usually joined for most of their length.

Zambia. N: Mpulungu–Mbala (Abercorn) road close to Chilongowelo turn, fl.& fr. 5.iv.1955, *Richards* 5306 (K). W: Kitwe, fl.& fr. 10.iii.1963, *Fanshawe* 7742 (K). C: Kabwe Dist., 10 km S of Kapiri Mposhi, fl.& fr. 27.iii.1955, *Exell, Mendonça & Wild* 1220 (LISC, SRGH). E: Chipata Dist., Chipangali area, fl.& fr. ii.1962, *Verboom* 483 (K, SRGH). S: Choma Dist., Mochipapa to Sinazongwe, mile 19, Mabwingombe Hills, fl.& fr. 2.iii.1960, *White* 7531 (K, SRGH). **Zimbabwe**. N: Mt Darwin Dist., Musengezi R., fl. 16.v.1955, *Watmough* 118 (K, SRGH). W: Hwange Dist., Zambezi R. between Matetsi and Deka rivers, fl.& fr. 28.ii.1963, *Wild* 6084 (SRGH). **Malawi**. N: Nkhata Bay Dist., Likoma Is., fl.& fr. 24.v.1901, *Kenyon* in *Riddelsdell* 51 (K). C: Nisasadzi, S Kasungu, fl. 11.iii.1953, *G. Jackson* 1137 (K, SRGH). S: Mulanje Dist., Likhubula Valley, fl.& imm.fr. 22.iii.1988, *J.D. & E.G. Chapman* 8999 (K, MAL, MO). **Mozambique**. N: Mueda Dist., Macondes, 83 km from Nairoto (Nantulo) towards Mueda, fl.& fr. 10.iv.1964, *Torre & Paiva* 11862 (LISC). Z: between Quelimane and Mocuba, fl.& imm.fr. 20.iii.1943, *Torre* 4969 (K, LISC).

Also in Congo (Katanga), Rwanda, Burundi, Tanzania and Kenya. In miombo and other woodland and grassland; 300–1830 m.

Var. **pilosa** Brummitt in Bol. Soc. Brot., sér.2, **41**: 302 (1968). —Gillett in F.T.E.A., Legum., Pap.: 198 (1971). Type: Tanzania, Ufipa Dist., ecarpment above Kasanga, 1050 m, fl.& fr. 30.iii.1959, *Richards* 11008 (K holotype, SRGH).

Usually rather shorter (up to 100 cm) and less branched than var. *stormsii*. Stems clothed with grey spreading hairs. Leaflets 2.5–9 × (0.3)0.6–0.9(1.1) mm, those of lower leaves often elliptic to oblong, those of upper leaves linear-oblong to linear; lower surface loosely or closely appressed-pubescent. Calyx ascending or spreading-pubescent; upper teeth often free for most of their length.

Zambia. N: Kaputa Dist., Nsama, fl.& fr. 4.iv.1957, *Richards* 9019 (K). C: Great North Road, 2 km NE of Serenje turnoff, fl.& fr. 29.iii.1984, *Brummitt, Chisumpa & Nshingo* 16955 (K). **Malawi**. N: Chitipa Dist., Kaseye Mission, 16 km E of Chitipa, fl.& fr. 5.iv.1969, *Pawek* 1949 (K).

Known only from N Zambia, Malawi, SW Tanzania and Burundi. Miombo woodland, often on disturbed ground, roadsides, etc; 1000–1400 m.

This species is superficially similar to *T. longipes*, but differs in its axillary rather than leaf-opposed racemes. Var. *pilosa* shows resemblance to *T. ringoetii*, which is distinguished by its capitulate inflorescences with conspicuous persistent bracts, and to *T. euprepes* which has an even shorter habit, usually branching near the base and not in the upper parts, brown hairs on stem, rachis and calyx , with those on the calyx longish and appressed, and pods with long appressed hairs.

47. **Tephrosia paradoxa** Brummitt in Bol. Soc. Brot., sér.2, **41**: 303 (1968). Type: Zambia, Mbala Dist., Ndundu, road outside driveway to house, 1740 m, fl.& fr. 22.ii.1959, *Richards* 10969 (K holotype).

> *Tephrosia longipes* as used by many.
> *Tephrosia paucijuga* Cronquist in F.C.B. **5**: 106 (1954) as regards *Lebrun* 9721.
> *Tephrosia stormsii* sensu Gillett in F.T.E.A., Legum., Pap.: 198 (1971).

Annual or short-lived perennial with erect, fairly stout, herbaceous, branched stems 60–140 cm high. Stems glabrous to sparsely appressed-pubescent in the lower and middle parts, becoming increasingly more pubescent in the upper parts towards the peduncles and racemes. Leaves with (1)2–6 pairs of leaflets; petiole 5–13 cm long, petiole and rachis together 8–26 cm long, subglabrous to rather thinly appressed-pubescent; leaflets 9–14(18) × 0.2–0.6(0.9) cm, linear, (sometimes, in Mozambique, the lowermost leaves with leaflets up to 1.6 cm broad and linear-oblong), leaflet apex acute to obtuse; upper surface glabrous, lower surface thinly appressed-pubescent; stipules (8)13–18(24) × 1–2 mm, linear, sometimes falcate. Flowers in lax terminal and axillary racemes up to 30 cm long, the axillary ones always subtended by a foliage leaf (not a bract) and forming a narrow angle of about 30°, their axis rather stout (1–2 mm diameter) and stiffly erect, densely pubescent to tomentose; bracts 5–8(11) × 0.7–1.5(2) mm, linear-triangular, fairly conspicuous, slow to fall; pedicels (4)5–8 mm long. Calyx (4)5–8(10) mm long, densely pubescent to villous with usually long whitish hairs interspersed among shorter brown hairs; two upper teeth free for $^1/_2$–$^3/_4$ their length, the lateral teeth approximately equalling the tube. Petals purple or reddish or brownish-purple, or the keel usually much paler; standard (14)15–19 mm long. Stamen tube joined above. Ovary pubescent; style pubescent. Pods 52–75 × 4.5–5 mm, shortly pubescent with usually irregular geniculate hairs. Seeds (14)16–19.

Zambia. N: Mbala Dist., near bridge over Lunzua R. on Kambole road c.32 km from Mbala (Abercorn), fl.& fr. 5.iv.1959, *McCallum Webster* 846 (K). **Zimbabwe**. N: Zvimba Dist., Trelawney, Tobacco Research Station, fl.& fr. 27.iii.1943, *Jack* 86 (K, PRE, SRGH). **Malawi**. N: Mzimba Dist., 32 km NE of Mzimba on M1, fl. 9.iii.1978, *Pawek* 13943 (K). **Mozambique**. N: Ribáuè Dist., 16 km from Ribáuè towards Nampula, fl. 31.i.1964, *Torre & Paiva* 10346 (LISC).

Also in Tanzania, Congo and Rwanda. In miombo woodland, grassland and on roadsides; 570–1740 m.

The specimen from Zimbabwe N: appears to be somewhat isolated from the main area of this species. It has rather shorter pods (14–15-seeded) than usual, but otherwise seems to be typical.

T. paradoxa is closely allied to *T. stormsii*, with which it is apparently sympatric over a wide area. In the Flora area the two seem quite distinct, *T. paradoxa* having a shorter and densely pubescent inflorescence axis and larger flowers and pods. However, several collections from Tanzania are difficult to place with certainty in either one or the other. It is possible that these intermediates could be the result of hybridization.

48. **Tephrosia laxiflora** R.E. Fr., Wiss. Ergebn. Schwed. Rhod.-Kongo-Exped. **1**: 83 (1914). —Baker f., Legum. Trop. Africa: 199 (1926). —Brummitt in Bol. Soc. Brot., sér.2, **41**: 324 (1968). Type: Zambia, Luwingu Dist., Malolo, near Luwingu (Luvingo), fl.& imm.fr. 25.x.1911, *Fries* 1115 (UPS holotype).

   Tephrosia lurida sensu R.E. Fries, Wiss. Ergebn. Schwed. Rhod.-Kongo-Exped. **1**: 83 (1914).

   Tephrosia paucijuga sensu Cronquist in F.C.B. **5**: 106 (1954), excl. *Lebrun* 9721.

   Tephrosia longipes var. *lurida* (Sond.) J.B. Gillett in Kew Bull. **13**: 125 (1958) in part, non *T. lurida* Sond.

Perennial suffrutex with a woody underground stock and herbaceous, erect, simple or 1(2)-branched stems up to 45(75) cm high with 2–5 leaves per stem. Stems glabrous to sparsely and very shortly appressed-pubescent. Leaves with (0)1–3(4) pairs of leaflets, the terminal leaflet subsessile; petiole 2–5(6.5) cm long, petiole and rachis together 2–9(12) cm long, glabrous or subglabrous; leaflets 3–8(11) × (0.2)0.3–0.8(1.4) cm, linear-elliptic to linear, ±cuneate at the base, acute to rounded at the apex, the veins ±conspicuously prominent on both surfaces; upper surface glabrous, lower surface sparsely and shortly appressed-pubescent to subglabrous; stipules 6–13 × 0.7–1.2 mm, linear-triangular, often falcate. Flowers in ±lax terminal racemes mostly 10–27 cm long; bracts 6–7 × 1 mm, quickly falling; pedicels 3–7(11) mm long. Calyx 3–5.5 mm long, fairly densely shortly appressed-pubescent; upper teeth connate for $^1/_2$–$^5/_6$ their length to form a rather broad triangle, the upper and lateral teeth usually slightly shorter than the tube, the lower tooth about equalling it. Petals pale purplish or pink, or yellow with pink markings; standard (10)12–18 mm long, cuneate to subcordate at the base. Staminal tube joined above. Ovary pubescent; style pubescent. Pods 35–60 × 3.5–4(5) mm, shortly pubescent with appressed or irregular hairs. Seeds 8–12.

Zambia. B: Kaoma Dist., 80 km E of Kaoma (Mankoya) on road to Kafue Hook, fl. 21.xi.1959, *Drummond & Cookson* 6719 (K, PRE, SRGH). N: Luwingu Dist., Malolo, near Luwingu (Luvingo), fl. 25.x.1911, *Fries* 1115 (UPS). W: Solwezi, boma environs, fl.& imm.fr. ix.1962, *Holmes* 1540 (K, SRGH). C: Kabwe Dist., 68 km from Mumbwa along road to Kabwe, fl. 20.x.1972, *Strid* 2339 (K). S: Mumbwa Dist., between Mumbwa and Chanobi, fl. 15.ix.1947, *Greenway & Brenan* 8085 (EA, K, PRE).

Also in Congo (Katanga) and Burundi. In *Brachystegia* woodland, dambo margins, etc., the aerial parts often burned off by fire; 1150–1400 m.

Closely related to *T. paucijuga* Harms from S Tanzania which is distinguished by its much broader (to 8 mm), 2–4 seeded pods.

49. **Tephrosia candida** DC., Prodr. **2**: 249 (1825). Type from NE India.

   Robinia candida Roxb., Fl. Ind. **3**: 327 (1832).

   Cracca candida (DC.) Kuntze, α *normalis*, ß *heterophylla* & γ *obtusifolia* Kuntze, Rev. Gen. Pl. **1**: 173 (1891).

Shrub 1–3 m high, usually much branched and bushy. Stems appressed or ascending, often deflexed, greyish or brownish tomentose. Leaves with (5)8–13 pairs of leaflets; petiole (1)1.2–1.8(2.6) cm long; petiole and rachis together (7)9–20(23) cm long, tomentose like the stem; leaflets (2)3–6(7.5) × (0.5)0.8–1.4(2) cm, elliptic-oblong, rounded to cuneate at the base, acute to subobtuse or rarely somewhat rounded at the apex, mucronate, the margins somewhat inrolled towards the lower surface; upper surface glabrous (rarely shortly pubescent but not in the Flora area) and dark green in colour, lower surface appressed; greyish-green pubescent and usually contrasting in colour with the upper surface; stipules 4–10 × 0.8–1.2 mm, linear-triangular. Flowers in fairly dense terminal and axillary racemes 4–24 cm long, or sometimes the lowermost fascicles of flowers in the axils of the uppermost foliage leaves; bracts 3–6 × 0.5–0.8 mm, linear-triangular, falling early; pedicels 9–17 mm long. Calyx 4–6 mm long, grey or brownish densely appressed-pubescent, the teeth usually darker than the tube; upper teeth short (1–2 mm), obtuse or rounded and fused for c.³/₄ their length, the lateral teeth about equalling the upper, obtuse or rounded, the lower tooth exceeding the others, about equalling the tube, channelled, obtuse. Petals white; standard (18)20–27 mm long, orbicular oblong, the wings and keel petals somewhat shorter. Upper stamen fused to the tube. Ovary appressed-tomentose; style pubescent. Pods 72–96 × 7–8(9) mm, closely appressed brown pubescent. Seeds 11–13, 4–5 × 3.5–4 × 1.5–2, grey or greyish-black, with a large white U-shaped aril c.1.5 mm long.

Zimbabwe. C: Greenlands Nursery, bud, 25.v.1963, *Corby* 1051 (SRGH). **Malawi**. S: Mulanje Dist., Swazi Estate, Mulanje, fl. 29.iii.1949, fr. 20.vi.1949, *Faulkner* 393 (K). **Mozambique**. N: Mogovolas, Posto Agrícola, fl.& fr. 30.iv.1934, *Ribeiro* s.n. (LISC).

Introduced. Native of NE India (from Dehra Dun east through the Himalayas to Assam and Orissa). Extensively cultivated in tropical countries as a fertilizing crop, ornamental or fish poison. Recorded naturalized in Indonesia (Java) and the West Indies.

50. **Tephrosia vogelii** Hook.f. in Hooker, Niger Fl.: 296 (1849). —Baker in F.T.A. **2**: 110 (1871). —Fries, Wiss. Ergebn. Schwed. Rhod.-Kongo-Exped. **1**: 81 (1914). — Baker f., Legum. Trop. Africa: 209 (1926); in J. Bot. **66**, Polypet. Suppl.: 106 (1928). —Cronquist in F.C.B. **5**: 108 (1954). —Hepper in F.W.T.A., ed.2, **1**: 528, 530 (1958). —White, F.F.N.R.: 166 (1962). —Torre in C.F.A. **3**: 163 (1962). — Gillett in F.T.E.A., Legum., Pap.: 210 (1971). —Burkill, Useful Pl. W. Trop.

Africa, ed.2, **3**: 460 (1995). Type: Nigeria, on Niger (Quorra) R., *Vogel* s.n. &
Bioko (Fernando Po), *Vogel* s.n. (K syntypes).

Tephrosia megalantha T. Durand & De Wild. in Bull. Soc. Roy. Bot. Belg. **36**: 57 (1897);
Ann. Mus. Congo Bot., sér.3, **1**: 60 (1901). —De Wildeman, Pl. Bequaert. **3**: 332 (1925).
Type from Congo.

Tephrosia periculosa Baker in Bull. Misc. Inform., Kew **1897**: 258 (1897). —De Wildeman,
Pl. Bequaert. **3**: 332 (1925). Type: Malawi, Kondowe to Karonga, *Whyte* s.n. (K holotype).

Shrub 1–3(4) m high, usually much branched and bushy. Stems brown tomentose with long
flexuous hairs intermixed among shorter and denser spreading hairs. Leaves with (6)8–13(15)
pairs of leaflets; petiole 9–28 mm long, petiole and rachis together (9)11–22(27) cm long,
tomentose like the stem; leaflets 2.5–5.5(7.5) × (0.6)0.9–1.7(2.3) cm, elliptic-oblong to
oblanceolate, rounded to cuneate at the base, rounded to emarginate at the apex, slightly
mucronate, the upper surface rather thinly appressed-pubescent, the lower surface densely
appressed-pubescent; stipules 11–20 × 2.5–4.5 mm, narrowly triangular or sometimes markedly
falcate, soon falling. Flowers in dense heads up to 10(20 in fruit) cm, or the lowermost
sometimes somewhat remote; bracts up to 16 × 13 mm, broadly ovate-acuminate to
suborbicular-acuminate, brown or greyish tomentose, conspicuous at bud stage but soon falling
as flowers open; pedicels 14–26 mm long, brown tomentose. Calyx 14–20(24) mm long, brown
or greyish tomentose; upper and lateral teeth about twice as long as the tube, oblong, ±
truncate at the apex, the lower tooth about 1.5 times as long as the lateral, strongly grooved and
upwardly curved distally into a keel-like shape. Petals white, rarely the standard purple;
standard 24–30(34) mm long, truncate to strongly cordate at the base, the wings and keel petals
somewhat shorter. Upper stamen loosely attached to, and easily detachable from, the adjacent
stamens about the middle of the filament. Ovary tomentose; style pubescent. Pods 9–14.5 ×
1.3–1.7 cm, light brown lanate-tomentose. Seeds numerous (more than 15), 6–8 × 4–4.5 × 2–2.5
mm, black, smooth, with a well developed white U-shaped aril c.2 mm long.

Zambia. B: Zambezi Dist., Kashiji (Kasisi) R., fl. 4.ix.1953, *Gilges* 223 (K, PRE,
SRGH). N: Mbala Dist., Chilongowelo, fl.& fr. 4.ii.1952, *Richards* 715 (K). W: Solwezi
Dist., Mutanda Bridge, fl. 25.v.1930, *Milne-Redhead* 600 (K). C: Kabwe Dist., Lukanga,
fl. 23.vii.1931, *W. Allan* 1500 (K). E: Isoka Dist., Nyika Plateau, fl. 24.xi.1955, *H.M.N.
Lees* 78 (K). S: Mumbwa Dist., near Mumbwa, st. 1911, *Macaulay* 394 (K). **Zimbabwe**.
C: Harare, fl.& fr. ii.1919, *Eyles* 1515 (PRE, SRGH). E: Nyanga Dist., Pungwe R., fl.&
fr. 7.vii.1948, *Chase* 838 (K, LISC, SRGH). **Malawi**. N: Nkhata Bay Dist., Mzuzu to
Nkhata Bay, c.1160 m, fl.& fr. 27.iv.1970, *Pawek* 3446 (K). S: Blantyre Dist., c.19 km N
of Limbe, fl.& fr. *Chase* 3871 (SRGH). **Mozambique**. N: Malema Dist., Mutuáli, foot
of Serra Cucuteia, fl.& fr. iv.1954, *A.J.S. Barbosa* 3718 (LMA). Z: "Zambesiland", fr.
ii.1866, *Kirk* (K). MS: Sussundenga Dist., between Dombe and Sanguene, Regedoria
Mepunga, fl. 28.x.1953, *Pedro* 4486 (LMA, PRE).

Probably introduced. Widespread in tropical Africa from Sierra Leone and
Ethiopia southwards to Angola, the Flora Zambesiaca area and the Comoro Islands;
also from Assam to Indonesia. The species has been extensively cultivated for use as
a fish poison and is clearly introduced in many of its present localities so that the
extent of its native area is now obscured. It is probably not native anywhere in the
Flora area. Usually found in or near cultivated land.

Specimens from the Flora area are consistently reported to have white flowers, apart
from two specimens from N Zambia (Kasama Dist., ix.1937, *Trapnell* 1825 (K) and
26.ii.1962, *Richards* 16183 (K)), which are described as mauve and purple respectively.

51. **Tephrosia aequilata** Baker in F.T.A. **2**: 113 (1871). —Baker f., Legum. Trop.
Africa **1**: 212 (1926). —Brenan in Mem. New York Bot. Gard. **8**: 252 (1953). —
Cronquist, F.C.B. **5**: 110 (1954). —White, F.F.N.R.: 166 (1962). Type: Tanzania,
Bukoba Dist., Karagwe Hills, *Grant* 401 (K holotype).

Tephrosia meyeri-johannis Taub. in Engler, Hochgebirgsfl. Trop. Afrikas: 260 (1892). Type: Tanzania, Kilimanjaro, *Meyer* 39 (B†).

Tephrosia nyasae Baker f. in Trans. Linn. Soc., London, Bot. **4**: 9 (1894); Legum. Trop. Africa **1**: 212 (1926). —Topham, Check List For. Trees Shrubs Nyasaland Prot.: 79 (1958). Type: Malawi, no locality, *Buchanan* 51 (K).

Tephrosia zombensis Baker in Bull. Misc. Inform., Kew **1897**: 257 (1897). —Burtt Davy, Fl. Pl. Ferns Transvaal **1**(2): 377 (1932). —Burtt-Davy & Hoyle, Check-list For. Trees Shrubs, Nyasaland: 62 (1936). —Forbes in Bothalia **4**: 990 (1948). Type: Malawi, Mt. Zomba, 1896, *Whyte* 394 (K).

Tephrosia aequilata subsp. *nyasae* (Baker f.) Brummitt in Bol. Soc. Brot., sér.2, **41**: 353 (1968). —Lock, Leg. Afr. Check-list: 368 (1989).

Tephrosia aequilata subsp. *mlanjeana* Brummitt in Bol. Soc. Brot., sér.2, **41**: 356 (1968). — Lock, Leg. Afr. Check-list: 368 (1989). Type: Malawi, Mt. Mulanje, Tuchila Plateau, 1830 m, fl.& fr. 25.vii.1956, *Newman & Whitmore* 198 (BM holotype, PRE, SRGH).

Tephrosia aequilata subsp. *namuliana* Brummitt in Bol. Soc. Brot., sér.2, **41**: 356 (1968). —Lock, Leg. Afr. Check-list: 368 (1989). Type: Mozambique, Namuli peaks, W face, 1525 m, fl. 26.vii.1962, *Leach & Schelpe* 11471 (K holotype, SRGH).

Tephrosia aequilata subsp. *gorongosana* Brummitt in Bol. Soc. Brot., sér.2, **41**: 357 (1968). —Lock, Leg. Afr. Check-list: 368 (1989). Type: Mozambique, Serra de Gorongosa, 1000 m, fl.& imm.fr. 6.v.1964, *Torre & Paiva* 12305 (LISC holotype).

Tephrosia aequilata subsp. *australis* Brummitt in Bol. Soc. Brot., sér.2, **41**: 358 (1968). — Lock, Leg. Afr. Check-list: 368 (1989). Type: Zimbabwe, Chimanimani Dist., 1675 m, fl.& fr. ix.1953, *Williams* 146 (K holotype, PRE, SRGH).

Shrub or rarely a small tree, 0.4–5 m high. Young stem densely appressed-pubescent, tomentose or villous. Leaves with (5)6–10(13) pairs of leaflets; petiole and rachis appressed-pubescent to villous; leaflets up to 52 × 20 mm, elliptic to oblong or oblanceolate, more or less rounded at the base, obtuse to rounded at the apex, more or less mucronate, the margins flat or inrolled towards the lower surface except at the base and apex, the upper surface glabrous, glabrescent or thinly appressed-pubescent with fine delicate hairs appearing almost glistening-silvery under a high-powered lens, the lower surface thinly to densely appressed-pubescent and sometimes subsericeous; stipules 1.5–5(8) mm broad, narrowly triangular to ovate. Flowers in dense heads not or scarcely elongating in fruit, usually more or less exceeded by the leaves, with bracts persisting through flowering; bracts 4–10 × 1–4 mm, narrowly triangular to lanceolate, acute; pedicels 2–20 mm, appressed-pubescent to villous. Calyx 5–12(14) mm, brown or grey appressed-pubescent to villous, the two upper teeth joined for not more than half their length. Petals 9–24 mm, purple; standard broadly rounded to emarginate at the apex, cuneate or truncate at the base. Ovary appressed pubescent; style pubescent; pods (20)24–40 × 5–7 mm, brown-villous or appressed pubescent. Seeds 3–5.

Zambia. N: Mbala Dist., rocks on old Sumbawanga road 8 km from Kawimbe, fl.& fr. 8.vi.1961, *Richards* 15229 (K, SRGH). **Malawi**. N: Viphya Plateau, above Rumphi (Rumpi) Drift, fl. 24.vi.1960, *Chapman* 783 (FHO, SRGH). S: Zomba Plateau, fl.& fr. 2.vi.1946, *Brass* 16154 (K, PRE, SRGH). **Zimbabwe**. E: Chimanimani Dist., Townlands, Sowerombi road, fl.& imm.fr. 12.viii.1950, *Crook* M68 (K, SRGH). **Mozambique**. N: Massangulo Mt, 66 km N of Mandimba, fl. 26.v.1961, *Leach & Rutherford-Smith* 11040 (K, LISC, SRGH). Z: Gurué Dist., near Namuli Peak, fl. n.d., *Mendonça* 2290 (LISC). MS: Gorongosa Mts., near Morombosi Falls, fr. 13.ix.1946, *Pedro & Pedrógão* 182 (LMU, PRE).

Southern Uganda and Kenya to N South Africa. In rocky and grassy places, forest margins and on open mountain slopes in upland areas; up to 2200 m.

This species is very variable, especially in the Flora Zambesiaca area. In my notes published in 1968, six subspecies and one variety were recognized, mostly characterisitic of different mountain massifs. However, in the plentiful material collected since then, especially in Malawi, the characters tend to break down, and here no infraspecific taxa are formally recognized. Nonetheless, marked differences between plants from different mountain areas can be demonstrated, and further investigation would be welcome. Significant variation is found in indumentum of stem, leaf rachis, leaflet surfaces, calyx and pods, in the number of leaflets, in stipule breadth, and in flower size. The two collections seen from Gorongosa Mountain (previously described as subsp. *gorongosana*) are markedly different from any others in having pods appressed-pubescent rather than villous. Since the flora of Namuli Mtn. in Mozambique is similar to that of Mt Mulanje in Malawi (the two subspecies of *T. whyteana* are confined to these massifs respectively), one might expect the plants of *Tephrosia aequilata* from these mountains to be similar, but this is not the case. The three collections seen from Namuli (previously subsp. *namuliana*) are larger in most parts, and have (7)10–13 pairs of leaflets compared with 5–8 pairs on Mulanje (previously subsp. *mlanjeana*), while the leaflets of specimens from Mt Mulanje are usually much more sericeous than those from Mt Namuli. The flowering pedicels on Namuli seem conspicuously longer than elsewhere. Plants from Zomba Plateau in Malawi (previously subsp. *nyasae*) are usually (but not always) distinguished from both the Mulanje and Namuli plants by having a grey appressed-pubescent (rather than golden and spreading) indumentum on the leaf rachis. Those from the Nyika Plateau, the Viphya Plateau (often), the Eastern Highlands of Zimbabwe and from the mountains in the former N Transvaal (previously susbp. *australis*) generally have a leaf rachis with spreading and often golden indumentum like those on Mulanje, but much more sparsely pubescent and non-sericeous leaflets. In East Africa the most conspicuous variant is on Mt Kilimanjaro where the stipules are up to 8 mm broad, about twice as broad as elsewhere.

52. **Tephrosia robinsoniana** Brummitt in Bol. Soc. Brot., sér.2, **41**: 359 (1968). Type: Zambia, Mkushi Dist., Fiwila, rocky hillside, 1340 m, fl. 3.i.1958, *E.A. Robinson* 2576 (K holotype, SRGH).

Shrub 0.7–1.5 m high. Young branches appressed or ascending greyish pubescent. Leaves with 4–8 pairs of leaflets; petiole and rachis together up to 6.5 cm long, appressed to ascending greyish pubescent; leaflets up to 25 × 15 mm, elliptic to obovate, rather narrowly rounded at the base, broadly rounded to subtruncate at the apex, not or very slightly mucronate, those on the lowermost leaf of each shoot usually densely appressed greyish pubescent and subsericeous on both surfaces, those on the other leaves appressed-pubescent beneath but glabrous abve with the margins tending to curve upwards; stipules 7–8 × 3–4 mm, triangular. Flowers in short compact racemes exceeded by the leaves; bracts up to 5 × 2 mm, not concealing the pedicels of the developing flowers; pedicels 4–10 mm long, appressed or ascending greyish pubescent. Calyx c.7 mm long, loosely appressed greyish pubescent, the teeth about equalling the campanulate part (the lower tooth slightly longer), the two upper teeth joined for most of their length to form a broad triangle with only the apices free as two closely adjacent prolongations c.0.7 mm long. Petals 16–18 mm long, bright pink; standard almost as broad as long, broadly rounded to emarginate at the apex, broadly cuneate to subtruncate at the base. Stamen tube c.10 mm long. Ovary appressed-pubescent, style pubescent. Pods c.35 × 9 mm, broadly oblong, finely and densely pubescent-villous with irregularly arranged whitish appressed hairs. Seeds 3–5, transversely elongate in the pod.

Zambia. C: Mkushi Dist., Mufulwe Hills, near Fiwila Mission, Munkokwe Peak, fl.& fr. 26.iii.1973, *Kornaś* 3561 (K).

Known only from these two collections, on rocky hillsides; 1340–1450 m.

53. **Tephrosia whyteana** Baker f. in Trans. Linn. Soc., London, Bot. **4**: 9 (1894); Legum. Trop. Africa: 212 (1926). —Burtt Davy & Hoyle, Check-list For. Trees Shrubs, Nyasaland: 62 (1936). —Chapman, Veg. Mlanje Mt.: 63, 67 (1962). —

Brummitt in Bol. Soc. Brot., sér.2, **41**: 360 (1968). —White, Dowsett-Lemaire & Chapman, Evergreen For. Fl. Malawi: 335 (2001). Type: Malawi, Mt. Mulanje, *Whyte* s.n. (K holotype).

Shrub up to 4 m high. Young stem with short light coloured spreading or appressed hairs interspersed with conspicuous long spreading or upwardly curved (rarely appressed) brown hairs. Leaves with (6)8–10(11) pairs of leaflets; pubescence of petiole and rachis as for young stem; leaflets up to 55(62) × 18 mm, narrowly elliptic to oblong, ± obtuse to rounded at base and apex, mucronate at apex, margins flat or slightly inrolled towards lower surface, the upper surface rather sparsely appressed-pubescent (rarely glabrous), the lower surface regularly but somewhat thinly appressed-pubescent; stipules 3–5 mm broad, triangular to ovate, soon falling. Inflorescence lax to fairly dense, the bracts falling before the opening of the subtended flower; bracts ovate-acuminate to linear-lanceolate; pedicels (7)10–20 mm long, indumentum as for young stem (longer hairs rarely absent). Calyx 6–13 mm long, appressed or ascending-pubescent, with or without conspicuously longer spreading hairs; teeth very variable, from 2 mm long and shorter than the tube to 7 mm and exceeding the tube, the upper two variously joined. Petals 16–24 mm long, purple; standard broadly cuneate at the base. Ovary ± appressed-tomentose or villous; style pubescent. Pods 4–6 × c.0.6 cm, variously brown or greyish villous (see subspecies). Seeds c.8, blackish, 6 × 3.2 mm, oblong with rounded ends, with a low rim aril in the middle of a long side.

Known only from Mt. Mulanje in Malawi and the Namuli Mtns some 130 km away in Mozambique.

The species is very similar to *T. aequilata* which occurs in the same areas. *T. whyteana* is distinguished by its early-falling stipules and bracts, generally laxer inflorescence, and longer pods with c.8 seeds.

Subsp. **whyteana** —Brummitt in Bol. Soc. Brot., sér.2, **41**: 360 (1968).

Shrub up to 3.5 m high. Inflorescence axis (from lowest flower) 3–8 cm long; pedicels (7)9–15 mm long, the long hairs usually light brown. Calyx 6–9(11) mm long, the teeth (1)3–4(6) mm, usually without or with few long conspicuously spreading hairs. Petals 16–23 mm long. Pods clothed with fairly dense ascending or spreading brown hairs similar to those on the stem.

Malawi. S: Mulanje (Mlanje) Mt., path to Lichenya (Luchenya) Hut, 1950 m, fl. 9.vi.1962, *Richards* 16611 (K, LISC).

Known only from Mt. Mulanje. Rocky slopes and forest margins; 1800–2000 m.

Plants referred to this subspecies are somewhat variable. A striking variant, possibly from lower altitudes (900–1200 m) than the other collections (Mt. Mulanje, hills above Ruo Gorge, 1200 m, fl. 18.vi.1962, *Richards* 16769 (K), and foot of Little Ruo path above Lujeri Estate, fl.& fr. 29.vii.1970, *Brummitt* 12349 (K)) has stem, leaf rachis, pedicel and calyx all shortly appressed-pubescent with few longer hairs which are also appressed, and a shorter calyx (6 mm) with shorter teeth (1–3 mm). This may represent a distinct taxon.

A specimen from Lichenya Plateau, Mt Mulanje, 1860 m, fl. 26.vi.1946, *Brass* 16445 (BM, K), has a longer calyx than others (11 mm, teeth c.6 mm) with copious long spreading hairs, and approaches subsp. *gemina*.

Subsp. **gemina** Brummitt in Bol. Soc. Brot., sér.2, **41**: 361 (1968). Type: Mozambique, Serra Namuli (montes do Gurué), fr. 20.ix.1944, *Mendonça* 2163 (LISC holotype).

Shrub to 6 m. Inflorescence axis c.1 cm long in flower, up to 3 cm long in fruit; pedicels c.20 mm long in flower, up to 28 mm in fruit, the long hairs blackish-brown. Calyx 12–13 mm long, the teeth 6–7 mm long, covered with long spreading blackish-brown hairs. Petals 20–24 mm long. Pods clothed with very dense appressed or ascending greyish hairs.

Mozambique. Z: Serra Namuli (montes do Gurué), fl.& fr. 20.ix.1944, *Mendonça* 2163 (LISC).

Known only from the type collection.

54. **Tephrosia montana** Brummitt in Bol. Soc. Brot., sér.2, **41**: 361 (1968). Type: Mozambique, Manica (Macequece) area, summit of Mt Vengo, 1300 m, fl.& fr. 23.xi.1943, *Torre* 6229 (LISC holotype, K).

Shrub up to 3 m or more high. Young branches shortly brown tomentose. Leaves with (8)10–17 pairs of leaflets; petiole and rachis shortly brown tomentose; leaflets up to 38(50) × 11 mm, elliptic to oblong, obtuse to rounded at the base, subacute to obtuse at the apex, usually fairly strongly mucronate, the margins incurved towards the upper surface, the upper surface glabrous, the lower surface rather thinly appressed-pubescent; stipules 6–12 × 1–2 mm, narrowly triangular to ± linear. Inflorescence compact in flower, exceeded by the upper leaves, elongating after flowering to 15 cm; bracts not conspicuous, falling early, 5–6 × 1–3 mm, ovate to lanceolate, acute, brown tomentose; pedicels 7–15 mm long, less shortly tomentose. Calyx 7–10 mm long, densely shortly greyish or brownish pubescent, teeth about equalling the campanulate part, the two upper teeth joined for most of their length to form a broad triangle with only the terminal 1 mm free. Petals 20–24 mm long, purple; standard rounded to emarginate, truncate at the base; lower margin of keel curved through c.90°. Stamen tube c.16 mm long. Ovary appressed-pubescent; style pubescent. Pods 70–80 × c.7 mm, densely brown tomentose when young, rather thinly tomentose when mature. Seeds 7–10.

Zimbabwe. E: Mutasa Dist., Stapleford, Nyamkombe (Nyam Kombi) R., fl. 8.v.1949, *Armitage* A13/49 (SRGH); Nyanga Dist., Nyazengu ridge, S of Inyangani peak, fl.& fr. 17.iii.1982, *Polhill, Müller & Pope* 4738 (K). **Mozambique**. MS: Gorongosa (Gorongoza), Mt Inhatete (Nhandete), fr. 15.x.1946, *Simão* 1107 (LISC).

Known only from the Nyanga/Mutare area of Zimbabwe and from Gorongosa 150 km further east.

Apparently closely related to *T. grandibracteata* and *T. festina* but differing from them markedly in its smaller acute bracts and in the short spreading brown tomentum. It may occur with both these species in the Nyanga area and further information is required as to its distribution, ecology and range of morphological variation.

55. **Tephrosia praecana** Brummitt in Bol. Soc. Brot., sér.2, **41**: 363 (1968). —K. Coates Palgrave, Trees Sthn. Africa, ed.2: 308 (1988). —M. Coates Palgrave, Trees Sthn. Africa: 375 (2002). Type: Zimbabwe, Mount Pene ('Singwekwe'), 2135 m, fl.& fr. 12–14.x.1908, *Swynnerton* 6176 (BM holotype).

Shrub or small tree up to 4.5 m high. Young branches shortly tomentose with grey or grey-brown hairs. Leaves pinnate with (4)6–8 pairs of pinnae; petiole and rachis together 6–9 cm long, with indumentum like that on the young branches; leaflets up to 38 × 10 mm, elliptic or oblong, ± rounded at the base, obtuse or subacute at the apex, with the margins somewhat inrolled towards the upper surface, upper surface glabrous, lower surface appressed-pubescent; stipules 8–10 × 1–3.5 mm, falcate-triangular. Inflorescence in flower compact, subglobose, exceeded by the upper leaves, elongating in fruit; bracts large and conspicuous but soon falling, up to 6 × 10 mm, broadly obtriangular and slightly apiculate, densely and shortly grey or brown tomentose. Calyx 6–9 mm long, tomentose like the bracts; teeth shorter than or about equalling the campanulate part, the two upper ones joined for all or most of their length forming a broad obtuse triangle which is entire or with the two apices free for only 0.7 mm. Petals 21–22 mm long, purple or bluish-purple; standard cuneate at the base. Stamen tube c.16 mm long, fused above. Ovary appressed-pubescent; style pubescent. Pods 72–78 × 7–8 mm, shortly brown tomentose. Seeds 7–8.

Zimbabwe. E: Chimanimani Dist., Tarka Forest Res., c.1160 m, fl.& fr. iii.1968, *Goldsmith* 10/68 (K, SRGH); Chimanimani, Orange Grove road, fl.& fr. 27.ii.1968, *Corby* 1993 (K, SRGH). **Mozambique**. MS: Sussundenga Dist., Monte Chiroso, Mavita, fl. 26.x.1944, *Mendonça* 2634 (K, LISC).

Known only from the Chimanimani area. Forest margins.

56. **Tephrosia festina** Brummitt in Bol. Soc. Brot., sér.2, **41**: 364 (1968). Type: Zimbabwe, Nyanga (Inyanga) road, kopje, fl.& fr. iii.1935, *Gilliland* 1684 (K holotype, BM, FHO, PRE).

Shrub to 5 m high. Young branches shortly grey or brown tomentose. Leaves pinnate with 7–13 pairs of leaflets; petiole and rachis shortly grey or brown tomentose; leaflets and stipules similar to those of *T. praecana* (see above). Inflorescence in flower compact and subglobose, exceeded by the upper leaves, elongated to up to 10 cm in fruit; bracts large and conspicuous but soon falling, up to 14 × 12 mm, ± orbicular-acuminate, densely and shortly grey or brown tomentose. Calyx 9–12 mm long; teeth equalling or longer than the campanulate part, the two upper ones joined for most of their length to form an acute triangle with their apices free for only 1 mm. Petals 13–17 mm long, purple; standard truncate to broadly cuneate at the base. Stamen tube 9–10 mm long, fused above. Ovary appressed-pubescent; style pubescent. Pods 52–60 × 7–8 mm, lanate-tomentose with the hairs grey at least at their tips. Seeds 5–6.

Zimbabwe. E: Nyanga Dist., old fort, Rhodes Estate, fl. 30.iii.1959, *Cleghorn* 491 (K, SRGH).

Known only from the mountains of Nyanga and Mutare Districts, on or among granite rocks in montane grassland, but more information required.

57. **Tephrosia chimanimaniana** Brummitt in Bol. Soc. Brot., sér.2, **41**: 365 (1968). Type: Zimbabwe, Chimanimani Dist., Chimanimani Mtns, among quartzite crags, fl. 20.viii.1954, *Wild* 4589 (K holotype, PRE, SRGH).
   Tephrosia nyasae sensu Baker f. in J. Linn. Soc., Bot. **40**: 54 (1911); Legum. Trop. Africa: 212 (1926) as regard *Swynnerton*. —sensu Eyles in Trans. Roy. Soc. S. Africa **5**: 375 (1916).
   Tephrosia aequilata sensu Goodier & Phipps in Kirkia **1**: 56 (1961).

Shrub 0.3–2 m high. Young branches densely brown or greyish spreading-pubescent to tomentose. Leaves with (5)7–12 pairs of leaflets; petiole and rachis ± appressed to spreading-tomentose, the hairs sometimes noticeably unequal; leaflets up to 25(30) × 8(10) mm, somewhat leathery, elliptic-oblong, rounded at the base, rounded to truncate at the apex, strongly mucronate, the margins somewhat incurved towards the upper surface, upper surface glabrous, lower surface densely appressed-pubescent and often subsericeous; stipules 4–5 × 1–1.5 mm, narrowly triangular. Flowers in subspherical terminal heads exceeded by the leaves; bracts large and usually conspicuous, up to 14 × 9 mm, suborbicular and long-acuminate, densely brown tomentose to villous, often interspersed with long whitish hairs; pedicels 5–7 mm long, whitish villous. Calyx 7–10 mm long, densely brown to whitish villous, the teeth longer than the campanulate part, the lower one very slightly longer than the others, the two upper teeth not or very slightly fused. Petals 13–15 mm long, purplish or sometimes blue; standard subrectangular to suborbicular, truncate to slightly cordate at the base; lower margin of keel curved ± through 90°. Stamen tube c.9 mm long. Ovary appressed-pubescent, style pubescent. Pod not seen.

Zimbabwe. E: Chimanimani Dist., Chimanimani Mts., Long Gully, fl. 29.v.1959, *Noel* 2023 (SRGH).

Endemic to the Chimanimani Mountains where it occurs from the foot of the first range to the quartzite crags; up to at least 2135 m.

58. **Tephrosia grandibracteata** Merxm. in Mitt. Bot. Staatssamml. München **6**: 200 (1953). —Brummitt in Bol. Soc. Brot., sér.2, **41**: 366 (1968). Type: Zimbabwe, Rusape, *Dehn* s.n. (M holotype). FIGURE 3.3.**27**.

Shrub up to 3(5) m high. Young branches shortly and closely appressed golden-brown pubescent. Leaves with (7)9–14(16) pairs of leaflets; pubescence of petiole and rachis similar to that of young branches; leaflets up to 40(48) × 12(15) mm, elliptic-oblong, ± rounded at base and apex, strongly mucronate, the margins somewhat incurved towards the upper surface, the upper surface glabrous, the lower surface shortly and closely appressed- brownish pubescent with hairs not or scarcely overlapping each other; stipules 5–15 × 1–1.5 mm, ± linear. Inflorescence at first an ovoid head exceeded by the upper leaves, elongating to 10 cm in fruit; bracts large and very conspicuous but soon falling, up to 20 × 13 mm, ovate to obovate and tapered gradually to a long-acute apex, shortly and closely appressed brown pubescent; pedicels 7–12 mm long (to 18 mm in fruit), appressed-pubescent. Calyx 9–13 mm long, the teeth longer than the campanulate part, the two upper teeth joined for most of their length to form a broad triangle with only the terminal 1–2 mm free; the whole shortly and rather thinly appressed-pubescent. Petals 20–27 mm long, purple; standard as broad as long, broadly rounded to emarginate at the apex, ± truncate at the base; lower margin of keel curved through c.90°. Stamen tube 15–18 mm long. Ovary shortly appressed-pubescent; style pubescent. Pods 7–8.5 × 0.6–0.8 cm, the sutures somewhat broadened, surfaces closely appressed dark brown pubescent. Seeds c.8.

Zimbabwe. C: Makoni Dist., Rusape, fl. viii.1952, *Dehn* s.n. (M). E: Nyanga Dist., top of road to Pungwe Falls, fl. 23.x.1946, *Rattray* 958 (K, SRGH).

Endemic to the broad Nyanga area. Often forming associations at forest margins or in open grassland; 1800–2250 m.

59. **Tephrosia rupicola** J.B. Gillett in Kew Bull. **13**: 131 (1958). —Brummitt in Bol. Soc. Brot., sér.2, **41**: 368 (1968). Type: Zimbabwe, Marondera town, Dombe-dombe, granite outcrop, *Corby* 410 (K holotype, SRGH).

Shrub up to 1 m high. Young stem densely pubescent to villous with grey or brown, spreading or deflexed hairs. Leaves with (3)5–8 pairs of leaflets; petiole c.5 mm long, petiole and rachis brownish-pubescent or villous, hairs spreading or some appressed and others spreading; leaflets to 30(34) × 11(13) mm, oblanceolate to oblong-cuneate, obtuse to truncate at apex, shortly mucronate, margins flat, the upper surface appressed-pubescent or glabrous, the lower surface appressed-pubescent; stipules 2–4 mm broad, ovate-acuminate. Inflorescence rather lax, up to 15(20) cm long, sometimes the lower flowers subtended by foliage leaves; bracts up to 11 × 4 mm, narrowly triangular to ovate-acuminate; pedicels 4–7(10 in fruit) mm long, brown or grey, pubescent or villous. Calyx 7–11 mm long, of which the campanulate part is only 2–3 mm long, the two upper teeth joined in the lower $^1/_3$, the whole grey or brown villous. Petals 12–22 mm long, purple; standard suborbicular-cuneate. Ovary densely villous along the sutures but the sides glabrous; style pubescent. Pod 27–35 × 8–9 mm, the margins pubescent but the sides glabrous. Seeds 7–9, transversely elongate, nearly oblong in outline, rounded at the ends, slightly broader at the lower end, with a small hilum near the upper narrower end with a small white rim aril.

Known only from Zimbabwe. On granite rocks in upland areas.

Subsp. **rupicola** —Brummitt in Bol. Soc. Brot., sér.2, **41**: 368 (1968).
 Tephrosia huillensis sensu Suessenguth & Merxmüller in Proc. & Trans. Rhod. Sci. Assoc. **43**: 25 (1951).

Hairs on young stem and inflorescence axis spreading or upwardly somewhat appressed, brown or grey-brown. Leaflets pubescent on upper surface. Calyx 9–11 mm long. Petals 15–22 mm long.

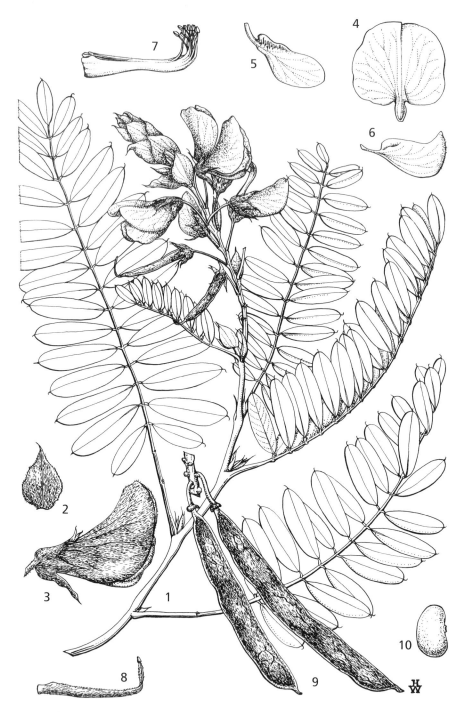

Fig. 3.3.**27**. TEPHROSIA GRANDIBRACTEATA. 1, flowering branch (× ²/₃), from *Methuen* 16;
2, bract (× 1¹/₂); 3, flower (× 1¹/₂); 4, standard (× 1); 5, wing (× 1); 6, keel (× 1); 7, stamens (×
1¹/₂); 8, gynoecium (× 1¹/₂), 2–8 from *Wild* 1484; 9, pods (× 1), from *Methuen* 16 and *Pardy*
31/36; 10, seed (× 2), from *Wild* 6722. Drawn by Heather Wood.

Zimbabwe. C: Marondera Dist., 12 km E of Marondera near Sanzara Road turn-off, fl.& fr. 8.ii.1997, *Brummitt & Pope* 19556 (BR, K, LISC, MAL, PRE, SRGH).

Known only from the adjacent Marondera and Makoni Districts of Zimbabwe.

Subsp. **dreweana** Brummitt in Bol. Soc. Brot., sér.2, **41**: 368 (1968). Type: Zimbabwe, Masvingo Dist., Great Zimbabwe Ruins (Zimbabwe), fl.& imm.fr. 5.ii.1961, *Drewe* 25 (SRGH).

Hairs on young stem and inflorescence axis deflexed, grey. Leaflets glabrous on upper surface. Calyx 7–8 mm long. Petals c.12 mm long.

Zimbabwe. S: Masvingo Dist., Great Zimbabwe Ruins, Zimbabwe Hotel, fr. 27.ii.1971, *Corby* 2179 (K, SRGH).

Known only from the above collections, from a rocky outcrop; c.1300 m.

60. **Tephrosia meisneri** Hutch. & Burtt Davy in Burtt Davy, Fl. Pl. Ferns Transvaal, pt.2: 377 (1932). Type: South Africa, 'Transvaal', no locality, *Sanderson* s.n. (K holotype).

> *Tephrosia shiluwanensis* sensu Goodier & Phipps in Kirkia **1**: 56 (1961), non Schinz (1907).
>
> *Tephrosia incarnata* Brummitt in Bol. Soc. Brot., sér.2, **41**: 370 (1968) excluding Mozambique specimens. Type: Zimbabwe, Mutare Dist., Himalaya Mts., Engwa Farm, 1940 m, in *Widdringtonia* scrub, fl.& fr. 2.iii.1954, *Wild* 4450 (SRGH holotype, K, LISC, PRE).
>
> *Tephrosia glomeruliflora* subsp. *meisneri* (Hutch. & Burtt Davy) Schrire in Bothalia **17**: 14 (1987).

Annual or short-lived perennial with erect branched stems up to 1.3 m high. Stems very sparsely pubescent above, glabrous below, smooth. Leaves pinnate with (3)4–6(7) pairs of leaflets; petiole (1.5)2.5–5.5 cm long, ¹/₃ as long or equalling the rachis, petiole and rachis together (3.5)5–12 cm long, with rather sparse appressed to spreading hairs; leaflets (1)1.5–3.5 × 0.5–1.4 cm, elliptic or oblong to obovate, ± cuneate at the base, obtuse to emarginate at the apex, the upper surface glabrous, the lower surface appressed-pubescent; stipules 7–13 mm from point of attachment to apex, 3–6(8) mm broad, long-acute at the apex, with a well developed auricle on the adaxial side, the abaxial side rounded or slightly auriculate, the whole reddish-brown, glabrous but the margins ciliate. Flowers in short dense racemes which are terminal or occasionally some in the axils of the uppermost leaves; peduncles mostly 9–18 cm long, bearing the flowers clear of the uppermost leaves; inflorescence axis densely brown or whitish pubescent; bracts large and conspicuous, 6–12 × 4–9 mm, broadly ovate to suborbicular, acuminate to rounded at the apex, concave and closely enfolding the buds but falling at flowering, sometimes splitting near the apex and appearing 2 or 3-dentate, reddish or purplish-brown, shortly pubescent; pedicels 4–10 mm long, densely appressed to spreading brown pubescent. Calyx 3.5–5(5.5) mm long, the campanulate part 1.5–2.5 mm long, appressed to spreading-pubescent, the two upper teeth joined for half their length with c.1.5 mm distinct, the lower tooth somewhat longer than the others. Petals 12–18 mm long, pink to purple; standard oblong to suborbicular, cuneate to subtruncate at the base. Stamen tube joined above or with the upper filament strongly or loosely joined to the others. Ovary glabrous or with few hairs on the sutures; style pubescent. Pod 50–62 × 7.5–9.5 mm, brown or blackish, glabrous. Seeds 12–15, transversely elongate.

Zimbabwe. E: Chimanimani Dist., near Skyline, Chimanimani–Chipinge (Melsetter–Chipinga) road, fl.& fr. 28.ii.1956, *Drummond* 5118 (COI, K, SRGH). S: Bikita Dist., S base of Mt. Hozvi (Horzi), fl. 9.v.1969, *Biegel* 3087 (K, SRGH).

Also in South Africa (Limpopo, Mpumalanga and KwaZulu-Natal Provinces). In Zimbabwe recorded from montane forest margins from Nyanga to Chirinda and Mt. Hozvi; 975–1950 m.

The specimens cited from S Mozambique as *T. incarnata* by Brummitt (1968) are now referred to *T. brummittii* (see below).

61. **Tephrosia brummittii** Schrire in Bothalia **17**: 12 (1987). Type: South Africa, KwaZulu-Natal, Hlabisa Dist., St Lucia Estuary Game Park, W of Vidal road, *Pooley* 1918e (NU holotype, K, PRE).

   Tephrosia incarnata sensu Brummitt in Bol. Soc. Brot., sér.2, **41**: 370 (1968) for Mozambique specimens.

Perennial herb with a woody rootstock, up to 60 cm tall. Stems 1–many, prostrate to ascending, branching from near the base, glabrous or sparsely pubescent to ascending-pilose. Leaves pinnate with 3–5(6) pairs of leaflets; petiole (1.5)2.5–5.5(7) cm long, petiole and rachis together 2.5–10 cm long; leaflets (1.2)1.8–3.5 × 0.3–0.7(1.2) cm, narrowly oblong-elliptic to oblanceolate, cuneate to rounded at the base, emarginate, obtuse or acute at the apex, the upper surface glabrous to puberulous, the lower surface moderately appressed to spreading greyish pubescent; stipules (6)7–17 × 3–11 mm, broadly ovate, base often cordate, apex acuminate, maturing conspicuously red or brown. Flowers in short dense pseudoracemes 1–3 cm long above a peduncle 7–17 cm long; bracts 5–15 × 4–10 mm, broadly ovate-acuminate, concave and enfolding the young buds, chestnut-brown, glabrous; pedicels 3–6 mm long in flower, elongating to 10 mm in fruit. Calyx 10 mm long, the campanulate part 4 mm long, sparsely to moderately pubescent, the teeth longer than the tube, lanceolate at the base, attenuate above and with long spreading hairs. Petals 11–21 mm long, pink to mauve-purple; standard broadly obovate to suborbicular, narrowing to the claw at the base. Staminal sheath open with the upper filament free. Ovary glabrous; style pubescent. Pods only known immature, glabrous, with 10–16 seeds set transversely.

Mozambique. M: Moamba Dist., Sábiè, Libombos Mts, M'Ponduine (Meponduine), near the border, fl. 25.iv.1947, *Pedro & Pedrógão* 735 (LMA); Namaacha, near Canada Dry factory, fl.& imm.fr. 27.iii.1957, *Barbosa & Lemos* 7534 (COI, LISC, LMA).

Also in South Africa (KwaZulu-Natal) and Swaziland. In South Africa known from streambanks and forest margins in the Lebombo Mts. and in grassland and dunes down to the coast; 10–450 m.

62. **Tephrosia cordata** Hutch. & Burtt Davy, Fl. Pl. Ferns Transvaal, pt.2: 375, 377 (1932). —Forbes in Bothalia **4**: 995 (1948). —Brummitt in Bol. Soc. Brot., sér.2, **41**: 370 (1968). —Schrire in Bothalia **17**: 14 (1987). Type: Swaziland, Mbabane, fr. i.1905, *Burtt Davy* 2886 (K holotype, BM, PRE).

?Annual or perennial, 1–1.8 m high, ?woody below. Stems glabrous, ± smooth. Leaves pinnate with 4–8 pairs of leaflets; petiole 2.5–6 cm long, petiole and rachis together 4.5–12 cm long, glabrous or with sparse appressed hairs; leaflets 1.5–3(4) × 0.6–1.2(1.5) cm, elliptic or oblong to obovate, obtuse to emarginate at the apex, ± cuneate at the base, the upper surface glabrous, the lower surface appressed-pubescent; stipules large and conspicuous, persistent, 9–17 mm long from point of attachment to apex, (6)7–14 mm broad, strongly cordate at the base with auricles on both upper and lower sides, the apex acute, the whole glabrous, not or scarcely ciliate, often turning blackish on drying. Racemes with c.6 flowers, short, up to 2 cm long, usually terminal on the main and lateral branches; peduncles 3–8 cm long, about equalling the upper leaves, glabrous except just below the flowers; inflorescence axis brown appressed-pubescent; bracts large, c.10 × 6 mm, ovate-acuminate, acute at the apex, concave and closely enclosing the buds but falling at flowering, brown, glabrous. Calyx 3–4.5 mm long, the campanulate part 1.5–2.5 mm, the two upper teeth joined for not more than half their length. Petals 17–20 mm long, pink to violet. Ovary glabrous; style pubescent. Pod 35–50 × 6–8 mm, glabrous; seeds 8–10, transversely elongate.

The species is recorded from Mozambique by Forbes (1948) on the basis of one specimen, without locality, collected by Schlechter, previously at Berlin but now destroyed. No other record of its occurrence in the Flora Zambesiaca area is known. However, it is a distinctive species unlikely to have been confused. The species occurs in the South Africa (Mpumalanga) and Swaziland close to the Mozambique border, at forest margins and in grassy places.

63. **Tephrosia gobensis** Brummitt in Bol. Soc. Brot., sér.2, **41**: 368 (1968). —Schrire in Bothalia **17**: 13 (1987). Type: Mozambique, Goba, by bridge over R. Umbeluzi, fl.& fr. 31.iii.1945, *A.E. Sousa* 132 (LISC holotype), and near spring 'Fonte-dos-Libombos', 9.5 km from Goba, bud & fr. 31.iii.1945, *A.E. Sousa* 132 (PRE).

A somewhat woody subshrub up to 1.5 m high. Young stems densely clothed with appressed to ascending grey hairs. Leaves pinnate with 2–5 pairs of leaflets; petiole 0.3–1 cm long, petiole and rachis together (0.6)1.2–2.4 cm long, with dense, grey, appressed hairs; leaflets (0.6)0.8–1.6 × (0.3)0.4–0.6 cm, obovate to oblanceolate, cuneate at the base, broadly rounded to emarginate at the apex; upper surface glabrous to sparsely and shortly pubescent, lower surface appressed-pubescent, greyish; stipules 3–5 × 0.4–0.8 mm, ± linear. Flowers in short terminal racemes 0.5–3(5 in fruit) cm long and in the axils of the upper 1–4 leaves; bracts 1.5–3 × 0.7–1 mm, ovate to narrowly triangular; pedicels 2–4 mm long. Calyx 5–6 mm long, appressed-pubescent; upper teeth joined for about ³/₄ their length, lateral teeth about equalling the tube. Petals c.15 mm long, pale rose-wine coloured to reddish-blue. Staminal tube connate above. Ovary glabrous on the surfaces but with long, rather stiffly spreading hairs on the margins; style pubescent, c.1 cm long in fruit. Pods 20–30 × 7–9 mm, boat-shaped, with the long style persisting almost to maturity, glabrous except at the margins, brown; seeds 3–5, elongated transversely.

Mozambique. M: Namaacha Dist., Boane–Changalane–Goba, fl.& fr. ii.1979, *de Koning* 7353 (K).

Also in Swaziland. Rocky places.

64. **Tephrosia radicans** Baker in F.T.A. **2**: 121 (1871). —Harms in Engler, Pflanzenw. Afrikas **3**(1): 589 (1915). —Eyles in Trans. Roy. Soc. S. Africa **5**: 376 (1916). — Baker f., Legum. Trop. Africa: 214 (1926). —Forbes in Bothalia **4**: 996 (1948). —Suessenguth & Merxmüller in Proc. & Trans. Rhod. Sci. Assoc. **43**: 26 (1951). —Cronquist in F.C.B. **5**: 106 (1954). —Hepper in F.W.T.A., ed.2, **1**: 530 (1958). —Torre in C.F.A. **3**: 165 (1962). —Schreiber in Merxmüller, Prodr. Fl. SW Afrika, fam. 60: 119 (1970). —Gillett in F.T.E.A., Legum., Pap.: 211 (1971). — Schrire in Bothalia **17**: 12 (1987). Type: Angola, Huíla, Varezeas and Catumba, *Welwitsch* 2082 (BM holotype, K).

   Cracca radicans (Baker) Kuntze, Rev. Gen. Pl. **1**: 175 (1891).
   Tephrosia radicans var. *rhodesica* Baker f. in J. Bot. **37**: 430 (1899); Legum. Trop. Africa: 214 (1926). —Eyles in Trans. Roy. Soc. S. Africa **5**: 376 (1916) as '*rhodesiaca*'. Type: Zimbabwe, Bulawayo, *Rand* 52 (BM).
   Tephrosia rensburgii Verdcourt in Kew Bull. **7**: 358 (1952). Type: Tanzania, Nzega Dist., Ndala, *van Rensberg* 56 (EA holotype).

Perennial, with numerous prostrate branches up to 160 cm long, the older parts often woody, arising from a woody vertical rootstock, the whole usually forming a dense mat. Stems usually conspicuously spreading-pubescent. Leaves pinnate with (2)3–4(5) pairs of leaflets; rachis and petiole together up to 3.5(5) cm long, the petiole up to 1.3(1.7) cm long, rather conspicuously spreading-pubescent; leaflets up to 22(33) × 12(17) mm, often becoming smaller towards the leaf base, obovate to obtriangular, rounded to truncate or emarginate at the apex, cuneate at the base, the upper surface glabrous, the lower surface appressed-pubescent; stipules 3–5(6) × 2–3 mm, ovate-acuminate to lanceolate. Flowers in terminal racemes up to 5(7) cm long, subsessile or with a peduncle up to 4 cm, rarely the lowermost flower in the axil of a foliage leaf; bracts ovate-lanceolate; pedicels (3)4–7(8) mm long. Calyx 4–6 mm long, appressed or spreading-pubescent, teeth about equalling the tube or the lower one somewhat longer, the two upper ones joined for about $^1/_2$–$^3/_4$ their length. Petals (7)10–14 mm long, mauve or pink; standard broadly rounded, cuneate or truncate at the base. Stamen tube open above, the upper stamen free. Ovary whitish pubescent; style pubescent. Pod (7)12–22 × 6–8 mm, oblong, shortly pubescent, or glabrescent at maturity. Seeds (1)2–4(5), 3–4 mm long, reniform to suborbicular-lenticular, greyish- or reddish-brown.

Zambia. B: Senanga Dist., Nangweshi, st. 28.vii.1952, *Codd* 7215 (K, PRE). N: Mbala (Abercorn), old road between the Uningi Pans, fl. 21.iii.1955, *Richards* 5047 (K, SRGH). W: Kitwe, fl.& fr. 4.vii.1963, *Mutimushi* 339 (K). C: Lusaka, fl.& imm.fr. 5.iii.1955, *Best* 61 (K, SRGH). S: Mazabuka, Vet. Res. Station, fl. 10.i.1963, *van Rensburg* KBS 1171 (K, SRGH). **Zimbabwe**. N: Zvimba Dist., Darwendale, Mpanda Farm, fl.& fr. 19.iii.1951, *Lennard* (SRGH). W: Matobo Dist., Matopos, fl. ii.1931, *Rattray* 271 (K, PRE, SRGH). C: Makoni Dist., c.8 km W of Rusape, fl. 22.i.1949, *Chase* 997 (K, LISC, PRE, SRGH). E: Mutare (Umtali), commonage, cricket grounds, fl. 6.iii.1956, *Chase* 6003 (K, LISC, PRE, SRGH). S: Bikita Reserve, fl.& imm.fr. 8.iv.1948, *D.A. Robinson* 318 (K, SRGH). **Malawi**. C: Lilongwe, fl.& imm.fr. 1.iv.1955, *Jackson* 1558 (K, LISC, SRGH). S: N.A. Nyambi, Mangochi (Fort Johnston), fl. 20.iv.1955, *Jackson* 1628 (K, SRGH).

Also in Nigeria, Congo, Tanzania, Angola, Namibia and South Africa (Limpopo and Mpumalanga Provinces). Open ground and grassy places, particularly on sandy soil, often a weed of roadsides or disturbed ground; 450–1700 m.

65. **Tephrosia zambiana** Brummitt in Bol. Soc. Brot., sér.2, **41**: 382 (1968). Type: Zambia, Kasama Dist., Mungwi, fl. 2.x.1960, *E.A. Robinson* 3899 (K holotype, SRGH).

Suffrutex with herbaceous unbranched stems up to 35 cm high. Stems densely whitish appressed-pubescent. Leaves predominantly 3-foliolate but some very small lower leaves unifoliolate or upper leaves 5-foliolate; petiole and rachis together up to 2 cm long, the rachis usually distinctly longer than the petiole, similar in pubescence to the stem; leaflets up to 5.5 × 1.1 cm, the terminal one slightly larger than the others, ± narrowly oblong, cuneate to rather strongly inrolled above at the base, obtuse to emarginate at the apex, the upper surface glabrous, the lower surface densely silky whitish appressed-pubescent and subsericeous when young; stipules c.6 × 0.8 mm, linear-triangular, pubescent on outer surface. Flowers in terminal racemes with at least 5–8 distinct nodes, clearly exceeding the uppermost leaves, or sometimes with the lowermost flowers in the axil of the one or two uppermost leaves; bracts resembling the stipules; pedicels 3–7 mm long, tomentose. Calyx 9–11 mm long, whitish appressed-tomentose, the teeth 5–7 mm long, subequal, 1.5–2 times as long as the tube, the two upper ones joined for about $^1/_3$ of their length, the lower one ± channelled and somewhat upwardly curved at its tip. Petals 18–20 mm long, probably purplish or the keel whitish; standard suborbicular, broadly cuneate towards the claw. Staminal tube not fused above, the upper filament free. Ovary ± tomentose; style pubescent. Pods not seen.

Zambia. N: Kasama Dist., Mungwi (Mungwe), fl. 2.x.1960, *E.A. Robinson* 3899 (K, SRGH).
Known only from the type collection, from miombo woodland.

66. **Tephrosia dasyphylla** Baker in F.T.A. **2**: 118 (1871). —Baker f., Legum. Trop. Africa: 210 (1926); in J. Bot. **66**, Polypet. Suppl.: 106 (1928). —Brummitt in Bol. Soc. Brot., sér.2, **41**: 375 (1968). —Gillett in F.T.E.A., Legum., Pap.: 193 (1971). Type: Angola, Huíla, morro de Lopolo, *Welwitsch* 2084 (LISU holotype, BM, K).

Suffrutex producing ± unbranched (except when damaged) herbaceous aerial stems up to 35(50) cm high. Stems spreading or sometimes ascending brown or rarely grey tomentose (in other areas closely appressed-pubescent). Leaves 1–3-foliolate or rarely 5-foliolate; petiole and rachis (if present) together 0.6–2.5(3) cm long; leaflets variable (see subspecies); stipules (6)9–15 × 1–2 mm (in other areas often smaller, 5–9 × 0.5–1.2 mm), narrowly triangular to linear. Flowers in dense terminal heads surrounded or exceeded by the uppermost leaves and often also in the axils of the uppermost leaves, altogether up to 15 in number; bracts resembling stipules; pedicels 5–10(13) mm long. Calyx (7)8–12(14) mm long (in other areas often 6–9 mm), appressed grey or brown villous, the teeth 2–4 times as long as the tube, subequal or the lower one longer than the others, the two upper teeth joined for $^1/_4$–$^1/_2$ their length, the lower teeth channelled and strongly curved. Petals (17)19–26(29) mm long (in other areas commonly 15–20 mm), pink to mauve or pale purple, rarely bluish, or the keel whitish, the standard brown pubescent outside. Staminal tube not fused above, the upper filament free. Ovary tomentose; style pubescent. Pods 35–55 × 8–10 mm, oblong, brown tomentose to subvillous. Seeds 8–10, lying transversely, c.5 × 3 × 2 mm.

Subsp. **dasyphylla** —Brummitt in Bol. Soc. Brot., sér.2, **41**: 375 (1968). —Gillett in F.T.E.A., Legum., Pap.: 193 (1971).

  Tephrosia luembensis De Wild. in Repert. Spec. Nov. Regni Veg. **13**: 103 (1914). —Baker f., Legum. Trop. Africa: 211 (1926). Type from Congo (Katanga).

  Tephrosia subfalcato-stipulata De Wild. in Bull. Soc. Roy. Bot. Belg. **57**: 126 (1925). — Baker f., Legum. Trop. Africa: 210 (1926). Type from Congo (Katanga).

  Tephrosia dasyphylla sensu Brenan, Check-list For. Trees Shrubs Tang. Terr.: 446 (1949). —sensu Suessenguth & Merxmüller in Proc. & Trans. Rhod. Sci. Assoc. **43**: 25 (1951). — sensu Cronquist in F.C.B. **5**: 104 (1954) in part excl. synonyms *T. butayei, T. argyrolampra* and specimen *Schmitz* 2812. —sensu Torre in C.F.A. **3**: 163 (1962) in part excl. *Exell & Mendonça* 1178.

Leaves predominantly 3-foliolate, though with occasional lower leaves unifoliolate or upper leaves 5-foliolate; petiole and rachis together 6–25(30) mm long, the petiole usually distinctly longer than the rachis which may be very short; terminal leaflets (4)5–10 × (0.6)0.8–1.7(2.3) cm, lateral leaflets 3.5–6.5(8) × 0.5–1.3(2) cm, all linear-oblong or very narrowly elliptic to narrowly oblanceolate, obtuse to emarginate at the apex, ± cuneate at the base; the upper surface glabrous, the lower surface grey or silver to white appressed-pubescent, stipules (6)8–15 × 1–1.5 mm, linear-triangular.

Zambia. N: Mbala Dist., Lake Chila, fl. 21.i.1955, *Richards* 4202 (K). W: Copperbelt, Mufulira Dist., fl. 17.i.1948, *Cruse* 250 (K). C: Walamba, fr. 23.v.1954, *Fanshawe* 1243 (K). **Zimbabwe**. C: Marondera Dist., Digglefold, fl.& fr. 24.xii.1948, *Corby* 296 (K, SRGH). E: Mutare Dist., Vumba, Norseland, fl. 9.ii.1950, *Chase* 1964 (K, LISC, SRGH). **Malawi**. N: Mzimba Dist., Perekezi Forest Res., c.42 km N of Katete, fl. 18.ii.1973, *Pawek* 6457 (K). C: Lilongwe Dist., Choulongwe (Chaulongwe) Falls, Dzalanyama Forest Res., fl.& fr. 14.i.1967, *Hilliard & Burtt* 4487 (F, K). **Mozambique**. N: Ngauma Dist., Massangulo, fr. 15.v.1948, *Pedro & Pedrõgão* 3507 (LMA).

Also in Congo (Katanga), Tanzania and Angola. In grassland, miombo woodland and fringing forest; 1200–1700 m.

Intermediates between subsp. *dasyphylla* and subsp. *amplissima*.

Intermediates have predominantly 3-foliolate leaves with elliptic leaflets, the terminal ones 1.6–2.2 cm broad.

Zambia. B: Zambezi (Balovale), fl. xii.1953, *Gilges* 301 (K, PRE, SRGH). W: Copperbelt, Mufulira Dist., fl. 16.xi.1947, *Cruse* 71 (K).

Subsp. **amplissima** Brummitt in Bol. Soc. Brot., sér.2, **41**: 380 (1968). —Gillett in F.T.E.A., Legum., Pap.: 193 (1971). Type: Zambia, Mufulira, fringing forest, fl. 11.i.1948, *Cruse* 249 (K holotype).

> *Tephrosia dasyphylla* sensu Cronquist in F.C.B. **5**: 104 (1954) as regard *Schmitz* 2812. — sensu Torre in C.F.A. **3**: 163 (1962) as regards *Exell & Mendonça* 1178.

Leaves unifoliolate or with 1 or 2 lateral leaflets much smaller than the terminal one; petiole and rachis (if present) together 8–20 mm long, the petiole usually longer than the rachis which may be very short; terminal (or solitary) leaflets at maturity 6–12 × (2.5)3–4.8 cm, ± elliptic, the apex obtuse to retuse or rarely subacute, the base rounded to subcuneate; lower surface densely covered with grey, curved or crisped, not strongly appressed hairs; stipules 10–15 × 1–2 mm.

Zambia. W: Mwinilunga Dist., 1 km S of Matonchi Farm, fl. 2.i.1938, *Milne-Redhead* 3924 (K).

Also in Congo (Katanga), Tanzania and E Angola. In miombo woodland and fringing forest.

This species should be reconsidered when more herbarium material and field observations are available. It may be more satisfactory to regard the two subspecies as separate species if the intermediates can be shown to be hybrids between them. For further discussion see Brummitt (1968), where subsp. *youngii* (Torre) Brummitt from Angola and subsp. *butayei* (De Wild. & T. Durand) Brummitt from western Congo are also recognized, both differing from the above in their silvery, often silky pubescence, shorter stipules and rather smaller flowers.

67. **Tephrosia muenzneri** Harms in Bot. Jahrb. Syst. **45**: 310 (1910). —Brummitt in Bol. Soc. Brot., sér.2, **41**: 381 (1968). —Gillett in F.T.E.A., Legum., Pap.: 194 (1971). Type: Tanzania, Ufipa Dist., Mbala–Muse road at Chapota turning, 1500 m, fl.& fr. 22.x.1960, *Richards* 13368a (K neotype).

Subsp. **pedalis** Brummitt in Bol. Soc. Brot., sér.2, **41**: 381 (1968). Type: Zambia, Lundazi Dist., Lundazi to Mzimba, mile 4, fr. 28.iv.1952, *White* 2495 (FHO holotype, K).

Suffrutex with unbranched procumbent or ?erect stems 30 cm long. Stems rather sparsely clothed with long 1.5 mm spreading hairs. Leaves simple, subsessile, the basal pulvinus 1–2 mm long; lamina up to 9.5 × 3.6 cm, oblanceolate, strongly cuneate at the base with the margins scarcely inrolled, broadly rounded at the apex, the upper surface glabrous, the lower surface rather thinly spreading-pubescent with stiffish straight or geniculate hairs; stipules 11–13 × 0.6–1 mm, linear, sparsely pubescent. Flowers in few-flowered terminal or subterminal heads overlapped by the uppermost leaves, or with one pair in the axil of an uppermost leaf; bracts resembling stipules; pedicels 7–10 mm long, rather thinly spreading-pubescent. Calyx 7–8 mm long, spreading-pubescent, the teeth c.3 times as long as the tube, the two upper teeth long-acute and joined for only about ¹/₃ their length. Petals, stamens and ovary not seen. Pods c.45 × 9 mm, thinly spreading-pubescent. Seeds c.6, widely spaced, oblique.

Zambia. E: Lundazi Dist., Lundazi to Mzimba, mile 4, fr. 28.iv.1952, *White* 2495 (FHO).

Known only from the type collection, from *Julbernardia paniculata*, *Brachystegia manga*, *B. spiciformis* woodland on pink sandy loam.

Subsp. *muenzneri* is known only from Ufipa District of SW Tanzania and differs mainly in its pubescence (stems grey to brown tomentose, the leaf lower surface with dense silvery-grey curved or crisped hairs) and its smaller stipules.

68. **Tephrosia cephalantha** Baker in F.T.A. **2**: 119 (1871). —Baker f., Legum. Trop. Africa: 210 (1926); in J. Bot. **66**, Polypet. suppl.: 106 (1928). —Cronquist in F.C.B. **5**: 102 (1954). —Schreiber in Mitt. Bot. Staatssamml. München **16**: 298 (1957). —Torre in C.F.A. **3**: 164 (1962). —Brummitt in Bol. Soc. Brot., sér.2, **41**: 383 (1968). —Schreiber in Merxmüller, Prodr. Fl. SW Afrika, fam. 60: 117 (1970). Type: Angola, Huíla, morro de Lopolo, *Welwitsch* 2087 (BM holotype).

> *Cracca cephalantha* (Baker) Kuntze, Rev. Gen. Pl. **1**: 174 (1891).

Var. **decumbens** Baker in F.T.A. **2**: 119 (1871), modified by Brummitt in Bol. Soc. Brot., sér.2, **41**: 384 (1968). —Baker f., Legum. Trop. Africa: 210 (1926). — Schrire in Bothalia **17**: 10 (1987). Type: Angola, Huíla, Ferrão da Sola, *Welwitsch* 2090 (K).

> *Tephrosia hypargyrea* Harms in Warburg, Kunene-Samb.-Exped. Baum: 259 (1903). — Baker f., Legum. Trop. Africa: 210 (1926). Type from Angola.

Annual or perhaps biennial, with a taproot and an erect or decumbent branched stem up to 80(100) cm high, sometimes woody in the lower part. Stems appressed or rarely shortly ascending-pubescent, brownish or greyish. Leaves predominantly trifoliolate, sometimes the lower leaves unifoliolate or an occasional upper leaf 5-foliolate, rarely most leaves 5-foliolate; petiole and rachis together 0.7–1.5 cm long, shortly pubescent to tomentose, the petiole shorter than the rachis or about equalling it; leaflets unequal with the terminal one 1.5–2 times as long as the lateral ones, the terminal one (25)35–55(65) × 10–20(26) mm, all leaflets obovate to oblanceolate, cuneate at the base, broadly rounded or more usually emarginate at the apex, the upper surface glabrous, the lower surface silvery appressed-pubescent; stipules 4–7(8) × 0.5–1.5(2) mm, narrowly triangular. Flowers in dense heads, up to 12 per head, surrounded or exceeded by the upper leaves, occasionally a pair of flowers remote from the head in the axil of a foliage leaf; bracts resembling the stipules, grey villous; pedicels 2–6 mm long, grey villous. Calyx (8)9–12 mm long, appressed grey villous, or brown villous on the teeth, the teeth 2–3 times as long as the campanulate part, long-acute, the two upper ones joined for about $^{1}/_{4}$–$^{1}/_{3}$ their length, the lower one slightly longer than the others. Petals 12–16 mm long, white or the standard and wings sometimes pinkish to purplish, the standard brown pubescent on the outside; standard cuneate. Stamen tube open above, the upper stem free. Ovary pubescent; style pubescent. Pods 2.5–4(4.5) × 0.6–0.7 cm, narrowly to linear-oblong, rather shortly brown villous. Seeds 9–13, lying transversely and almost separated from each other by inward projections of the endocarp.

Caprivi Strip. About 69 km from Singalamwe on W.N.L.A. road to Katima Mulilo, fl. 3.i.1959, *Killick & Leistner* 3278 (K, SRGH). **Botswana**. N: Ngamiland Dist., island off Kwando R. mainstream, fl. 29.i.1978, *P.A. Smith* 2311 (K, SRGH). SE: Tamasanka, E Bamangwato Reserve (Territory), imm.fr. 1883, *Holub* (K). **Zambia**. B: Mongu Dist., Bulozi Plain, fl. 7.i.1960, *Gilges* 850 (SRGH). N: Mbala Dist., bathing place, Lake Chila, fr. 30.iv.1955, *Richards* 5468 (K). W: Copperbelt, Mufulira Dist., fr. 20.iii.1948, *Cruse* 303 (K), C: Serenje Dist , Great North Road 5 km from Serenje turning, fr. 5.iv.1961, *Richards* 14941 (K). S: Choma SE, fl.& fr. 19.iii.1958, *Robinson* 2807 (K, PRE, SRGH). **Zimbabwe**. N: Gokwe, fl.& fr. 5.iii.1962, *Bingham* 148 (K, SRGH). W: Gwayi (Gwaai), S.N.A., fr. iv.1953, *R.M. Davies* 547 (SRGH). C: Chirumanzu Dist., Mtao Forest Res., fl. 24.iii.1947, *Goldsmith* 1/47 (SRGH). S: Masvingo Dist., Makaholi Expt. Station, fr. 16.iii.1978, *Senderayi* 282 (K, SRGH).

Also in Angola and Namibia. Usually in woodland or bushland on sandy ground; 900–1450 m.

Var. *cephalantha*, found in Congo and Angola, differs in having much smaller and narrower leaflets, the terminal one 3–7(10) mm broad and 5–8 times as long as broad, and the leaves often 5-foliolate.

A specimen from Zambia N: (Samfya, Lake Bangweulu, *Watmough* 200 (K, LISC, PRE, SRGH)), resembles var. *decumbens* in the size and shape of its leaflets, but differs markedly in several other characters – the leaves are mostly 5-foliolate, with the petiole and rachis together up to 2.5(3) cm long; the flowers have much larger petals, up to 22 mm long, though the calyx is fairly short, 9–10 mm long, and so only about half as long as the petals; and the pods are larger, 5–5.5 × 0.7–0.8 cm, and rather closely appressed-pubescent.

69. **Tephrosia tanganicensis** De Wild. in Ann. Mus. Congo, Bot., sér.5, **1**: 262 (1906). —Gillett in F.T.E.A., Legum., Pap.: 195 (1971). —Lock, Leg. Afr. Check-list: 384 (1989) as "*tanganyikensis*". Type: Tanzania, Mpanda Dist., Karema, *Storms* s.n. (BR holotype).

Robust single-stemmed annual or short-lived perennial herb, 0.5–2 m tall, erect from a taproot, shortly branched above. Stem with a dense indumentum of brownish hairs, persisting and becoming paler on older parts. Leaves with (1)2–3(4) pairs of leaflets; petiole and rachis (if present) together 1–5 cm long; leaflets 2–6 × 0.5–1.5 cm, oblanceolate, cuneate at the base, rounded to more commonly emarginate at the apex, glabrous to shortly and sparsely pubescent above, densely covered with silvery appressed hairs beneath; lateral nerves closely set and very numerous; stipules 3–8 mm long, narrowly triangular. Flowers 6–12 in terminal sessile heads; pedicels short, up to 5 mm long in fruit; bracts resembling the stipules. Calyx 9–12 mm long, brown tomentose; lateral teeth 2–3 times as long as the campanulate part, narrowly triangular, the lower one longer. Standard c.12 mm long, orange- to apricot-yellow inside, brownish tomentose outside. Upper filament free or shortly attached to the sheath for 2–4 mm. Ovary brownish tomentose; style pubescent. Pods 4–6 × 0.7–1 cm, oblong, style base pointing slightly downwards, densely brown tomentose, the hairs darkest along the edges, becoming silvery on the sides. Seeds 6–8, lying transversely and separated from each other by inward projections of the endocarp, 4–5 × 3–3.5 mm, oblong-reniform, the hilum towards the narrower basal end of the seed and covered by a large conspicuous wedge-shaped aril.

Malawi. N: Chitipa Dist., Misuku Hills, fr. 8.vii.1973, *Pawek* 7132 (K, MAL, MO); Nkhata Bay Dist., North Viphya, 3 km NW of Chikwina, fl. 22.v.1970, *Brummitt* 11033 (K).

Also in Tanzania. Miombo woodland; 1150–1450 m.

For notes on similarity to *T. aurantiaca* (species 24) from the same area see under that species.

70. **Tephrosia chisumpae** Brummitt in Kew Bull. **35**: 465 (1980). Type: Angola, Cuito, Rio Sobe, in open grassy woodlands at Rio Sobe, 13.iii.1906, *Gossweiler* 2768 (BM holotype). FIGURE 3.3.**28**.

    Caulocarpus gossweileri Baker f., Legum. Trop. Africa: 170 (1926) non *Tephrosia gossweileri* Baker f. Type as above.

Erect much branched shrub to 1.5 m tall, with numerous stems from a woody rootstock, upper parts herbaceous. Stems slightly ribbed, densely pubescent, the hairs silvery and mostly appressed, but some longer spreading hairs. Leaves sessile, digitately (1)3–5-foliolate; leaflets 10–50 × 2–8 mm, linear-oblanceolate to oblanceolate, sometimes folded lengthwise, long-cuneate to the base, obtuse to rounded or emarginate at the apex, with a minute ± recurved mucro, inconspicuously puberulous above, densely pubescent to tomentose beneath with appressed silvery-white hairs; stipules 2–5 mm long, subulate to narrowly triangular, pubescent outside, brown. Flowers in axillary clusters, but leaves lacking at uppermost nodes so inflorescences continued as a short pseudoraceme; bracts similar to the stipules; pedicels 3–4(8 in fruit) mm long. Calyx 4–5 mm long, densely covered with short white appressed hairs; teeth broadly triangular, shorter than the tube. Petals 12–16 mm long; standard elliptic-obovate, yellow-green turning purple, with brown hairs on the back; wings purple; keel white. Upper filament separate or only lightly attached to the staminal sheath. Ovary tomentose; style pubescent. Pod 30–40 × 6–8 mm, oblong, with a stipe 3–4 mm long, densely pubescent,

Fig. 3.3.**28**. TEPHROSIA CHISUMPAE. 1, habit (× ²⁄₃); 2, flower (× 2); 3, standard (× 2); 4, wing (× 2); 5, keel (× 2); 6, stamens (× 2); 7, gynoecium (× 2); 8, pod (× 1 1/3), 1–8 from *Gossweiler* 2768; 9, seed (× 2), from *Fanshawe* 6968. Drawn by Heather Wood. From Kew Bulletin (1980).

somewhat glabrescent at maturity. Seeds 6–9, 4–4.5 × 3–3.5 mm, ovoid-oblong, ± transversely elongate, with a small hilum on the narrower end.

Zambia. B: 64 km S of Mongu on road to Senanga, fl.& imm.fr. 31.i.1975, *Brummitt, Chisumpa & Polhill* 14190 (BR, K, LISC, MO, NDO, PRE, SRGH); Mongu Dist., Lukolo plain near Namilia village, 35.1 km E of Mongu–Senanga road, fl.& fr. 18.111.1996, *Harder, Bingham, Zimba & Luwiika* 3739 (K, MO).

Also in Congo (Katanga) and S Angola. Woodland on Kalahari sands; c.1000 m.

20. REQUIENIA DC.

by R.K. Brummitt

Requienia DC. in Ann. Sci. Nat. (Paris) **4**: 91 (1825). —Brummitt in Kew Bull. **35**: 469–473 (1980).

Perennials, woody at base, with prostrate to erect branches up to 1.5 m high. Young stems variously densely pubescent. Leaves unifoliolate; petioles up to 5(6) mm long; leaflet from 7 × 7 mm, suborbicular, to obovate or obcordate, to 4(4.5) × 2(2.4) cm, elliptic, sometimes coriaceous with raised venation on upper surface, both surfaces pubescent or rarely glabrescent; stipules up to 3(4) mm long, linear-triangular, pubescent. Flowers in groups of up to c.6 in axils of leaves; pedicels up to 1(2) mm long. Calyx up to 5 mm long, pubescent, the teeth ± equal, subacute, equalling or a third to half as long as the tube, the upper two connate for a ¹/₂ to ²/₃ of their length. Petals purple or pinkish; standard up to 8 mm long, broadly spathulate to subpandurate, hairy outside; wings slightly shorter than the standard, the limb about equalling the very slender claw; keel petals with a very slender claw equalling that of the wings, the limb slightly shorter. Stamens uniterd into a tube. Ovary pubescent; style glabrous; pod 1-seeded, shortly falcate up to 10 × 4 mm, pubescent to villous.

A genus of three species, found in dry country, two in southern Africa and one from W Africa to the Sudan.

Calyx 4–5 mm long; leaflet ovate to oblong, broadest below the middle, mostly c.3 times as long as broad, with hairs ± obscuring the leaflet surface above and beneath; low shrub ·······················**1.** *pseudosphaerosperma*
Calyx 1.5–3 mm long; leaflet variably round to elliptic or elliptic-oblong to slightly oblanceolate, broadest about the middle or a little above, usually 1–2 times as long as broad, appressed-pubescent but hairs not obscuring the surface; woody herb, stems numerous from a branching rootstock, decumbent or ascending ·· ·····················**2.** *sphaerosperma*

1. **Requienia pseudosphaerosperma** (Schinz) Brummitt in Bol. Soc. Brot., sér.2, **41**: 220 (1968). —Schreiber in Merxmüller, Prodr. Fl. SW Afrika, fam. 60: 99 (1970). —Brummitt in Kew Bull. **35**: 492 (1980). Type: Namibia, "Kalahari", Uschi, *Fleck* 334 (Z holotype).

    Tephrosia pseudosphaerosperma Schinz in Vierteljahrsschr. Naturf. Ges. Zürich **57**: 557 (1912). —Harms in Engler, Pflanzenw. Afrikas **3**(1): 591 (1915). —Baker f., Legum. Trop. Africa: 215 (1926).

An undershrub or shrub up to 1.5 m high. Young stems shortly appressed- or spreading-tomentose, whitish. Leaves unifoliolate; petiole 1–4 mm long, pubescent; leaflet up to 5.5 × 1.3 cm, ovate to oblong, ± cuneate at the base, rounded at the apex, both surfaces about equally densely whitish or greyish appressed-pubescent to sublanate. Calyx 4–5 mm long, whitish appressed-pubescent to tomentose. Petals purple; standard 6–8 mm long; wing petals 6–7 mm long; keel petals 5–6 mm long. Pod c.10 × 4 mm, appressed- or spreading-tomentose or villous, whitish.

Botswana. N: between Odiakwe and Kanyu (Francistown–Maun road), fl.& fr. 9.iii.1965, *Wild & Drummond* 6830 (K, SRGH). SW: c.108 km N of Kang (Kan) on road to Ghanzi, fl. 19.ii.1960, *de Winter* 7390 (K, SRGH). SE: Mahalapye Dist., W Lephephe (Lephaphe), fl. 13.i.1958, *de Beer* 578 (SRGH). **Zimbabwe**. W: Hwange (Wankie) Nat. Park, Libuti Camp, c.128 km S of Main Camp, fl.& fr. 28.ii.1971, *Rushworth* 2541 (K, SRGH).

Known also from Namibia and South Africa (Limpopo Province). Wooded grassland on Kalahari sands; c.1000 m.

2. **Requienia sphaerosperma** DC. in Ann. Sci. Nat. (Paris) **4**: 91 (1825). —Harvey in F.C. **2**: 231 (1862). —Schreiber in Merxmüller, Prodr. Fl. SW Afrika, fam. 60: 100 (1970). —Brummitt in Kew Bull. **35**: 472, fig. 4 (1980). Types: South Africa, Northern Cape, Hay Divison, bewteen Asbestos Mtns and Witte Water, *Burchell* 1693 (G-DC holotype, K). FIGURE 3.3.**29**.

    Tephrosia sphaerosperma (DC.) Baker in F.T.A. **2**: 125 (1871). —Harms in Engler, Pflanzenw. Afrikas **3**(1): 590, fig.292 (1915). —Baker f., Legum. Trop. Africa: 215 (1926). —Wilman, Check List Griqualand West: 66 (1946). —Forbes in Bothalia **4**: 955 (1948). — Schreiber in Mitt. Bot. Staatssamml. München **16**: 299 (1957) excl. syn. *T. pseudosphaerosperma*. —Leistner in Koedoe **2**: 164 (1959).

    Tephrosia pseudosphaerosperma sensu Forbes in Bothalia **4**: 956 (1948).

Perennial herb with a ± woody rootstock with numerous decumbent or ascending branches up to 40 cm long. Young stem spreading- or sometimes appressed-tomentose, occasionally whitish. Leaves unifoliolate; petiole 2–5(6) mm long, pubescent; leaflet varying from c.7 × 7 cm and orbicular, to 4(4.5) × 2(2.4) cm and elliptic, rounded to cuneate at the base, obtuse to truncate at the apex, both surfaces closely appressed-pubescent to somewhat sparsely

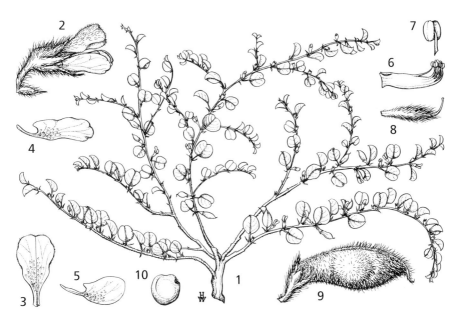

Fig. 3.3.**29**. REQUIENIA SPHAEROSPERMA. 1, habit (× ¹/₂), from *Dinter* 315 and *Leistner* 2022; 2, flower (× 3); 3, standard (× 3); 4, wing (× 3); 5, keel petal (× 3); 6, stamens (× 3); 7, anther (× 16); 8, gynoecium (× 3), 2–8 from *Leistner* 2022; 9, pod (× 3); 10, seed (× 3), 9 & 10 from *Dinter* 315. Drawn by Heather Wood. From Kew Bulletin (1980).

spreading-pubescent. Calyx (1.5)2–3 mm long, appressed-pubescent. Petals purple or pinkish; standard (4)5–6 mm long. Pod (6)7–9(10) × 2.5–3.5 mm, fairly densely spreading pubescent or sometimes appressed-pubescent or villous.

Caprivi Strip. Mashi, fl.& fr. 4.xi.1962, *Fanshawe* 7122 (K, SRGH). **Botswana**. N: Central Dist., near Odiakwe, N of Makgadikgadi Pans, fl.& fr. 9.iii.1965, *Wild & Drummond* 6824 (K, SRGH). SW: c.72 km N of Kang, fl.& fr. 18.ii.1960, *Wild* 5045 (K, PRE, SRGH). SE: Kweneng Dist., c.9.5 km S of Matlolakgang Ranch, fl.& fr. 25.iii.1977, *Hansen* 3093 (GAB, K, PRE, SRGH). **Zambia**. B: Sesheke Dist., Ngweze (Ngwezi), fl.& fr. 29.x.1962, *Fanshawe* 7101 (K, SRGH).

Also in Namibia and N South Africa (Northern Cape, North-West and Limpopo Provinces). Dry country, on loose sand or in grassland on Kalahari sand; 800–1100 m.

A very variable species, particularly in leaf shape and size; the two extremes look very different. Plants with long elliptic leaves were regarded as specifically distinct, and apparently mistakenly referred to *T. pseudosphaerosperma* by Forbes (1948), but all intermediates occur. Of specimens seen from the Flora area those from Zambia, the Caprivi Strip and SW Botswana tend to have small suborbicular leaves and those from SE Botswana have long elliptic leaves. However, over the whole range of the species there does not appear to be any close correlation between leaf shape and distribution and it does not seem possible to recognize infraspecific taxa.

21. PTYCHOLOBIUM Harms

by R.K. Brummitt

Ptycholobium Harms in Engler, Pflanzenw. Afrikas **3**(1): 591 (1915). —Brummitt in Kew Bull. **35**: 460–464 (1980).

Undershrubs up to 40 cm high. Young stems variously densely pubescent. Leaves digitately trifoliolate (i.e. the terminal one sessile, lacking a rachis) or sometimes unifoliolate, rarely with occasional leaves 5-foliolate with a short rachis; petiole short, up to 7(9) mm long; leaflets ± oblanceolate, usually glabrous above and appressed-pubescent beneath; stipules 1–4 mm long, linear-triangular. Flowers in groups of (1)2–4(7) in axils of foliage leaves; pedicels not exceeding 2 mm; bracteoles absent. Calyx 3–5 mm long, tubular to funnel-shaped with 5 subequal acute teeth slightly shorter than the tube, the two upper ones slightly connate at the base, pubescent. Petals up to 7 mm long, usually bluish or purplish; standard spathulate, oblong cuneate or subpandurate, gradually narrowed to the claw, hairy outside; wings with a very slender claw, the limb ± oblong; keel petals with a very slender claw equalling those of the wings, the limb shorter than the claw, $^1/_2$ to $^2/_3$ as long as the wing limb. Stamens united into a tube, the upper one easily detached. Ovary pubescent; style glabrous. Pod thin-walled, variously contorted, indehiscent, variously pubescent.

A genus of three species restricted to the drier parts of Africa and SW Asia. Very similar to *Tephrosia* but distinguished by the thin-walled, contorted, indehiscent pods.

1. Pods very conspicuously transversely plicate with 3–4 folds at each side, 4–6 mm broad · **1.** *plicatum*
– Pods variously contorted but not conspicuously plicate, if slightly plicate then 8–13 mm broad · 2
2. Pods (4)5–6 mm broad, coiled or spiralled, rather conspicuously greyish spreading-pubescent or villous · **2.** *contortum*
– Pods 8–13 mm broad, undulate or coiled, finely appressed-pubescent or rarely rather shortly spreading-pubescent · **3.** *biflorum*

1. **Ptycholobium plicatum** (Oliv.) Harms in Engler, Pflanzenw. Afrikas **3**(1): 591 (1915). —Baker f., Legum. Trop. Africa: 216 (1929). —Brummitt in Kew Bull. **35**: 462, fig.1L (1980). Type: South Africa, Mpumalanga, Eland R., *Rehmann* 4922 (K lectotype, selected by Brummitt 1980).

> *Tephrosia plicata* Oliv. in Hooker's Icon. Pl. **15**: 36, t.1445 (1883). —Bews, Fl. Natal Zululand: 111 (1921). —Burtt Davy, Fl. Pl. Ferns Transvaal, pt.2: 378 (1932). —Forbes in Bothalia **4**: 959 (1948).

Subsp. **plicatum** —Brummitt in Kew Bull. **35**: 462, fig.1L (1980). FIGURE 3.3.**30**/11.

An undershrub up to 30 cm high, erect or ascending. Young stem densely spreading-pubescent, not conspicuously whitish. Leaves digitately trifoliolate or sometimes occasional leaves unifoliolate; petioles 2–5 mm long, appressed- or ascending-pubescent; leaflets 15–38 × 3–7(9) mm, oblanceolate to narrowly oblanceolate, the upper surface glabrous, the lower surface closely appressed to rarely shortly spreading-pubescent. Calyx 3–4 mm long, appressed-pubescent. Petal colour – see note below; standard 4–5 mm long, spathulate or subpandurate. Pods conspicuously deeply plicate with 3–4 folds showing on each side, 8–12 × 4–6 mm, rather densely whitish to brownish spreading-pubescent.

Zimbabwe. W: Umguza Dist., Nyamandhlovu Pasture Res. Station, fl.& fr. 25.i.1963, *Denny* in NRSH 407 (SRGH). **Mozambique**. M: between Umbelúzi and Namaacha, fl.& fr. 16.x.1940, *Torre* 1802 (LISC).

Mainly in South Africa (Limpopo, Mpumalanga and KwaZulu-Natal Provinces), known in the Flora Zambesiaca area from only a few collections from S Mozambique and W Zimbabwe. River banks and disturbed places in *Acacia* woodland; 50–600 m.

Petal colour is noted on only two specimens seen – *Thorncroft* 111 from South Africa, which is said to be pink, and *Corby* 1550, which is said to be ?yellow. Further information is required.

Subsp. *arabicum* Brummitt, which is much less hairy and with less plicate curved pods, occurs in Yemen and Oman.

2. **Ptycholobium contortum** (N.E. Br.) Brummitt in Regnum Veg. **40**: 24 (1965). — Brummitt in Kew Bull. **35**: 462, fig.1A–J (1980). Type: Botswana, Kwebe, *E.J. Lugard* 132 (K holotype). FIGURE 3.3.**30**/1–9.

> *Tephrosia contorta* N.E. Br. in Bull. Misc. Inform., Kew **1909**: 103 (1909). —Forbes in Bothalia **4**: 960 (1948).
> *Sylitra contorta* (N.E. Br.) Baker f., Legum. Trop. Africa: 168 (1926).

An undershrub up to 40 cm high, erect or ascending. Young stem whitish appressed-pubescent to spreading-tomentose. Leaves digitately trifoliolate or sometimes occasional leaves unifoliolate; petiole 1–4(6) mm long, ± appressed-pubescent; leaflets (12)15–40 × (2)3–7(9) mm, oblanceolate to narrowly oblanceolate, the upper surface glabrous, the lower surface appressed- to sometimes almost spreading-pubescent. Flowers (1)2–4(6) per axil. Calyx 3–5 mm long, appressed- to spreading-pubescent. Petals blue or purple or perhaps rarely yellow; standard 5–7 mm long, spathulate or oblong-cuneate. Pods strongly coiled or spiralled, ± longitudinally channelled, (4)5–6 mm broad, densely greyish spreading-pubescent or villous.

Botswana. N: Central Dist., 16 km NE of Thabatshukudu, fl.& fr. 26.iv.1957, *Drummond & Seagrief* 5237 (K, SRGH). SW: Ghanzi Dist., 10 km SW of Kuke, fl.& fr. 7.iii.1987, *Long & Rae* 140 (K). SE: Kweneng Dist., Molepolole, Letlhakeng Valley, fl.& fr. 15.ii.1960, *Yalala* 7 (K, SRGH). **Zimbabwe**. S: Gwanda Dist., near junction of Mzingwane (Umzingwane) and Limpopo rivers, fl.& fr. i.1956, *Rattray* 1718 (K, SRGH). E: Birchenough Bridge, Save R., fl.& fr. i.1938, *Obermeyer* 2455 (K, SRGH).

Also in South Africa (Limpopo Province). In sandy grassland, river alluvium, mopane and *Acacia-Terminalia* woodland.

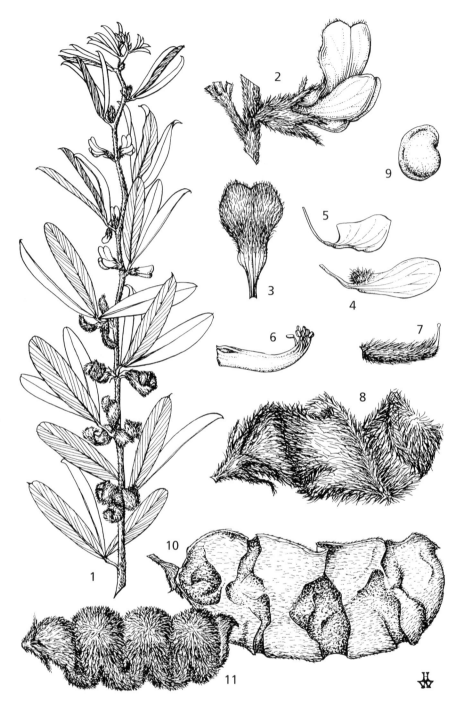

Fig. 3.3.**30**. PTYCHOLOBIUM CONTORTUM. 1, flowering and fruiting branch (× ²/₃); 2, flower (× 4); 3, standard, outer side (× 4); 4, wing (× 4); 5, keel petal (× 4); 6, stamens (× 4); 7, gynoecium (× 4); 8, pod (× 4); 9, seed (× 6), 1–9 from *Acocks* 16796. PTYCHOLOBIUM BIFLORUM subsp. BIFLORUM. 10, pod (× 4), from *Pearson* 4404. PTYCHOLOBIUM PLICATUM subsp. PLICATUM. 11, pod (× 4), from *Rehmann* 4922. Drawn by Heather Wood. From Kew Bulletin (1980).

3. **Ptycholobium biflorum** (E. Mey.) Brummitt in Regnum Veg. **40**: 24 (1965); in Kew
    Bull. **35**: 462, fig.1K (1980). Type: South Africa, Northern Cape, Bitterwater,
    Gamka R., *Drège* s.n. (K).

> *Sylitra biflora* E. Mey., Comment. Pl. Afr. Austr.: 114 (1836). —Harvey, Thes. Cap. **1**: 50,
> t.78 (1859); in F.C. **2**: 224 (1862). —Harms in Engler, Pflanzenw. Afrikas **3**(1): 585 excl.
> fig. (1915). —Baker f., Legum. Trop. Africa: 168 (1926). —Hutchinson, Gen. Fl. Pl. **1**:
> 397 (1964).

An undershrub up to 40 cm high, erect or ascending or rarely ± prostrate. Young stem
appressed- white pubescent, sometimes subsericeous. Leaves unifoliolate to digitately
trifoliolate, or rarely with occasional leaves 5-foliolate; petiole 1–7(9) mm long, appressed-
pubescent; leaflets 20–50(60) × 2–6(9) mm, oblanceolate to linear-elliptic, the upper surface
glabrous or rarely very sparsely or minutely appressed-pubescent, the lower surface shortly
and regularly closely appressed-pubescent. Calyx 2.5–4 mm long, appressed-pubescent.
Petals bluish or purple, or perhaps rarely white; standard 4–7 mm long, spathulate or
subpandurate. Pods 20–36 × 8–13 mm, somewhat shallowly plicate, flat or coiled (see
subspecies), finely and inconspicuously appressed-pubescent or rarely very shortly
spreading-pubescent.

Subsp. **biflorum** —Brummitt in Regnum Veg. **40**: 24 (1965). —Schreiber in
    Merxmüller, Prodr. Fl. SW Afrika, fam. 60: 99 (1970). —Brummitt in Kew Bull.
    35: 462, fig.1K (1980). FIGURE 3.3.**30**/10.

Stems generally more rigid than in subsp. *angolense*, often conspicuously zigzag, the lower
parts often forming an intertwining defoliate network. Leaves unifoliolate or an occasional leaf
trifoliolate; petiole 1–5(7) mm long. Pods with the walls usually shallowly transversely plicate
but the margins flat, the whole not or only slightly folded longitudinally, often not or only
weakly coiled.

Botswana. SW: Mabuasehube (Mobua Shobea) Pan, c.117 km S of Tshane, fl.& fr.
24.ii.1960, *Wild* 5139 (K, SRGH).

Also in S Namibia, South Africa (Northern Cape Province) and in Somalia (see
Thulin in Nordic J. Bot. **8**: 465, 1989). Dry country calcicole, often on beds or banks
of seasonal rivers.

Subsp. **angolense** (Baker) Brummitt in Regnum Veg. **40**: 24 (1965). —Schreiber in
    Merxmüller, Prodr. Fl. SW Afrika, fam. 60: 98 (1970). —Brummitt in Kew Bull.
    35: 462 (1980). Type: Angola, Mossamedes, Benguela, near Maiombo, *Welwitsch*
    4124 (BM holotype).

> *Sylitra angolensis* Baker in F.T.A. **2**: 103 (1871). —Harms in Engler, Pflanzenw. Afrikas
> **3**(1): 586, fig.290 (1915). Baker f., Legum. Trop. Africa: 168 (1926). —Bremekamp &
> Obermeyer in Ann. Transv. Mus. **16**: 419 (1935). —Torre in C.F.A. **3**: 141 (1962).

Stems more herbaceous than in subsp. *biflorum*, not conspicuously zigzag nor forming an
intertwining network. Leaves trifoliolate, or occasional leaves unifoliolate or rarely 5-foliolate;
petiole (1)3–7(9) mm long. Pods folded longitudinally so that the two sutures lie adjacent to
each other, ± strongly coiled backwards, and also slightly transversely plicate.

Botswana. SW: Ghanzi Dist., Damara Pan, 20.iv.1930, *van Son* in *Tvl. Mus.* 28905 (K).
Also in N Namibia and S Angola. Habitat probably similar to that of subsp. *biflorum*.
The only other known collection of this subspecies from Botswana – c.35 km S
of Takatshwaane Pan, fl.& fr. 21.ii.1960, *Wild* 5108 (K, SRGH) – is very conspicuously
spreading-pubescent on stems and pods and at first sight has more the aspect of
P. contortum.

Tribe 5. **ABREAE**

by D.K. Harder

Abreae (Endl.) Hutch., Gen. Fl. Pl. **1**: 451 (1964).

Woody subshrubs or lianes. Leaves paripinnate; leaflets numerous, opposite; stipels present, minute. Flowers in pseudoracemes or rarely sessile in the axils; bracts and bracteoles small. Calyx shortly toothed to subtruncate, the two upper lobes connate higher than the others. Standard shortly clawed, with small inflexed auricles; keel petals shallowly falcate, joined along the lower side, interlocked with the wings. Stamens 9, joined into a sheath and split at the apex; anthers uniform or 4 slightly smaller. Style short, incurved. Pods mostly compressed, ± septate, dehiscent. Seeds subglobose or compressed-ellipsoid; hilum short; radicle incurved.

A monotypic tribe.

22. **ABRUS** Adans.

Abrus Adans., Fam. Pl. 2: 327 (1763). —Breteler in Blumea **10**: 607 (1960). —Verdcourt in Kew Bull. **24**: 235–253 (1970).
Hoepfneria Vatke in Oesterr. Bot. Z. **29**: 222 (1879).

Stipules small, usually persistent; stipels filiform; rachis projecting beyond the last pair of leaflets. Inflorescence axillary or terminal, the flowers in fascicles on short reduced branchlets, often arranged unilaterally on the rachis, or rarely sessile in the axils; bracts and bracteoles present. Flowers small, white, yellow, pink or dark purple. Standard ovate to round, with a short broad claw, glabrous. Upper stamen absent. Ovary subsessile, many-ovuled; style not bearded; stigma capitate. Pods linear or oblong, subturgid or compressed, or ± septate. Seeds subglobose or ellipsoid and compressed, sometimes bright red and black; rim aril sometimes developed, usually minute or absent.

A pantropical genus of 14 species, 4 of these divided into 2–5 subspecies giving a total of 22 taxa (Berhaut in Adansonia **5**: 359, 1965; Verdcourt in Kew Bull. **24**: 235, 1970). Breteler in Blumea **10**: 607 (1960) recognized only four taxa. For consistency and accuracy this treatment closely follows the revision by Verdcourt (1970) and in F.T.E.A. (1971). Four species occur in the Flora Zambesiaca region. Since all of the taxa considered for the present treatment have already been covered previously, the present author simply verified the accuracy of this work. During preparation of this treatment, an earlier name for *Abrus pulchellus* Thwaites was uncovered; the new combinations for *Abrus melanospermus* (Hassk.) D.K. Harder and its typification are presented here.

1. Seeds almost always red and black (rarely light brown to yellowish to completely black), shiny, spherical; inflorescence rachis less than half as long as the peduncle, thickening (1.5–3 mm in diameter from early fruiting stage) and generally curved; bracts and bracteoles small, much shorter than calyx; vegetative parts of plant sparsely pubescent; fruit densely pubescent to velutinous ··· **1.** *precatorius*
   – Seeds never black and red, laterally compressed; inflorescence rachis very short to longer than the peduncle, not markedly thickening (generally less than 1 mm in diameter), mostly ± straight ······································· 2
2. Bracts and bracteoles as long as calyx or exceeding it; corolla deep red or purple ··· **2.** *canescens*
   – Bracts and bracteoles much shorter than the calyx; corolla cream to pale purple or pink ·· 3

3. Pods woody, tapering at the apex, conspicuously tuberculate; leaflets mostly elliptic-oblong, sometimes oblong (subsp. *oblongus*); inflorescence rachis elongate, much longer than the short peduncle; shrub · · · · · · · · **3.** *schimperi*
− Pods thinner, rounded at the apex, not tuberculate; leaflets mostly oblong or rectangular, but often a little broader near the apex; inflorescence rachis (excluding sterile bracts below flowers) shorter to a little longer than the peduncle and not much elongated; subshrub or liane · · · · · · **4.** *melanospermus*

1. **Abrus precatorius** L., Syst. Nat., ed.12, **2**: 472 (1767). —Breteler in Blumea **10**: 617, fig.5 (1960). —Verdcourt in Kew Bull. **24**: 240 (1970). Type: Sri Lanka, *Hermann* Vol. 2: 6 (BM lectotype). FIGURE 3.3.**31**.

Subsp. **africanus** Verdc. in Mitt. Bot. Staatssamml. München **7**: 328 (Mar. 1970); in Kew Bull. **24**: 241 (Apr. 1970). —Schreiber in Merxmüller, Prodr. Fl. SW Afrika, fam. 60: 11 (1970). —Verdcourt in F.T.E.A., Legum., Pap.: 114 (1971). Type: Kenya, Tana River Dist., 48 km S of Garsen, Kurawa, 7.x.1961, *Polhill & Paulo* 628 (K holotype, EA).

    Abrus tunguensis Pires de Lima in Brotéria, sér.Bot. **19**: 127 (1921). Types: Mozambique, vicinity of Palma, *Pires de Lima* 94 & 134 (PO syntypes).

    Abrus precatorius as used by many incl. Klotzsch in Peters, Naturw. Reise Mossamb. **6**(1): 29 (1861). —Baker in F.T.A. **2**: 175 (1871). —Harms in Engler, Pflanzenw. Afrikas **3**(1): 648 (1915). —Eyles in Trans. Roy. Soc. S. Africa **5**: 380 (1916). —Baker f., Legum. Trop. Africa: 351 (1929). —Miller in J. S. Afr. Bot. **18**: 17 (1952). —Boutique in F.C.B. **6**: 85 (1954). —Hepper in F.W.T.A., ed.2, **1**: 574 (1958) non L. sensu stricto. —White, F.F.N.R.: 138 (1962). —Mogg in Macnae & Kalk, Nat. Hist. Inhaca Is.: 146 (1969) non L. sensu stricto.

Woody twining climber, 1–4.5 m tall, sometimes procumbent or supported on low shrubs. Stems greenish to greenish-brown, often more than 1.5 cm in diameter, sometimes with warty excrescences, sparsely appressed pubescent to somewhat spreading pubescent or glabrous. Leaves 16–36-foliolate; petiole 0.5–1.8 cm long; leaflets deciduous, 5–27 × 2–10 mm, oblong, obovate-oblong or ovate, obtuse to acuminate at the apex, rounded or subcordate at the base, glabrous to glabrescent on upper surface, sparsely appressed-pubescent or puberulous beneath; stipules caducous, 3–5 mm long. Inflorescence axis 2–7 cm long, thickening and robust in fruit; peduncles 1.5–6 cm long; flowers subsessile in dense fascicles; bracts and bracteoles much shorter than calyx, 0.5–1 mm long. Calyx 3 mm long, sparsely puberulous, denticulate. Corolla yellow, white, mauve, pale purple or pink, 9–15 mm long. Infructescence often falcate. Pods 2–4(5) × 1–1.5 cm, oblong, somewhat swollen, with a hooked reflexed beak, covered with dense short rust-coloured appressed hairs, densely covered with tubercles forming obliquely transverse ridges. Seeds (1)3–7, scarlet with black area around the hilum or very rarely entirely black or yellowish, 4–7 × 4–5 mm, ovoid, shining, usually remaining attached to the edges of the open valves for some time.

Caprivi Strip. Singalamwe, c.1100 m, fr. 1.i.1959, *Killick & Leistner* 3236 (K). **Botswana**. N: Chobe Dist., Chobe Nat. Park, Gubatsa Hills near Savute R., fr. 24.x.1972, *Pope, Biegel & Russell* 849 (K). **Zambia**. B: Sesheke Dist., Masese, *Acacia–Terminalia* thickets on Kalahari sand, fr. 8.ix.1969, *Mutimushi* 3584 (K). N: Mbala Dist., Nkamba (Kamba) Bay, L. Tanganyika, 15.iv.1957, *Richards* 9200 (K, UZL). C: Luangwa Valley, South Luangwa Nat. Park, Mfuwe, c.670 m, fr. 17.vii.1969, *Astle* 5709 (K, UZL). E: Petauke (Old Boma), c.700 m, fl.& imm.fr. 15.xii.1958, *Robson* 954 (BM, K). **Zimbabwe**. N: Gokwe Dist., Sasame (Sesami) R., fr. 3.iv.1964, *Bingham* 1217 (K). W: Hwange Dist., Victoria Falls, Kandahar Is., fr. iv.1918, *Eyles* 1314 (BM, K). E: Nyanga Dist., Nyamaropa Communal Land (T.T.L.), Regina Coeli Mission, 1030 m, fr. 17.i.1967, *Biegel* 1768 (K, MO). S: Chiredzi Dist., Gonarezhou Nat. Park, W bank of Save R. at Save-Runde (Sabi-Lundi) R. junction, fr. 31.v.1971, *Ngoni* 143

(K). **Malawi**. N: Karonga Dist., Vinthukutu, c.3 km N of Chilumba, c.530 m, fr. 17.ix.1977, *Pawek* 13107 (K, MAL, MO, SRGH, UC). C: Kasungu Dist., Kasungu Nat. Park (Game Reserve), near Lifupa Camp, 1020 m, fr. 23.vi.1970, *Brummitt* 11649 (K). S: Machinga Dist., Chiuta Lake, near Namigongo harbour, fr. 9.vii.1983, *Seyani &* *Tawakali* 1138 (K, MO). **Mozambique**. N: Marrupa Dist., margin of Rio Messenguecé

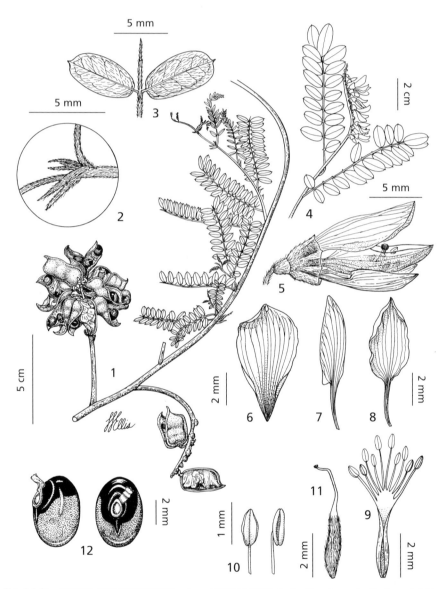

Fig. 3.3.**31**. ABRUS PRECATORIUS subsp. AFRICANUS. 1, fruiting branch; 2, stem detail with stipules; 3, leaf apex with adaxial leaflets. 1–3 from *Richards* 25908. 4, flowering branchlet; 5, flower, parts partly exposed; 6, standard; 7, wing; 8, keel petal; 9, stamens, spread out; 10, anther, front and side; 11, gynoecium; 12, seed, front and side. 4–12 from Frontier Tanzania CFRP 39. Drawn by Linda Ellis.

(Massanguezi), 18.ii.1981, *Nuvunga* 614 (MO). Z: Mocuba Dist., between Mocuba and Mucharro, 1.2 km from Macuia market, fr. 17.v.1949, *Barbosa & Carvalho* 2696 (K). T: Magoe Dist., Msusa, Zambezi R., 100 m, 27.vii.1950, *Chase* 2828 (K, MO). MS: Sussundenga Dist., Dombe, c.2 km ENE of Mission, 22.xi.1965, *Pereira & Marques* 820 (E, MO). GI: Massinga Dist., Pomene, 22.ix.1980, *de Koning, Jansen & Zunguze* 8464 (MO). M: Maputo Dist., Inhaca Is., near Marine Biology Station towards Ponta Rasa, fr. 13.xii.1984, *Groenendijk & Dungo* 1577 (K, MO).

Also distributed and naturalized in many regions of tropical Africa, South Africa, Seychelles, Mauritius, Madagascar, and introduced into N, S and C America, the Caribbean, Hawaii, Society Islands, Australia, Sri Lanka, China, India, Singapore, Philippines, Taiwan and Thailand. Widespread but not common, in riverine and lake shore vegetation often in thickets, in coastal littoral thicket vegetation and thickets on Kalahari Sand; also in mixed deciduous woodland; sea level to 1200 m.

Commonly called "Lucky bean". The leaves have a sweet taste and are eaten as a salad. The seeds are extremely poisonous, but only harmful if eaten when the seed coat is broken. The seeds are widely used for decoration.

2. **Abrus canescens** Baker in F.T.A. **2**: 175 (1871). —Baker f., Legum. Trop. Africa: 351 (1929). —Boutique in F.C.B. **6**: 83, fig.1A, B (1954). —Hepper in F.W.T.A., ed.2, **1**: 575 (1958). —Torre in C.F.A. **3**: 239 (1966). —Verdcourt in Kew Bull. **24**: 251 (1970); in F.T.E.A., Legum., Pap.: 117 (1971). Type: Angola, Cuanza Norte, Pungo Andongo, *Welwitsch* 2250 (LISU holotype, BM, K).

Twining woody climber or subshrub 0.5–3 m tall from a woody rootstock. Stems pubescent with spreading or appressed rust-coloured hairs, later glabrescent. Leaves 18–28-foliolate; petiole short, 0.1–2 mm long; leaflets (0.6)1–2(3) cm × 2–7 mm, narrowly oblong or elliptic-oblong, rounded and mucronulate or emarginate at the apex, rounded or subcordate at the base, drying to dull brown on upper surface, dull grey beneath, densely pubescent, more so beneath, with somewhat long, curved, ± appressed, greyish hairs; stipules 2–3 × 0.5–1 mm at base, lanceolate, caducous. Inflorescences terminal and axillary; flowers sessile in usually well-separated verticillate fascicles; peduncle 3–8 cm long; bracts and bracteoles equal to or exceeding the calyx in length, 3–6 mm long. Calyx pubescent, 2–3 mm long, denticulate. Corolla dark red or purple, 8–15 mm long. Pods (2.4)4–5.5(6) × 0.8–1.2 cm, oblong or broadly linear, pubescent, occasionally with some glands. Seeds dark brown, grey-brown or black, glossy, 4–5.5 × 3–4 mm, compressed-ellipsoid; rim aril cream, ellipsoid with a narrow cartilaginous funicle remnant at one end.

Zambia. N: Kaputa Dist., Kundabwika Falls on the Kalungwishi R., c.90 km W of Mporokoso, 29°19'E, 9°12'S, c.1000 m, fl.& imm.fr. 17.iv.1989, *Goyder, Pope & Radcliffe-Smith* 3071 (K, UZL).

A widespread species in tropical Africa but rare in the Flora area. Elsewhere found in Nigeria, Cameroon, Central African Republic, Ethiopia, Uganda, Burundi, Tanzania, Congo, Congo (Brazzaville), Gabon and Angola. In riverine and dambo vegetation, often in wet or marshy areas.

Easily recognized by the long bracteole length, dark flower colour and the well developed rim aril.

3. **Abrus schimperi** Baker in F.T.A. **2**: 175 (1871). —Baker f., Legum. Trop. Africa: 352 (1929). —Verdcourt in Kew Bull. **24**: 242 (1970); in F.T.E.A., Legum., Pap.: 115 (1971). Type: Ethiopia, Tigray, near Djeladjeranne, *Schimper* 1552 (K holotype, BM, BR, FI, G, L, M, P, W).

Woody shrub 0.9–3.6 m tall, virgately branched. Stems appressed-pubescent, later glabrous; bark smooth, red to red-brown, lenticels light beige-grey. Leaves 14–34-foliolate; petioles

0–1(2.8) cm long; leaflets 5–33 × 4.5–15 mm, oblong-elliptic, distinctly rounded at the sides, or oblong with parallel sides in one subspecies, dimensions increase distally, rounded or mucronulate at the apex, rounded to subcuneate at the base, glabrous on upper surface, finely appressed puberulous beneath; stipules 3–6 mm long, subulate. Inflorescences terminal and in upper axils, 13–30 cm long; peduncle very short; flowers subsessile in numerous fascicles on short shoots; pedicels 2.5 mm long; bracts and bracteoles very short. Calyx 3 mm long, obconic, truncate or undulate, appressed-pubescent, with hairs ± oriented in one direction. Corolla cream, yellow, pale purple, mauve, blue to light violet, wings usually darker than the standard, 8–14 mm long. Pods 5.4–7(7.9) × 0.7–1.3 cm, oblong to oblanceolate, the valves somewhat thick and woody, appressed-pubescent sometimes velutinous and usually markedly tuberculate. Seeds brown with darker brown mottling especially near hilum, dull, arranged in the pod with the longest dimension at right angles to the pod's long axis, 5–7.5 × 5–6 × 2–3 mm, rounded-oblong or rhombic, compressed; hilum small; rim aril slightly developed with cartilaginous funicle remnant present.

Subsp. **africanus** (Vatke) Verdc. in Kew Bull. **24**: 243 (1970); in F.T.E.A., Legum., Pap.: 115 (1971). Type: Kenya, Ukamba, Kitui, *Hildebrandt* 2797 (B† holotype, BM, K).

    Hoepfnera africana Vatke in Oesterr. Bot. Z. **29**: 222 (1879).

    Abrus schimperi sensu auctt. incl. Taubert in Engler & Prantl, Pflanzenfam. III, **3**: 355 (1894). —Baker f., Legum. Trop. Africa: 352 (1929) in part. —Brenan, Check-list For. Trees Shrubs Tang. Terr.: 404 (1949) non Baker sensu stricto.

Leaflets mostly elliptic-oblong, 7–9 on each side of the rachis. Inflorescence 13–26 cm long and often well exceeding the length of subtending leaves. Tubercles on pod well developed and conspicuous.

Zambia. C: Lusaka Dist., above Chinyunyu Hot Springs, 85 km from Lusaka towards Petauke, fl.& fr. 12.ii.1975, *Brummitt & Lewis* 14326 (K). E: Chipata Dist., Machinje Hills, map ref. UA9428, fr. 15.v.1965, *B.L. Mitchell* 2954 (K). S: Choma Dist., between Mutama R. and Mapanza on road to Pemba, fr. 14.i.1964, *van Rensburg* 2738 (K). **Zimbabwe**. N: Hurungwe Dist., Urungwe Safari Area, 10 km from Makuti on Kariba road, banks of R. Nyakasanga, 980 m, fl. 18.ii.1981, *Philcox, Leppard & Dini* 8701 (K, MO, SRGH). W: Nkayi Dist., Shangani, Mbahze Dam, c.1000 m, fr. x.1952, *R.M. Davies* in *GHS* 41356 (MO, SRGH) C: Kwekwe Dist., c.11 km S of Kwekwe (Que Que), c.1260 m, fr. 11.v.1966, *Biegel* 1182 (K, MO, SRGH). **Malawi**. N: Rumphi Dist., 16 km NW of Rumphi on M1, *Combretum* thicket, 1250 m, fl. 11.iii.1978, *Pawek* 14035 (K, MA, MO, SRGH, UC). C: Lilongwe Dist., Lilongwe Nature Sanctuary, Zone B, 5.ii.1985, *Patel & Banda* 2071 (MAL). **Mozambique**. T: Cahora Bassa Dist., Marueira towards Songo, near Marueira, fl.& fr. 5.ii.1972, *Macêdo* 4781 (K, LMA, MO).

Also in Kenya and Tanzania. On rocky hillsides, riverine vegetation, deciduous thicket, open savanna woodland, and in *Commiphora*–mopane savanna; copper-tolerant and frost sensitive; 300–1260 m.

Stems used for tooth brushes and fish traps (*Davies* 406).

Subsp. **oblongus** Verdc. in Kew Bull. **24**: 244 (1970). Type: Malawi, Rumphi Dist., near Njakwa, on rocky banks of the South Rukuru R, 30.iv.1952, *White* 2539 (K holotype, FHO).

Leaflets oblong, 10–17 on each side of the rachis. Inflorescence shorter, 2–8 cm in length, not exceeding the length of subtending leaves by more than one-half. Tubercles on pod less well developed and inconspicuous.

Zimbabwe. N: Shamva Dist., Mazowe R., Golden Star, Shamva, fl.& fr. 29.xii.1963, *Masterson* 298 (K). E: Mutare (Umtali), above customs border station, commonage, c.1080 m, fl. 7.i.1954, *Chase* 5178 (BM, K, MO, SRGH). S: Mwenezi Dist., Malangwe R., SW of Mateke Hills, stream banks, c.670 m, fr. 5.v.1958, *Drummond* 5570 (K, SRGH). **Malawi**. N: Mzimba Dist., Njakwa (N'Jacqua), fl.& fr. 27.iii.1976, *Phillips* 1552 (K, MAL, MO). **Mozambique**. MS: Manica Dist., on road to Manica (Macequece) in wooded gully, c.1100 m, 22.xi.1951, *Chase* 4182 (BM) – see note below.

Only known from the Flora Zambesiaca area. Stony river banks, hill slopes and escarpments, in riverine thickets and *Euphorbia–Brachystegia* scrub, often scrambling over surrounding vegetation; 600–1250 m.

There is some confusion over the origin of *Chase* 4182. Specimens at K and MO indicate that the specimen was collected in Mutare (Umtali) commonage in Zimbabwe, while the duplicate at BM indicates an origin in Mozambique. However, these localities are close together and the taxon would be expected to occur in adjacent Mozambique.

Subsp. *schimperi* is 20–32-foliolate with golden indumentum on the inflorescence axes and calyx. It occurs in the Central Africa Republic, Sudan, Ethiopia and N Uganda.

4. **Abrus melanospermus** Hassk., Cat. Hort. Bot. Bogor: 282 (1844). —Harder in Novon **10**: 124 (2000). Type from Java.

> *Abrus pulchellus* Thwaites, Enum. Pl. Zeyl.: 91 (1859). —Taubert in Engler & Prantl, Pflanzenfam. III, **3**: 355 (1894). —White, F.F.N.R.: 138 (1962). —Verdcourt in Kew Bull. **24**: 246 (1970); in F.T.E.A., Legum., Pap.: 115 (1971). Type: Sri Lanka, Belanger, *Thwaites* 1467 (G, K, P).

Subshrub to liane prostrate to twining ascending, with woody stems up to 6 m long. Rootstock usually woody, fibrous or rhizomatous. Stems glabrous to densely spreading-pubescent to appressed-pubescent, the indumentum greyish or rust-coloured. Leaves (10)12–38-foliolate; petiole short to more than 2 cm long; leaflets 0.35–5 × 0.1–2 cm, oblong to oblong-obovate, truncate or broadly rounded and with a short stiff point at the apex, truncate to rounded or subcordate at the base, glabrous to densely silky pilose on upper surface, glabrescent to densely silky pilose beneath; stipels minute, c.1 mm long; stipules 2–4 mm long, subulate, caducous. Inflorescences axillary or terminal, 2.5–24 cm long; flowers subsessile in fascicles on cushion-like short shoots; bracts and bracteoles minute, under 1 mm long. Calyx 1.5–3 mm long, subtruncate, shallowly lobed or distinctly toothed, appressed-pubescent, indumentum rust-coloured to silvery. Corolla white, cream, yellow, orange, carmine-red, to purple, usually pale pink, drying yellowish, 8–15 mm long. Pods 2.2–7(9) × 0.8–1.5 cm, oblong or linear-oblong, compressed, smooth, sparsely to densely appressed-pubescent, the indumentum greyish or rust-coloured. Seeds greyish- or deep red-brown, sometimes speckled with violet or darker brown, 4–6 × (2)3–4.5 × 2–2.5 mm, oblong or pyriform, usually strongly compressed, arranged with longest dimension across the pod; rim aril minute with a conspicuous remnant cartilaginous funicle.

Across its range of distribution *A. melanospermus* is highly variable in leaflet number and leaflet size. Five subspecies occur – three from Asia and two African. Typical plants of *A. melanospermus* occur only in India, China and the Malesian Islands. Two subspecies are found in the Flora area: subsp. s*uffruticosus* has consistent and distinctive characters of low habit, small, very hairy leaves, small pods, and often distinctly lobed calyx from the centre of its distributional range in East Africa. In the western parts of its range, the characters become more similar to subsp. *tenuiflorus*. Intermediate specimens have been collected in West Africa from Sierra Leone, Mali and Guinea. In addition to the localities cited below, the less widespread subsp. *suffruticosus* occurs in Sierra Leone, N Nigeria, Central African Republic, Congo, Tanzania and Angola. Subsp. *tenuiflorus* has been collected in most parts of tropical, humid West Africa including Senegal, Guinea, Sierra Leone, Liberia, Ghana, Nigeria, Cameroon, Equatorial Guinea, Congo, Sudan, Uganda, Tanzania and Angola.

Subsp. **tenuiflorus** (Benth.) D.K. Harder in Novon **10**: 124 (2000). Type: Brazil, Santarem, *Spruce* 786 (K holotype, BM, C, G, NY).

> *Abrus tenuiflorus* Benth. in Martius, Fl. Bras. **15**(1): 216 (1859).
> *Abrus pulchellus* as used by many incl. Baker in F.T.A. **2**: 175 (1871). —Boutique in F.C.B. **6**: 84 (1954). —Hepper in F.W.T.A., ed.2, **1**: 574 (1958) non Thwaites sensu stricto.
> *Abrus pulchellus* subsp. *tenuiflorus* Verdc. in Kew Bull. **24**: 250 (1970); in F.T.E.A., Legum., Pap.: 116 (1971).

Climbing woody vine 1.8–6 m tall. Leaflets 6–13 on each side of the rachis, 1.2–4.5 × 0.7–1.8 cm, the leaflets generally increase proportionally in size distally along the rachis. Indumentum on vegetative portions usually sparse, more appressed. Pods 2.2–4.5(6) × 0.8–1.3 cm.

Zambia. N: Mansa (Fort Rosebery), streamside thicket, fl. 3.iii.1964, *Fanshawe* 8511 (K, NDO). W: Mwinilunga Dist., Kalene Hill, mushitu, fr. 17.v.1969, *Mutimushi* 3204 (K). **Mozambique**. N: vicinity of Palma, fl. 8.v.1917, *Pires de Lima* 259 (K photo, PO). MS: Muanza Dist., Cheringoma Plateau, upper Chinizíua R. catchment, *Brachystegia* thicket, fl. vii.1972, *Tinley* 2650 (K, MO, P, SRGH). M: Maputo, Catembe, along path to R. Tembe, c.6 km from Catembe, fr. 6.v.1981, *de Koning & Boane* 8684 (K).

Also in Senegal, Guinea, Sierra Leone, Liberia, Ghana, Nigeria, Cameroon, Equatorial Guinea, Sudan, Uganda, Tanzania, Congo and Angola; and in Bolivia, Brazil, Venezuela and New Caledonia. In streamside thickets, mushitu fringes and in coastal bush; 0–1500 m.

Subsp. **suffruticosus** (Boutique) D.K. Harder in Novon **10**: 124 (2000). Type: Congo, Lubumbashi, *de Georgi* (BR holotype).

> *Abrus suffruticosus* Boutique in F.C.B. **6**: 84, fig. 1C, D (1954), without a Latin description, in Bull. Jard. Bot. État **25**: 127 (1955).
> *Abrus pulchellus* sensu Hepper in F.W.T.A., ed.2, **1**: 574 (1958) in part, non Thwaites.
> *Abrus pulchellus* subsp. *suffruticosus* (Boutique) Verdc. in Kew Bull. **24**: 249 (1970); in F.T.E.A., Legum., Pap.: 116 (1971).

Procumbent or ascending woody herb or subshrub 0.2–1 m long, with a woody rhizomatous rootstock. Leaflets 11–17 on each side of the rachis, smaller, 1–1.5 cm × 3–4.5 mm. Indumentum usually dense, the stem and rachis with appressed or spreading rusty-brown hairs contrasting with sometimes dense silky grey hairs on the leaflets. Pods 2.2–3(4) × 0.9–1.2 cm.

Zambia. B: Kaoma (Mankoya) to Mongu, mile 31, edge of Luampa R., near Luampa Mission, *Cryptosepalum/Brachystegia* woodland on Kalahari sand, fl.& fr. 21.xi.1952, *White* 2105 & 2105A (FHO, K). N: Mbala Dist., old road to Kasakalawe (Cascalawa) from Chembe Village, at L.Tanganyika, 1090 m, fl.& fr. 16.ii.1960, *Richards* 12492 (K, MO, UZL). W: Copperbelt, Mufulira Dist., *Brachystegia* woodland, c.1300 m, fl. 31.iii.1949, *Cruse* 512 (K). C: Mkushi Dist., Mulungushi Hill near Mulungushi Dam, 1090 m, fl.& fr. 18.iii.1973, *Kornaś* 3509 (K). **Malawi**. N: Chitipa Dist., 16 km E of Chitipa at Kaseye Mission, c.1400 m, fl.& fr. 18.iv.1976, *Pawek* 11090 (K, MA, MO, UC). S: L. Malawi (Nyasa), Tumbi, fr. 28.iv.1902, *W.P. Johnson* 348 (K).

Also in West Africa from Sierra Leone to the Central African Republic, and in Congo, Tanzania and Angola. Miombo and Kalahari sand woodlands; 400–1800 m.

According to Verdcourt (1970), this taxon maintains consistency in the characters of low habit, small, very hairy leaves and small pods within the central parts of its range. These characters gradually converge with those of subsp. *tenuiflorus*. Intermediate forms of these taxa can be found in Sierra Leone, Mali and Guinea.

Tribe 6. **SESBANIEAE**

by G.P. Lewis

Sesbanieae (Rydb.) Hutch., Gen. Fl. Pl. **1**: 401 (1964). —Lavin & Schrire in Lewis et al., Legumes of World: 452–453 (2005).
Galegeae subtribe *Sesbaniinae* Rydb., Amer. J. Bot. **10**: 495.
Galegeae subtribe *Robiniinae* sensu Benth. in Bentham & Hooker, Gen. Pl. **1**(2): 445 (1865) in part.

Herbs, rarely tall and shrubby. Leaves paripinnate, leaflets opposite; stipules present; stipels inconspicuous. Flowers with a well-developed hypanthium, arranged in axillary racemes in the leaf axils; bracts and bracteoles present. Calyx campanulate, generally attenuate at the base, with 5 lobes shorter than the tube. Standard orbicular to oblong, emarginate, with two prominent callosities, one on either side of the midrib in the region of the nectar guide. Staminal tube diadelphous, with basal openings; anthers uniform, included in the keel tip, basally attached to the filaments. Styles glabrous or puberulous. Fruits mostly transversely septate between the seeds but not breaking up into articles, stipitate, glabrate, modified into linear pods, sometimes inflated or with four wings.

A tribe consisting of the single pantropical genus *Sesbania*, with c.60 species, about half of which occur in Africa and Madagascar.

The tribe Sesbanieae was formerly included in the tribe Robinieae, but recent findings have suggested it should be separate. There are no native species of Robinieae found in the Flora Zambesiaca area, but three exotic species have been recorded. They are described below, even though they do not form part of the Sesbanieae.

Gliricidia sepium (Jacq.) Steud., from Central America, is grown as an ornamental in Zimbabwe. Shrub or small tree. Leaves imparipinnate 15–35 cm long; leaflets (3)6–10(12) on each side of the rachis, 2–8 × 1.2–4.5 cm, elliptic-oblong. Flowers in rather short racemes at old nodes, rose-pink with a yellow blotch at the base of the standard, 1.5–2.3 cm long, the calyx truncate and without bracteoles. Pods 7.5–18 × 1.4–2.2 cm, glabrous, with 3–10 lenticular seeds. **Zimbabwe**. Harare, Enterprise Road, fl. 14.x.1974, *Biegel* 4663 (K, SRGH); Chipinge Dist., Zona Tea Estate, fl. x.1967, *Goldsmith* 108/67 (K, SRGH). It has considerable potential in agroforestry, see Hughes in Commonwealth Forestry Review **66**: 31–45 (1987).

Robinia pseudoacacia L., native of eastern USA but now widely distributed, is grown as an ornamental in Zimbabwe. Tree with rough bark. Leaves pinnate, similar to *Millettia* spp. (but without stipels). Racemes of white to pinkish flowers; standard 1.5–2.3 cm long. Pods flattened, c.5–10 × 1–1.5 cm, narrowly winged along the upper margin, dehiscent. **Zimbabwe**. Harare, Highlands, Kia Ora Nurseries, 21.xii.1968, *Biegel* 2721 (K, SRGH).

Robinia hispida L. is recorded by Biegel (Check-list Ornam. Pl. Rhod. Parks & Gard.: 93, 1977), as a cultivated ornamental in Zimbabwe. It has characteristic pink flowers and hispid twigs.

Key to Sesbanieae and exotic Robinieae in the Flora area

1. Leaves paripinnate; pods transversely septate; racemes numerous, axillary; branches herbaceous or becoming slightly woody ·············· **23. Sesbania**
－ Leaves imparipinnate; pods not septate; racemes on young shoots or precocious on old nodes; branches soon becoming woody (introduced trees) ········· 2
2. Style glabrous; calyx truncate, the lobes very reduced; standard reflexed through 180° ·· **Gliricidia**
－ Style hairy on the upper part; calyx lobes ± as long as the tube; standard reflexed through 90° ·· **Robinia**

23. SESBANIA Adans.

Sesbania Adans., Fam. Pl. **2**: 326 (1763), as *"Sesban"*; orthographic change by
  Scopoli, Introd.: 308 (1777) nom. conserv. —Gillett in Kew Bull. **17**: 91–159
  (1963); in F.T.E.A., Legum., Pap.: 330–351 (1971). —Lewis in Kirkia **13**: 11–51
  (1988). —Lavin & Sousa in Syst. Bot. Monogr. **45**: 39–45 (1995).
    Daubentonia DC., Prodr. **2**: 267 (1825); Mém. Lég.: 285 (1826).

Erect annual or briefly perennial herbs or softly woody shrubs or small trees, often producing a dark gummy juice when the bark is cut. Hairs simple, white or golden. Leaves paripinnate, the rachis channelled above; leaflets often in more than 10 pairs, oblong, entire; stipels usually present at all petiolules; stipules truncate at the base, occasionally persistent. Flowers in axillary racemes; bracts and bracteoles present but often caducous. Calyx campanulate, usually sparsely woolly at the margin; teeth subequal, shorter than the tube. Corolla glabrous, blue, mauve, white, red or orange, or more commonly yellow with the standard usually streaked and spotted or continuously veined with purple, the claw with two vertical parallel or divergent variously shaped appendages (these rarely lacking, as in *S. grandiflora*); blade of wing with transverse lamellate sculpturing (except in *S. grandiflora*), usually toothed or hooked at the base, the claw much shorter than the blade and shorter than that of the keel; blade of keel rounded below, rounded or broadly pointed at the tip, usually toothed at the base, not, or little, longer than the claw. Upper stamen free, sharply bent near the base corresponding to the conspicuous auricles at the base of the filament sheath, the sheath longer than the free parts of the filaments which are curved upwards; anthers all alike, oblong-elliptic (much longer than wide in *S. grandiflora*), dorsifixed. Ovary glabrous or rarely pilose with soft spreading hairs; style glabrous or, less often, pubescent near the tip; stigma small, globose or ovoid. Pod usually long, dehiscent, rostrate, usually shortly stipitate, sometimes winged, transversely septate, (2)4–51-seeded. Seeds usually ellipsoid or cylindrical, rarely subreniform; hilum circular or, rarely, broadly elliptical, often surrounded by a narrow rim aril.

A genus of about 60 species, widely distributed in the tropics and subtropics, most species being found in seasonally wet habitats.

Dimensions of leaves and leaflets refer to those found on flowering and fruiting branches normally seen on herbarium specimens; leaves from the lower parts of plants may be much larger than recorded.

1. Filament sheath 35–100 mm long, curved for much of its length; wing petals without sculpturing; standard without appendages · · · · · · · · · · · · **1.** *grandiflora*
– Filament sheath less than 30 mm long, only upcurved near the tip (except in *S. punicea* with bright red or orange flowers); wing petals with lamellate sculpturing (Fig. 3.3.**33**/21) · 2
2. Pod with four wings, each more than 2 mm broad · · · · · · · · · · · · · · · · · · · 3
– Pod not winged, or the wing rudimentary and much less than 2 mm wide; ovary unwinged · 4
3. Flowers red or bright orange; plant not aculeate; ovary arcuate, unwinged; ovules 10 or less; style glabrous; keel and wing petals without basal teeth; standard appendages with minute free tips; pod usually less than 9 cm long · **2.** *punicea*
– Flowers yellow, standard speckled purple; peduncle, at least, aculeate; ovary only upturned at the tip, winged; ovules 15 or more; style pubescent; keel and wing petals with basal teeth; standard appendages rounded, without free tips; pod usually more than 12 cm long · **18.** *tetraptera*
4. Appendages of standard with free tips 2 mm or more long (Fig. 3.3.**32**/2 & 3) · 5
– Appendages of standard without free tips or with tips less than 2 mm long · · 6
5. Filament sheath 16–24 mm long; racemes 1–2-flowered; most leaves with less than 8 pairs of leaflets · **3.** *goetzei*
– Filament sheath 9–15 mm long; racemes (2)3–20-flowered; most leaves with more than 8 pairs of leaflets · **4.** *sesban*
6. Septa of pod 8–11 mm apart; stipules sometimes rather persistent, erect; style glabrous; appendages of standard rounded at the top without any free tip (Fig. 3.3.**32**/4 & 5) · 7
– Septa of pod less than 8 mm apart, usually 4–6 mm apart · · · · · · · · · · · · · · 8
7. Corolla blue or purple (rarely white); filament sheath 17–23 mm long; blade of standard elliptic, longer than wide; sutures of pod angular with the lower suture wider than the upper one (Fig. 3.3.**35**/10); nowhere aculeate · · · **5.** *coerulescens*
– Corolla yellow (the standard spotted purple on the outer face); filament sheath 9–15 mm long; blade of standard broadly ovate, wider than long; sutures of pod not angular, of similar width · **6.** *macrantha*
8. Filament sheath 16–21 mm long; appendages of standard with blunt, strongly incurved, free tips of up to 1.5 mm long (Fig. 3.3.**32**/6) · · · · · · · · **7.** *cinerascens*
– Filament sheath less than 16 mm long · 9
9. Standard yellow with continuous purple veins and sometimes also a few flecks near the margin; appendages reduced to slight ridges low down on the claw; filament sheath 10–12 mm long; pod up to 20-seeded · · · · · · · · · · · · · · · · · 10
– Standard yellow, usually mottled with purple but not with conspicuous continuous purple veins; appendages present at the top of the claw, except in *S. brevipeduncula* where they are lower down and minute · · · · · · · · · · · · · · · 12
10. Axis of inflorescence less than 10 mm long, including a peduncle up to 3 mm long; bract and bracteoles not or hardly persistent; pod hardly constricted between the seeds (Fig. 3.3.**36**/9); plant sometimes aculeate · · · · · · · · · · · · **15.** *macowaniana*
– Axis of inflorescence more than 20 mm long, including a peduncle of more than 10 mm in length; bracts and bracteoles ± persistent; pod constricted between the seeds (particularly when immature) (Fig. 3.3.**34**/4 & 5) · · · · · · · · · · · · · · · 11
11. Ovules more than 16; pod slightly torulose; raceme 1–5-flowered; plant usually aculeate · **16.** *notialis*
– Ovules less than 13; pod moniliform; raceme usually more than 5-flowered; plant not aculeate · **17.** *transvaalensis*

12. Filament sheath more than 9 mm long · 13
– Filament sheath less than 9 mm long; appendages of standard without free tips
· 19
13. Appendages of standard with free tips 0.5–2 mm long (Fig. 3.3.**32**/7); vertical
rows of spots, developing into glandular outgrowths, on stems above leaf-axils;
upper margins of wing petals inrolled (Fig. 3.3.**33**/21); persistent pocket on
inner face of keel petals (Fig. 3.3.**33**/7); seeds cube-shaped (Fig. 3.3.**34**/3b & c);
never aculeate · **8.** *rostrata*
– Appendages of standard without free tips; stems without vertical rows of spots or
outgrowths; wing petal margins not inrolled; keel petals without pocket · · · · 14
14. Leaf rachis sparsely aculeate, especially towards the base (although prickles
sometimes rather obscure) · 15
– Leaf rachis not aculeate · 17
15. Blade of keel longer than wide, the basal tooth horizontal (Fig. 3.3.**33**/8);
mature pod not torulose; peduncle usually aculeate at the base · · · · **9.** *bispinosa*
– Blade of keel as wide as, or wider than, long; mature pod clearly torulose (Fig.
3.3.**34**/1 & 2); peduncle rarely aculeate · 16
16. Basal tooth of keel at an angle of 40–70° to the claw, no second tooth between it
and the keel apex (Fig. 3.3.**33**/12); style pubescent; seeds 4–4.5 mm long · · · ·
· **13.** *greenwayi*
– Basal tooth of keel vertical at 90° to the claw, a broad second tooth between it
and the keel apex (Fig. 3.3.**33**/13); style glabrous; seeds less than 3.5 mm long
· **14.** *brevipeduncula*
17. Style pubescent · **13.** *greenwayi*
– Style glabrous · 18
18. Basal tooth of keel not erect, no broad second tooth between it and keel apex;
blade of standard usually rather longer than wide; ovules 25 or more; peduncle
usually more than 10 mm long · **11.** *microphylla*
– Basal tooth of keel at 90° to claw, a broad second tooth between it and the keel
apex; blade of standard wider than long; ovules usually fewer than 25
(occasionally to 32); peduncle usually less than 9 mm long · · · **14.** *brevipeduncula*
19. Leaflet lower surface and leaf rachis pubescent; plant not aculeate (outside the
Flora area, sometimes with a few minute obscure prickles on the stem); pod not
torulose · **12.** *sericea*
– Leaflets glabrous, sometimes a few hairs at the base of the petiolules · · · · · · 20
20. Blade of the standard, nearly always, longer than wide, widest above the middle;
calyx with 5 red bands (although these sometimes obscure) each running into a
tooth; standard appendages rounded (Fig. 3.3.**32**/10); ovary slightly thickened
below the style; plant not aculeate or very rarely with a few obscure prickles · · ·
· **11.** *microphylla*
– Blade of the standard usually as wide as, or wider than, long; calyx without red
bands; standard appendages wedge-shaped, truncate or slightly emarginate at
the apex (Fig. 3.3.**32**/8 & 9); ovary not thickened below the style; plant usually
aculeate · 21
21. Unopened standard angular, truncate at the apex; blade of keel petal longer
than wide with a short blunt tooth (Fig. 3.3.**33**/8); pod usually straight-sided,
never truly torulose, brownish mottled with red-brown; ovules usually 30 or
more; seeds 3 mm long · **9.** *bispinosa*
– Unopened standard rounded; blade of keel petal usually as wide as, or wider than
long, with a usually long, narrow acuminate tooth (Fig. 3.3.**33**/9); pod torulose,
pale green or straw-coloured, often with a well-marked darker blotch above the
septal area; ovules usually fewer than 30; seeds 3.5–4 mm long · · · · **10.** *leptocarpa*

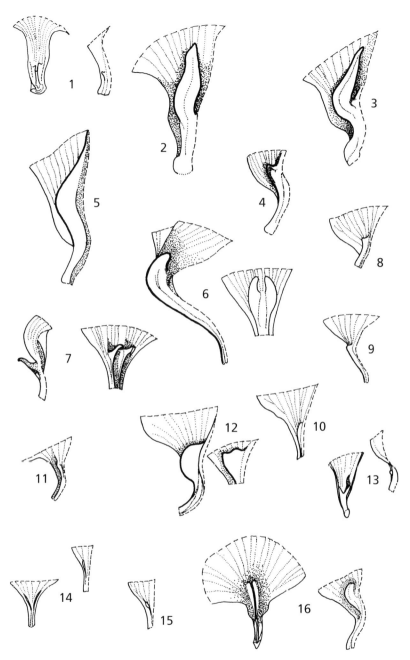

Fig. 3.3.**32**. SESBANIA SPECIES. Standard bases cut lengthways to show one appendage. For some species both appendages are also shown in front view (all × 4). 1, *S. punicea* (*van Rensburg* 3111); 2, *S. goetzei* (*Brummitt & Williams* 11212); 3, *S. sesban* var. *nubica* (*Chase* 2812); 4, *S. coerulescens* (*Richards* 23203); 5, *S. macrantha* var. *levis* (*Leach & Schelpe* 11452); 6, *S. cinerascens* (*Angus* 2754); 7, *S. rostrata* (*P.A. Smith* 1905); 8, *S. bispinosa* (*Philcox et al.* 8655); 9, *S. leptocarpa* var. *leptocarpa* (*Gomes e Sousa* 3662); 10, *S. microphylla* (*P.A. Smith* 2402); 11, *S. sericea* (*Brummitt* 15473); 12, *S. greenwayi*, rounded appendage (*Verboom* 691), blunt appendage (*Torre* 987); 13, *S. brevipeduncula* (*van Rensburg* 1374); 14, *S. notialis* (*Mitchison* 118); 15, *S. transvaalensis* (*Holub* s.n.); 16, *S. tetraptera* subsp. *rogersii* (*Brummitt* 8849). Drawn by G.P. Lewis. From Kirkia (1988).

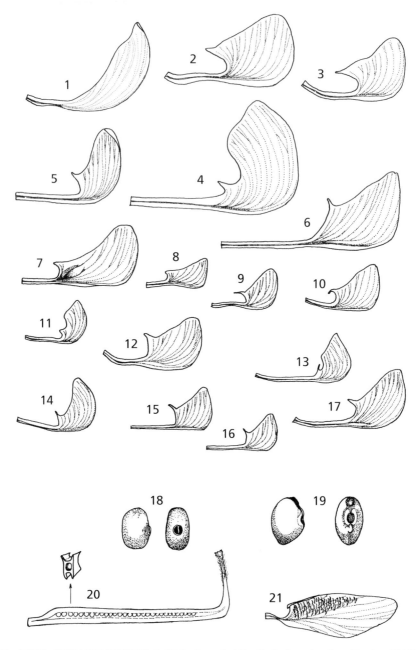

Fig. 3.3.**33**. SESBANIA SPECIES. 1–17, keel petals (all × 2). 1, *S. punicea* (*van Rensburg* 3111); 2, *S. goetzei* (*Brummitt & Williams* 11212); 3, *S. sesban* var. *nubica* (*Chase* 2812); 4, *S. coerulescens* (*Richards* 23203); 5, *S. macrantha* var. *levis* (*Leach & Schelpe* 11452); 6, *S. cinerascens* (*Angus* 2754); 7, *S. rostrata* (*P.A. Smith* 1905); 8, *S. bispinosa* (*Philcox et al.* 8655); 9, *S. leptocarpa* var. *leptocarpa* (*Gomes e Sousa* 3662); 10, *S. microphylla* (*P.A. Smith* 2402); 11, *S. sericea* (*Brummitt* 15473); 12, *S. greenwayi* (*Verboom* 691); 13, *S. brevipeduncula* (*van Rensburg* 1374); 14, *S. macowaniana* (*Wild & Drummond* 6925); 15, *S. notialis* (*Mitchison* 118); 16, *S. transvaalensis* (*Holub* s.n.); 17, *S. tetraptera* subsp. *rogersii* (*Brummitt* 8849); 18 & 19, seeds (× 2). 18, *S. macrantha* var. *macrantha* (*Pawek* 2536); 19, *S. punicea* (*Chase* 6781); 20, winged ovary (× 4), *S. tetraptera* subsp. *rogersii* (*Brummitt* 8849); 21, wing petal (× 2), *S. rostrata* (*P.A. Smith* 1905). Drawn by G.P. Lewis. From Kirkia (1988).

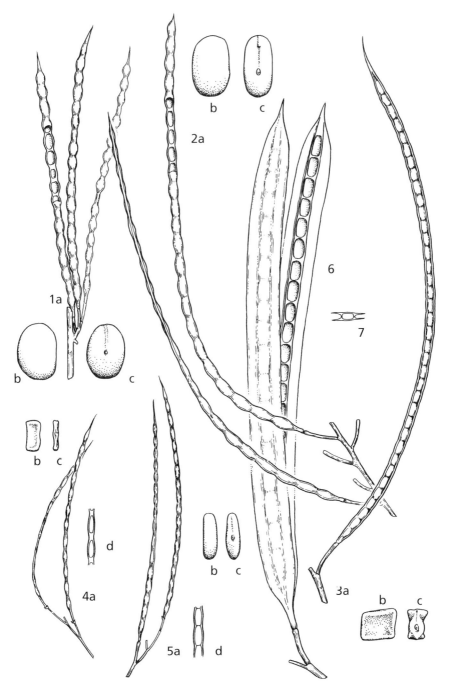

Fig. 3.3.**34**. SESBANIA SPECIES. Fruits and seeds: a, pods (all × $^2/_3$); b, seeds, side view; c, seeds, front view (all × 4); d, longitudinal section of pod (× 1). 1, *S. brevipeduncula* (*Zimbabwe Dept. Agric.* 3316); 2, *S. greenwayi* (a, *Verboom* 691, b & c, *Pedro & Pedrógão* 5022); 3, *S. rostrata* (*Savory* 40 S); 4, *S. transvaalensis* (*Coetzer* 33); 5, *S. notialis* (*Mitchison* 118); 6, *S. tetraptera* subsp. *rogersii* (*Angus* 1592); 7, *S. tetraptera* subsp. *rogersii*, transverse section of pod (× $^2/_3$) (*Angus* 1592). Drawn by Eleanor Catherine. From Kirkia (1988).

1. **Sesbania grandiflora** (L.) Poir. in Lamarck, Encycl. Méth. Bot. **7**: 127 (1806). —
   Phillips & Hutchinson in Bothalia **1**: 46 (1921). —Baker f., Legum. Trop. Africa:
   260 (1929). —Hepper in F.W.T.A., ed.2, **1**: 532 (1958). —Gillett in Kew Bull. **17**:
   105 (1963); in F.T.E.A., Legum., Pap.: 331 (1971). —Lewis in Kirkia **13**: 19
   (1988). Type: India, Herb. Linn. no.922.1 (LINN lectotype).
   Robinia grandiflora L., Sp. Pl.: 722 (1753).

A fast-growing tree to 7 m tall. Stems tomentose, unarmed. Leaves up to 30 cm long; rachis
slightly pubescent or glabrous; petioles 0.7–1.5 cm long; leaflets in 15–21 pairs, opposite to
alternate on the same leaf, 1.2–4.4 cm long (including a petiolule of c.2 mm), 0.5–1.5 cm wide,
oblong to oblong-elliptic, rounded to obtuse to slightly emarginate at the apex, slightly
asymmetrical at the base, glabrous or with small scattered, appressed hairs on both surfaces
(just visible with a ×10 lens); stipels 0.75–1 mm long, filiform, pubescent, persistent; stipules 8
mm long, broadly lanceolate, caducous. Raceme axillary, 2–6.5 cm long, 2–3-flowered;
peduncle 1.5–3.5 cm long, tomentose; pedicels 1.5–1.8 cm long, pubescent; bracts 3–6 mm
long, linear-lanceolate to lanceolate, caducous; bracteoles 4–6 mm long, broadly lanceolate,
caducous. Calyx 1.5–2.2 × 1.6–2 cm, ± two-lipped with a rather indeterminate margin due to the
mode of splitting as the bud develops, abscising together with the nectarial region as the fruit
develops but tending to persist around the stipe of the pod, the lips broadly rounded and
terminating in slightly pubescent subulate tips. Flowers white, flesh-pink or crimson. Standard
up to 10.5 × 6 cm, the claw without appendages; wings up to 10.5 × 3 cm, without a basal tooth
and lacking petal sculpturing; keel up to 10.5 × 4.5 cm with a basal, broadly triangular tooth.
Anthers up to 6 times as long as wide. Ovary glabrous, stipitate; style glabrous, flattened,
terminating in a slightly lobed stigma. Pod up to 52.5 cm × 8 mm, linear to slightly falcate,
occasionally slightly torulose, stipitate, apex long-acuminate, glabrous, up to c.50-seeded; septa
7.5–10 mm apart; lower suture angled, wider than the rounded upper suture; valve surface
rather warty with prominent raised venation. Seeds dark brown, 6.5 × 5 mm, 2.5–3 mm thick,
subreniform, the hilum in a small pit.

Malawi. S: Machinga Dist., Balaka, beside G.R.Patel's Bar, cult., fr. 24.x.1961,
Chapman 1481 (K, SRGH).

A native of parts of Asia but cultivated in many parts of the world both as an
ornamental and for forage, firewood, pulp and paper, food and green manure. It
appears to be a recent introduction to the Flora Zambesiaca area and is only known
from the Chapman Malawi collection.

2. **Sesbania punicea** (Cav.) Benth. in Martius, Fl. Bras. **15**(1): 43 (1859). —Biegel,
   Check-list Ornam. Pl. Rhod. Parks & Gard.: 98 (1977). —Pienaar in Stirton, Plant
   Invaders: 136, photos, figs.1–3 (1978). —Lewis in Kirkia **13**: 20 (1988). Type from
   tropical America. FIGURES 3.3.**32**/1, 3.3.**33**/1 & 19.
   Piscidia punicea Cav., Ic. Pl. **4**: 8, t.316 (1797).

Shrub or small tree, 1–4(5)m tall. Young stems glaucous, green to dark-brown, often
longitudinally ribbed, glabrous to slightly appressed-pubescent, older stems with cracked
greyish-black bark. Leaves 5–16 cm long; petioles 1–1.8 cm long; leaflets in (6)10–18(21) pairs,
1.1–2.9 cm × 3–8 mm, linear-oblong to oblong, rounded at the apex, apiculate, entire, slightly
pubescent beneath especially on the margins and also punctate with minute, scattered black
dots, the venation very evident beneath; stipules 3–7 mm long, lanceolate, ± persistent. Raceme
axillary, 4–11.3 cm long, 12–24-flowered; peduncle 1–3.3 cm long, slightly appressed-pubescent
especially near the base; pedicels 5–10 mm long, glabrous; bracts 2 mm long, lanceolate,
marginally pubescent, caducous; bracteoles 0.5–1 mm long, filiform to narrowly lanceolate,
caducous. Calyx a deep red colour in bud, 4.5–6 × 5–7 mm, apically rather oblique, almost
truncate, the teeth inconspicuous, 0.5 mm long, broadly triangular, slightly pubescent at the
tips. Petals red or bright orange. Standard 1.7–2 × 1.9–2.1 cm, broad, rounded, apically
emarginate, the appendages minute with acute free tips and located at the very base of the claw;

wings 1.6–1.9 cm × 4.5–6.5 mm, slightly falcate, without a basal tooth; lamellate sculpturing weakly present in upper basal two-thirds on the outer face of the petals; keel 1.6–1.9 cm × 8–10 mm, arcuate, without a basal tooth, the limbs of the two petals fused apically and slightly overlapping (but not fused) basally. Ovary arcuate, glabrous, not winged; style glabrous. Pod pale brown, darker brown in central area over the seed cavities, 4.5–9.2 × 1.3–1.8 cm, linear to slightly falcate, 4-winged, stipitate, beaked, glabrous, (3)4–9(10)-seeded; septa 6.5–8 mm apart, slightly oblique; the main part of the pod (i.e. excluding the wings) torulose with prominent reticulate venation on the valves above the seed cavities. Seeds greyish-green to straw-coloured, 6.5 × 5 mm, 4 mm thick (quite distinct in shape from all the African species of the genus) without a rim aril but with a paler coloured raised area around the elliptic (not circular) hilum which is not centrally located. Seedlings with first foliage leaf simple, much larger than the persistent cotyledons; second foliage leaf paripinnate with four pairs of leaflets.

Zambia. W: Kitwe Dist., fl. & fr. 22.v.1967, *Fanshawe* 'K' (K, SRGH). C: Lusaka Dist., Mount Makulu, fl. 29.ix.1966, *van Rensburg* 3111 (K, SRGH). **Zimbabwe**. C: Harare Dist., Umwindsidale, 25 km E of Harare (cult.), fl.& imm.fr. 20.ix.1966, *Müller* 472 (K, SRGH); Harare, Mount Pleasant, seedlings cultivated in garden, 19.iii.1972, *Biegel* 3932 (SRGH). E: Mutare Dist., Mutare (Umtali) garden, fr. 8.xii.1957, *Chase* 6781 (K, SRGH).

A native to South America found in Argentina, Uruguay and S Brazil, where it has also been cultivated as an ornamental for over 100 years. It is now widely cultivated as a garden plant in many parts of the world but has, in many cases, escaped to become a problematic weed invading indigenous vegetation. It is now a particular problem in the USA and South Africa. In the Flora Zambesiaca area most records are from cultivated specimens but it has undoubtedly escaped from gardens and is also recorded as a weed of aquatic areas. Although a handsome plant with its distinctive red flowers, the seeds, flowers and leaves are, nevertheless, very poisonous to many forms of livestock.

The record from Malawi in Lock (Leg. Afr. Check-list, 1989) of *S. punicea* is untraced.

Gillett in Kew Bulletin **17**: 93 (1963) comments that Rydberg's proposal (1923) "to uphold all the segregate genera which had previously been separated from *Sesbania*", including *Daubentonia* DC. (1825), seems less acceptable than Rydberg's views on the subtribal status of the *Sesbaniinae*. Gillett supports the views of both Bentham and Taubert as to the generic limits of *Sesbania*, except that he recognizes *Glottidium* as a distinct genus based mainly on differing fruit and seed characters. He suggests that all other segregates are best treated as subgenera. In Kew Bull. **17**: 149 (1963) he describes a new subgenus, *Pterosesbania*, removing the two African species *S. tetraptera* and *S. rogersii* (the latter considered as a subspecies of the first in this treatment) from the otherwise American group *Daubentonia*. The flowers of *S. punicea* have the wings and keel petals without basal teeth and the standard has only minute appendages near the base of the claw (Rydberg incorrectly states under the generic description of *Daubentonia* that "the banner is without callosities"). The first character, in particular, separates this species from all the African Sesbanias. In addition the seeds are very distinct, being reniform and not cylindrical, and are without a central hilum or rim aril. When considered together with the winged pods and oblique calyx with inconspicuous teeth, the overall picture suggests that it would be as sensible, on morphological grounds, to recognize *Daubentonia* DC. as a genus distinct from *Sesbania* as it is to recognize *Glottidium*. Lavin and Schrire in Lewis et al., Legumes of the World (2005), reinstate tribe *Sesbanieae* as molecular data (Wojciechowski et al., unpublished) clearly shows that *Sesbania sensu lato* is as closely related to tribe *Loteae* as to the *Robinieae*. In addition, the putative position of *Sesbania* as the sister clade to the rest of *Robinieae* (Lavin & Sousa in Syst. Bot.

Monogr. **45**: 1–165, 1995) reinforces its segregation as a distinct group. Lavin & Schrire (2005) recognize no segregate genera from within *Sesbania sensu lato*, and this is followed here, pending further detailed studies of the segregate genera accepted by Rydberg and Gillett.

3. **Sesbania goetzei** Harms in Bot. Jahrb. Syst. **30**: 327 (1901). —Phillips & Hutchinson in Bothalia **1**: 42 (1921). —Baker f., Legum. Trop. Africa: 258 (1929). —White, F.F.N.R.: 165 (1962). —Gillett in Kew Bull. **17**: 111 (1963); in F.T.E.A., Legum., Pap.: 338 (1971). —Lewis in Kirkia **13**: 22 (1988). Type: Tanzania, L. Rukwa, *Goetze* 1115 (B† holotype, BM, E, K). FIGURES 3.3.**32**/2, 3.3.**33**/2.

An erect, softly woody shrub to small tree 3–6 m tall, the trunk up to 12 cm in diameter. Young stems densely and softly grey pubescent, not aculeate; older ones glabrescent and striate. Leaves (0.7)1.5–5.6 cm long; rachis pubescent; petioles 1.5–5 mm long, pubescent; leaflets in (3)4–8(12) pairs, (0.5)1–1.8 cm × (2)3–5 mm, oblong to oblong-elliptic, rounded to obtuse to emarginate at the apex, minutely apiculate, tapering to an asymmetrical base, pubescent on both surfaces; stipules 2–4 mm long, triangular-lanceolate to triangular-ovate, pubescent, caducous. Raceme axillary, 0.5–2.7 cm long, 1–2(rarely 4)-flowered, the rachis pubescent; peduncle 0.5–2 cm long, pubescent; pedicels (3)5–6(9) mm long, pubescent; bracts and bracteoles linear-lanceolate, caducous to rarely persistent, the bracts 3–4 mm long; bracteoles 1.5–2 mm long. Calyx 7–8 × 8 mm, the tube sparsely pubescent or occasionally glabrous; teeth 0.5–1 mm long, subulate, sparsely pubescent. Standard dark yellow with purple flecks and spots on outer face (sometimes without the purple markings on plants outside the Flora Zambesiaca area), 1.6–2.8 × 1.6–2.4 cm, suborbicular, deeply emarginate, appendages with free tips 3–4 mm long; wings yellow, 1.6–2.4 cm × 7–9 mm, with an obscure broadly-triangular upper basal tooth and prominent lamellate sculpturing in upper basal half on outer face; keel greenish-yellow, 1.6–2.4 cm × 8–10 mm, the basal tooth triangular, ± horizontal. Ovary sometimes speckled reddish-purple, strongly to sparsely pubescent to glabrescent along the upper margin; style glabrous to pubescent. Pod 14–21.5(26) cm × (2)4 mm, ± linear, slightly torulose, thicker at the centre than the margin, sometimes with prominent fracturing at right angles to the sutures when mature, glabrous, (26)30–46-seeded; septa 5–6 mm apart; sutures slightly thickened, 1 mm wide. Seeds greenish, olive-brown or orangish-brown, 2.5–3.5 × 1.5–2 mm, 1–1.75 mm thick, subovoid to oblong, the hilum subcentral in a small circular pit surrounded by a rim aril.

Subsp. **goetzei**

Leaflets mainly in 4–8 pairs. Racemes 1–2-flowered. Calyx sparsely pubescent. Back of standard with purple flecks and spots.

Zambia. N: Kaputa Dist., shores of L. Mweru Wantipa, fl.& fr. 22.x.1949, *Bullock* 1354 (K). **Malawi**. S: Zomba Dist., Lake Chilwa, Chaone Is., fl.& imm.fr. 1.vi.1970, *Brummitt & Williams* 11212 (K, PRE). **Mozambique**. GI: Inharrime Dist., next to Lagoa Poelela, fl.& imm.fr. 13.ix.1948, *Myre & Carvalho* 169 (COI, LISC).

Also in Ethiopia, Kenya and Tanzania. Lake shores and more rarely coastal sandy beaches; 0–900 m.

Gillett in F.T.E.A., Legum., Pap.: 338 (1971) presents habitat data as: "Land liable to flood beside alkaline lakes, sometimes forming pure stands. 900–1900 m; rainfall 500–1000 mm per annum."

Myre & Carvalho 169 is the most southerly example of the species and, coming from the coast, greatly extends its altitudinal range. Flowers from this specimen display unusually hairy styles, although on young fruit these persistent organs are glabrescent to glabrous.

Subsp. *multiflora* J.B. Gillett, with up to 12 pairs of leaflets, up to 5-flowered racemes, a calyx glabrous except at the margin and a wholly yellow standard, is only known from one specimen, namely *Dames* 62 (EA) from Tanzania. However, *Rykebusch* 17238, also from Tanzania, is intermediate between subsp. *goetzei* and subsp. *multiflora* and throws some doubt on to the acceptance of the latter.

4. **Sesbania sesban** (L.) Merr. in Philipp. J. Sci., C, **7**: 235 (1912). —Cronquist in F.C.B. **5**: 76, t.5 (1954). —Hepper in F.W.T.A., ed.2, **1**: 532 (1958). —Sousa in C.F.A. **3**: 181 (1962) in part. —White, F.F.N.R.: 165 (1962). —Gillett in Kew Bull. **17**: 112 (1963). —Schreiber in Merxmüller, Prodr. Fl. SW Afrika, fam. 60: 111 (1970). —Gillett in F.T.E.A., Legum., Pap.: 339 (1971). —Lewis in Kirkia **13**: 23 (1988). —K. Coates Palgrave, Trees Sthn. Africa, ed.2: 314 (1988). —M. Coates Palgrave, Trees Sthn. Africa: 381 (2002). Type: Egypt, *Hasselquist* (LINN no. 922.12 lectotype).

> *Aeschynomene sesban* L., Sp. Pl.: 714 (1753).
> *Sesbania aegyptiaca* Poir. in Lamarck, Encycl. Méth. Bot. **7**: 128 (1806) as *"Sesban aegyptiacus"*. —Baker in F.T.A. **2**: 134 (1871). —Phillips & Hutchinson in Bothalia **1**: 44 (1921). —Baker f., Legum. Trop. Africa: 259 (1929). —Burtt Davy, Fl. Pl. Ferns Transvaal, pt.2: 380 (1932). —Hutchinson, Botanist Sthn. Africa: 462 (1946). Type as for *Sesbania sesban*.

Shrub or small short-lived tree 1–8 m tall with stem up to 12 cm in diameter. Stems sometimes tinged pinkish or bluish, laxly to densely spreading pubescent, sometimes glabrescent or, less often, almost totally glabrous. Leaves 2–18.5 cm long; rachis usually pubescent; petioles 0.3–1.6 cm long; leaflets in 6–27 pairs, 0.5–2.6 cm × 2–5.5 mm, oblong, obtuse to emarginate at apex, apiculate, glabrous or nearly so above, often pilose at the margins, midvein beneath usually pubescent, sometimes pubescent over the whole lower surface, occasionally densely so; stipules up to 7 mm long, narrowly triangular, usually inrolled or twisted, caducous or persistent, pubescent. Raceme (2)3–20-flowered, 2.8–14.5(19.7) cm long, glabrous or sparsely pilose; peduncle 1.4–4.4(5.7) cm long, glabrous or pubescent; pedicels 4–12 mm long, glabrous or less often sparsely pubescent; bracts and bracteoles (1)3–5 mm long, linear-lanceolate, ± glabrous or pilose, falling very early. Calyx 4–6.5 mm long (including receptacle), 4–8 mm wide, the tube glabrous; teeth 0.5–1 mm long, broadly triangular with a minute acuminate point, marginally pubescent. Standard uniformly yellow (sometimes with a pale green tinge) or more commonly speckled or flecked purplish on outer face, or sometimes suffused with purple (not in the Flora area, e.g. var. *bicolor*), 1.1–2 cm long including a claw of 2.5–4 mm, 1.3–2.1 cm wide (up to 2.5 cm wide in West Africa), the blade $1^1/_4$–$1^1/_2$ times as wide as long, widest below the middle, cordate at the base, appendages with acuminate free tips c.2–5 mm long; wings yellow, 1.5–1.9 cm long including a claw of 4–6 mm, 4.5–7 mm wide, with a broad tooth or short hook at the base and lamellate sculpturing in the upper basal two-thirds; keel yellow or creamish sometimes tinged green, the blade 11–16 mm long (up to 21 mm long in West Africa) including a claw of 6–9 mm, 6.5–9 mm wide, subtriangular with a broad, acute basal tooth at 0–20° to the claw. Ovary 30–50-ovulate, glabrous or rarely somewhat pilose along the upper suture or near the base of the style. Pod straw-coloured, often with a brown blotch over each septum, or reddish-brown, (3.7)15–28(31) cm × 2–5 mm, straight or slightly curved, sometimes slightly twisted, glabrous, (3)11–49-seeded; septa 4–8 mm apart; sutures cylindric, very rarely with swollen portions at intervals; valves sometimes fractured at right angles to the sutures. Seeds olive-green or brown, usually mottled dark purple or black, 3–4.5 × 2–2.25 mm, 2 mm thick, subcylindrical; hilum in a subcentral circular pit surrounded by an obscure rim aril.

Var. **nubica** Chiov., Fl. Somala **1**: 145 (1929); **2**: 163 (1932). —Gillett in Kew Bull. **17**: 112, fig. 3/4 (1963); in F.T.E.A., Legum., Pap.: 339 (1971). —Palmer & Pitman, Trees Sthn. Africa: 931 (1972). —Corby in Kirkia **9**: 327 (1974). —Lewis in Kirkia **13**: 24 (1989). Type: S Somalia, Belet Uen, *Puccioni & Stefanini* 185 (FT holotype). FIGURES 3.3.**32**/3, 3.3.**33**/3.

> *Sesbania punctata* sensu Taubert in Engler, Pflanzenw. Ost-Afrikas **C**: 213 (1895), in part, non DC.

Sesbania pubescens as used by many, e.g. Harms in Warburg, Kunene-Samb.-Exped. Baum: 261 (1903) and R.E. Fries, Wiss. Ergebn. Schwed. Rhod.-Kongo Exped. **1**: 84 (1914), non DC.

Sesbania of Eyles in Trans. Roy. Soc. S. Africa **5**: 377 (1916) in part as regards specimen *Eyles* 144.

Sesbania sesban var. *zambesiaca* J.B. Gillett in Kew Bull. **17**: 113 (1963). —Drummond & Corby, Legum. Indig. & Nat. S. Rhodesia: 25 (1964). —Drummond in Kirkia **8**: 226 (1972). —Palmer & Pitman, Trees Sthn. Africa: 931 (1972). —Corby in Kirkia **9**: 327 (1974). —Gonçalves in Garcia de Orta, sér.Bot. **5**: 112 (1982). Type: Zambia, Livingstone Dist., Zambezi R., a little below Katombora, *Brenan & Greenway* 7757 (K holotype, EA, FHO).

Caprivi Strip. Schuckmannsburg, fr. 20.x.1970, *Vahrmeijer* 2202 (PRE). **Botswana**. N: Ngamiland Dist., Chobe Nat. Park, fl.& fr. 28.viii.1970, *Mavi* 1130 (K, LISC, PRE, SRGH). **Zambia**. B: Senanga Dist., Zambezi R., Nangweshi Resthouse S of Senanga, fl.& fr. 24.vii.1961, *Angus* 3035 (K, LISC, SRGH). N: Mbala Dist., Kumbula (Nmbulu) Is., L. Tanganyika, fl. 11.iv.1955, *Richards* 5403 (K). C: Lusaka Dist., Luangwa R., short distance above Luangwa Beit Bridge, fl.& fr. 5.ix.1947, *Brenan & Greenway* 7815 (K). E: Chipata Dist., Lutembwe R. gorge E of Machinje Hills, fr. 13.x.1958, *Robson & Angus* 95 (K, LISC, PRE, SRGH). S: Mazabuka Dist., Kafue Pilot Polder, fl.& fr. 6.ii.1963, *van Rensburg* 1322 (K, SRGH). **Zimbabwe**. N: Gokwe North Dist., Copper Queen, Sanyati R., fl.& fr. 19.vii.1962, *Bingham* 304 (K, SRGH). W: Hwange Dist., Victoria Falls, Rainforest, fr. 2.ix.1966, *Crozier* 80/66 (K, SRGH). C: Harare, Salisbury Expt. Station, ?cult., fl.& fr. 10.vi.1937, *Arnold* 6542 (K, SRGH). E: Chipinge Dist., Birchenough Bridge, fl.& imm.fr. 26.ix.1949, *Chase* 1781 (K, LISC, SRGH). S: Mwenezi Dist., Runde (Lundi) R., fl.& imm.fr. 30.vi.1930, *Hutchinson & Gillett* 3257 (K). **Malawi**. N: Rumphi Dist., Njakwa Gorge (S Rukuru Gorge), 5 km E of Rumphi, fl.& imm.fr. 21.v.1970, *Brummitt* 10965 (K, PRE). C: Dedza Dist., Chongoni Forestry School turn-off, fl.& fr. 8.ix.1965, *Jeke* 18 (K, SRGH). S: Blantyre Dist., Nkula Falls, E bank of Shire R., fl.& fr. 2.vii.1970, *Brummitt* 11767 (K, PRE, SRGH). **Mozambique**. N: Lichinga Dist., Meponda, next to L. Niassa, fr. 4.ix.1958, *R. Monteiro* 37 (LISC). T: Magoe Dist., Msusa, Zambezi R., fl.& fr. 27.vii.1950, *Chase* 2812 (COI, K, LISC, SRGH). MS: between Gorongosa (Vila Paiva) and Amatongas, Pungue R., fl.& imm.fr. 9.ix.1942, *Mendonça* 182 (LISC). GI: Save (Sabi) R., Massangena (Massengena), fl.& imm.fr. vii.1932, *Smuts* P.366 (COI, K, PRE). M: Magude Dist., Motaze, R. Mazimechopes, fl. 16.vii.1948, *Torre* 8077 (LISC).

Also in the Yemen, and from Lake Chad to Somalia, Uganda, Kenya, Tanzania, Angola, Namibia, eastern parts of South Africa and Swaziland. The commonest member of the genus in the Flora area. At low altitudes on river or stream margins and lake shores, in wet or swampy ground, only away from water in areas of high and frequent rainfall; 30–2200 m.

Differs from subsp. *punctata* (DC.) J.B. Gillett (Senegal to Sudan) chiefly in its shorter flowers (filament sheath 9–13 mm long instead of 15–17 mm long).

It is with some hesitation that Gillett's var. *zambesiaca* is here placed in synonymy. The type, together with several other specimens, particularly from the Upper Zambezi, Barotseland and the Caprivi Strip, show some correlation between yellow flowers (i.e. those with standards not mottled purple) and wider (and usually shorter) pods. However, these characters cannot be consistently used to separate it from var. *nubica* and, as a result, more pubescent forms of the latter have, in the past, often been incorrectly identified as var. *zambesiaca*. Furthermore, many morphologically intermediate specimens have proved impossible to place with certainty in either var. *nubica* or var. *zambesiaca*. For these reasons, only one taxon is recognized in this treatment, although further study may necessitate a reappraisal of the *sesban* 'complex'. The other two varieties of subsp. *sesban*, namely var. *bicolor* and var. *sesban,* neither of which occurs in the Flora Zambesiaca area, both require a closer look. Given the degree of variation found in var. *nubica,* these two varieties seem to be separated from it by rather weak, unstable characters. The var. *bicolor* has a standard suffused with purple and specimens of var. *sesban* are almost or completely glabrous and, as such, are very close to a few specimens of var. *nubica* in the Flora area. Pods of var. *nubica* are especially variable in length, width, thickness and surface coloration. Those of *Brummitt* 12156 (from Malawi) are somewhat twisted but this occasional feature appears to be of little taxonomic significance. *Corby* 1923A, a cultivated specimen from Zimbabwe, has very long, broad fruits and large inflorescences. Some fruiting specimens with long, narrow fruits can be difficult to separate from *S. cinerascens* and flowers are usually needed to be certain of correct identification. *Lemos & Balsinhas* 82 from Mozambique is a specimen with a sparsely hairy ovary and style.

5. **Sesbania coerulescens** Harms in Warburg, Kunene-Samb.-Exped. Baum: 260 (1903). —Phillips & Hutchinson in Bothalia **1**: 46 (1921) as *"caerulescens"*. —Baker f., Legum. Trop. Africa: 260 (1929). —Cronquist in F.C.B. **5**: 78 (1954). —Sousa in C.F.A. **3**: 183 (1962) as *"caerulescens"*. —White, F.F.N.R.: 165 (1962) as *"caerulescens"*. —Gillett in Kew Bull. **17**: 117 (1963). —Schreiber in Merxmüller, Prodr. Fl. SW Afrika, fam. 60: 110 (1970). —Corby in Kirkia **9**: 326 (1974). —Lewis in Kirkia **13**: 25 (1988). —K. Coates Palgrave, Trees Sthn. Africa, ed.2: 313 (1988). —M. Coates Palgrave, Trees Sthn. Africa: 381 (2002). Type: Angola, Miané R. near Kavamba, c.16°10'S, 19°10'E, 14.iii.1900, *Baum* 782 (EA, K). FIGURES 3.3.**32**/4, 3.3.**33**/4, 3.3.**35**.

    Sesbania hockii De Wild. in Repert. Spec. Nov. Regni Veg. **11**: 544 (1913). Type from Congo.

    Sesbania of Eyles in Trans. Roy. Soc. S. Africa **5**: 377 (1916) in part as to specimen *Flanagan* 3121.

Annual or biennial herb to subshrub 2.5–5 m tall. Stems pithy, glabrous, unarmed, marked with dark green or brown vertical lines or dashes. Leaves (6)7.5–40 cm long (with some lower ones to 60 cm); petioles (0.8)1.1–3.7(4.6) cm long; leaflets in (10)14–45 pairs, (0.7)1–5.2 × 0.2–1.5 cm, oblong to oblong-elliptic, obtuse to slightly emarginate at the apex, apiculate, slightly narrowed and asymmetrical at the base, entire, glabrous, punctate above with minute black dots, obscurely mottled dirty-red beneath; petiolules 0.25–2 mm long; stipels 0.5–2 mm long, usually persistent; stipules 0.7–1.5 cm × 2–6 mm, ovate-lanceolate to oblanceolate, long acuminate, caducous to persistent. Raceme (2.5)5–23 cm long, 2–15-flowered; peduncle (1.8)4–9.5 cm long, glabrous; pedicels (0.8)1–3.5 cm long, glabrous; bracts 0.6–1.2 cm × 1–3.5 mm, linear-lanceolate to ovate-lanceolate, long acuminate, basally striped reddish-brown, glabrous, caducous to occasionally persistent; bracteoles (3)4–9 × 0.5–1.75 mm, linear-lanceolate, glabrous. Calyx (0.6)0.7–1.3 × 0.5–1.1 cm; tube glabrous; teeth acuminate, 1.5–4 mm long, shortly woolly on margins and inner surface. Petals usually blue, less frequently mauve or purple, rarely white (e.g. *R.M. Davies* 2879). Standard (1.9)2.1–3.4 × (1.7)1.8–3.2 cm, broad, rounded, apically emarginate, the appendages small and without free tips; wings (1.7)2.1–2.8 × 0.5–1 cm, with an upper basal tooth and with lamellate sculpturing in upper basal section on outer face; keel 1.9–2.4 × 1–1.5 cm, limb subquadrate with basal tooth projecting upwards. Ovary and style glabrous. Pod (20)25–36 cm × 3.5–6 mm, linear, long acuminate, glabrous, 25–37-seeded; septa 8–9 mm apart; lower suture angular, up to 4 mm wide, broader than the keeled to somewhat angular upper suture which opens first. Seeds greenish-black, heavily mottled black, 5–5.5 × 2.5 mm, 2.5 mm thick, cylindrical; hilum in small central circular pit surrounded by obscure rim aril. Seedlings with cotyledons shortly petiolulate; first foliage leaf simple, larger than cotyledons; second foliage leaf paripinnate.

Zambia. B: Kaoma Dist., Kamuni Dambo, Kasempa road, fl.& imm.fr. 3.iv.1964, *Verboom* 1207 (K, SRGH). N: Mbala Dist., Lake Chila shore, fl.& fr. 13.v.1955, *Richards*

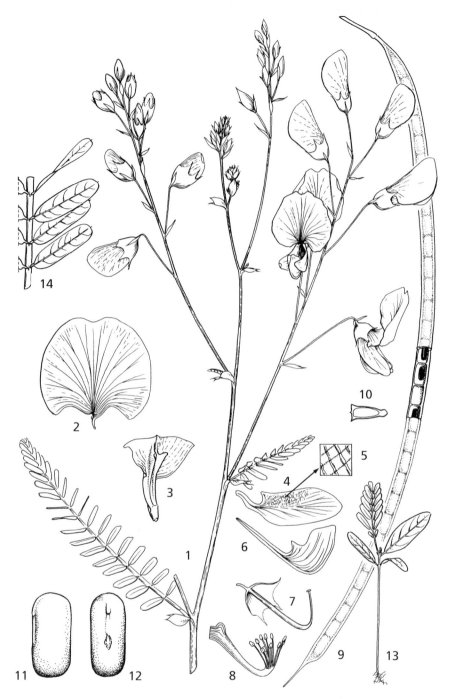

Fig. 3.3.**35**. SESBANIA COERULESCENS. 1, flowering branch (× ²/₃); 2, standard (× 1); 3, appendages at base of standard (× 4); 4, wing (× 1); 5, detail of sculpturing on wing (× 8); 6, keel (× 1); 7, calyx (part cut away) and gynoecium (× 1); 8, staminal tube (× 1), 1–8 from *Richards* 23203; 9, pod (× ²/₃), from *Richards* 5695; 10, transverse section of pod (× 2); 11 & 12, seed, side and front views (× 4), 9–12 from *Stohr* 810; 13, seedling (× ²/₃), from *Fanshawe* s.n.; 14, portion of mature leaf (× ²/₃), from *Richards* 23203. Drawn by Eleanor Catherine. From Kirkia (1988).

5695 (K, SRGH). W: Ndola, fl.& imm.fr. 11.iv.1954, *Fanshawe* 1077 (K, SRGH); Kitwe, seedling 23.ii.1970, *Fanshawe* 10767 (K). C: Mkushi Dist., Munshiwemba, fl.& imm.fr. iii.1943, *Stohr* 819 (K, PRE). **Zimbabwe**. N: Gokwe South Dist., Lutope R., fl. 18.iii.1963, *Bingham* 532 (K, LISC, SRGH); Murehwa Dist., Headlands road, fl. (white-flowered form) ii.1961, *R.M. Davies* 2879 (SRGH). C: Harare Dist., below Cleveland Dam, fl.& imm.fr. 14.iii.1969, *Biegel* 2886 (K, PRE, SRGH); same locality, fr. 3.vii.1969, *Biegel* 3167 (K, PRE, SRGH).

Also in Congo, Angola and Namibia. Gregarious on river or stream margins, lake shores and beside dams, in wet or swampy ground on edges of dambos and in floodplain grassland, occasionally in miombo woodland; 1100–1750 m.

The large blue flowers readily distinguish this from all other species. Given fruits only, particularly when immature, there is a tendency for confusion with *S. macrantha* var. *levis*. However, the angled sutures of the mature pods of *S. coerulescens* (the lower wider suture may be up to 4 mm wide) at once separate it from the former. The seeds of *S. coerulescens* are shorter (up to 5.5 mm long) than those of *S. macrantha* var. *levis*, which are 6–8 mm in length.

Verboom 1207 is both morphologically and geographically closer to the Angolan type material than any other specimen in the Flora Zambesiaca area. *R.M. Davies* 2879 is a form with white flowers but is not unique. Both *Fanshawe* (9619) and *Shepherd* (32) refer to this variation in flower colour.

6. **Sesbania macrantha** E. Phillips & Hutch. in Bothalia **1**: 47 (1921). —Baker f., Legum. Trop. Africa: 261 (1929). —Cronquist in F.C.B. **5**: 79 (1954). —Hepper in F.W.T.A., ed.2, **1**: 532 (1958). —White, F.F.N.R.: 165 (1962). —Sousa in C.F.A. **3**: 184, t.15 (1962). —Gillett in Kew Bull. **17**: 117 (1963). —Gillett in F.T.E.A., Legum., Pap.: 341(1971). —Corby in Kirkia **9**: 326 (1974). —Lewis in Kirkia **13**: 28 (1988). Type: Angola, Cuanza Norte, Pungo Andongo, *Welwitsch* 1997 (BM holotype).

Softly woody herb, slender shrub or small tree, (1)2–6 m tall; glabrous throughout except for the calyx margin and inner surface of the calyx teeth, a few hairs on the stipules and often on the rachis and leaflet margins of juvenile leaves. Stems greyish-green often tinged purplish-black, aculeate or completely without prickles. Leaves (6.5)10–33(45) cm long, the rachis aculeate or not; petioles (0.3)0.5–2.9(3.5) cm long; leaflets in (9)15–55(64) pairs, (0.7)1.4–3.1(4.4) cm × (2.5)3–8(13) mm (the basal pair sometimes much smaller than the others), oblong, obtuse to slightly emarginate at the apex, apiculate, slightly asymmetrical at the base, entire; stipules (0.9)1.4–2.5(2.9) cm × (2)3–5 mm, lanceolate to ovate-lanceolate, acuminate, slightly falcate, erect, caducous or occasionally rather persistent. Raceme (6.3)11–26.5(27.2) cm long (especially long in E Zimbabwe), (5)6–20(21)-flowered; peduncle (1.6)3–9(10) cm long, aculeate (sometimes so densely so that the prickles fuse to form dentate ridges) or completely without prickles; pedicels 0.9–2.2(2.4) cm long; bracts 5–10 mm long, linear to linear-lanceolate, acuminate, caducous; bracteoles 3–7 mm long, linear, caducous. Calyx 5–12 × 6–10 mm, the teeth 1–2 mm long, broadly acuminate. Standard deep yellow, flecked or spotted with violet or purple on the outer face, (1.4)1.6–2.6 × (1.4)1.7–2.3(2.5) cm, the blade first widening gradually from the claw and then very abruptly so that it is subcordate just below the middle, apically emarginate, appendages semicircular or broadly triangular and without free tips; wings yellow, 1.3–2.1 cm × 4–9 mm, with upper basal acuminate tooth slightly hooked or ± horizontal and with lamellate sculpturing in upper basal and central sections on outer face; keel greenish-yellow, slightly speckled with purple, 1.3–1.6 cm × 7.5–11 mm, the basal tooth usually large, horizontal, ±

parallel to the petal-claw. Ovary 22–32-ovulate; style glabrous. Pod 14.5–31 cm × 3.5–6.5(8) mm, curved, long acuminate, thicker at the centre than at the sutures which are sometimes slightly constricted between the seeds particularly along the upper margin, 12–30-seeded; septa 9–12(13) mm apart; sutures of younger fruits drying green in contrast to the brown central area. Seeds dark olive-brown to almost black, not mottled, 6–8 × 2.5–3.5 mm, 2–3 mm thick, subcylindrical to cylindrical; hilum in ± central circular pit surrounded by a white rim aril.

Var. **macrantha** —Gillett in Kew Bull. **17**: 117, figs. 2/13, 2/14 and 3/5 (1963). —Gillett in F.T.E.A., Legum., Pap.: 341 (1971). —Corby in Kirkia **9**: 326 (1974). —Lewis in Kirkia **13**: 28 & map 1 (1988). FIGURE 3.3.**33**/18.

> *Sesbania punctata* sensu Baker in F.T.A. **2**: 133 (1871) in lesser part, non DC.
> *Sesbania aculeata* sensu Baker f. in J. Linn. Soc., Bot. **37**: 143 (1905) non Poir.
> *Sesbania bispinosa* sensu Brenan, Check-list For. Trees Shrubs Tang. Terr.: 440 (1949) in part, non (Jacq.) W.F.Wight.

Peduncle, and usually stem and leaf rachis, aculeate, although prickles sometimes very few and not immediately evident.

Zambia. N: Mbala Dist., Mwengo, fl. & fr. 10.vi.1951, *Bullock* 3954 (K). W: Solwezi Dist., Solwezi Dambo, fr. 15.vi.1930, *Milne-Redhead* 518 (K). E: Chipata Dist., Nkokwa Dambo, fl.& fr. ii.1962, *Verboom* 462 (K, LISC). **Zimbabwe**. C: Marondera Dist., Middlesex road N of Marondera (Marandellas), fl. 20.i.1965, *Corby* 1213 (K, SRGH). E: Mutare (Umtali), fl.& fr. 23.iv.1945, *Chase* 133 (COI, K, LISC, SRGH). **Malawi**. N: Mzimba R., fl. 26.ii.1959, *Robson* 1723 (K, LISC, PRE, SRGH); Chitipa Dist., Nyika Plateau, NE foot of Nganda, fr. 1.viii.1972, *Brummitt & Synge* WC121 (K). C: Dedza Dist., foot of Dedza Mt., 2 km NW of town, fl.& fr. 1.iv.1970, *Brummitt* 9574 (K, PRE, SRGH). S: Blantyre Dist., Matenje road, 1–2 km N of Limbe, fl.& imm.fr. 10.iii.1970, *Brummitt* 9011 (K, SRGH). **Mozambique**. N: Lichinga Dist., R. Licuir, c.15 km from Litunde, fr. 6.x.1983, *Groenendijk* 705 (K). Z: between Gurué and Alto Molócuè (Moloqué), fl.& fr. 28.v.1937, *Torre* 1510 (COI, LISC, both specimens with very few minute prickles, close to var. *levis*). T: Moatize Dist., Zóbuè, fr. 17.vi.1941, *Torre* 2864 (LISC). MS: Sussundenga Dist., Serra Zuira, Tsetserra, 15 km on Vacaria road to Chimoio (Vila Pery), fl.& fr. 5.iv.1966, *Torre & Correia* 15721 (LISC).

Also in West Africa, Cameroon, Congo, Burundi, Uganda, Kenya, Tanzania, Angola and possibly the former Transvaal in South Africa (Burtt Davy, Fl. Pl. Ferns Transvaal, pt.2: 381, 1932). River or stream margins, lake shores, in wet or swampy ground in dambos and on mushitu margins; also in wet areas in miombo woodland; 750–2300 m.

Corby 1213 from Zimbabwe is particularly prickly, with the prickles fused into dentate ridges on the peduncles; *Arnold* 12580 is unusual in having persistent bracts on the well-developed inflorescences.

Var. **levis** J.B. Gillett in Kew Bull. **17**: 118 (1963). —Gillett in F.T.E.A., Legum., Pap.: 341 (1971). —Corby in Kirkia **9**: 326 (1974). —Ross in Fl. Pl. Africa **47**: t.1856 (1982). —Lewis in Kirkia **13**: 30 & map 1 (1988). Type: Zimbabwe, Harare and Marondera Districts, *Eyles* 1514 (BM holotype, PRE, SRGH). FIGURES 3.3.**32**/5, 3.3.**33**/5.

> *Sesbania cinerascens* sensu Baker in F.T.A. **2**: 134 (1871) in part, not the type. —sensu Phillips & Hutchinson in Bothalia **1**: 47 (1921). —sensu Baker f., Legum. Trop. Africa: 260 (1929). —sensu Cronquist in F.C.B. **5**: 79 (1954) for most part, excepting the type, non Baker emend E.P. Sousa. —sensu White, F.F.N.R.: 164 (1962).

Completely non-aculeate.

Zambia. N: Luapula Prov., Upper Pambashye dambo, fl.& fr. 14.iv.1961, *Verboom* LK228 (K, PRE, SRGH). C: Lusaka Dist., Mt. Makulu Res. Station, 20 km S of Lusaka, fl.& fr. 2.iv.1957, *Angus* 1535 (K, PRE). S: Choma Dist., Mapanza, fl.& fr. 29.v.1953, *Robinson* 284 (K). **Zimbabwe**. N: Mazowe Dist., main road through Henderson Res. Station, fr. 11.vi.1970, *Biegel* 3327 (K, SRGH). W: Matobo Dist., Matopos, fl. v.1915, *Rogers* 7924 (PRE). C: Harare Dist., Christon Bank, Botanic Garden Extension, fl. 6.iv.1981, *Philcox & Drummond* 9074 (K). S: Masvingo Dist., Mushandike Sanctuary (Nat. Park), Shatu R., fl. vii.1971, *Wright* 5271 (K, SRGH). **Malawi**. C: Lilongwe Dist., beside Ngala Mt. on SW side, 23 km SE of Lilongwe, fl.& fr. 28.iv.1970, *Brummitt* 10232 (K, PRE). S: Zomba Dist., S side of Mwinje Hill Forest, fr. 6.vi.1987, *Kaunda & Kwatha* 621 (K, MAL). **Mozambique**. Z: Gurué Dist., Namuli Peaks, lower slopes near Gurué (Vila Junqueiro), fl.& fr. 25.vii.1962, *Leach & Schelpe* 11452 (K, LISC, PRE, SRGH). MS: Sussundenga Dist., Serra Mocuta (Serra de Mavita), fl.& imm.fr. 10.iv.1948, *Barbosa* 1421 (LISC).

Also in West Africa, Cameroon, Congo, Tanzania, Angola and South Africa (Limpopo, Mpumalanga and KwaZulu-Natal Provinces). River or stream margins and in wet or swampy ground in dambos, in tall grassland beside rivers and in wet areas in miombo woodland, on damp sandy soils or deep ferrugineous clays; 750–1600 m.

Biegel, Check-list Ornam. Pl. Rhod. Parks & Gard.: 98 (1977), records this species as cultivated in Zimbabwe.

A few specimens of var. *macrantha* possess very few prickles (e.g. *Salubeni* 1458 and *E. Phillips* 3832 & 3833 from Malawi; *Verboom* 462 from Zambia and *Norlindh & Weimarck* 2451 & 4443 from Zimbabwe) and are difficult to separate from var. *levis*. But plants are usually either clearly aculeate in the former or completely without prickles in the latter. Gillett's comments on the geography of the two taxa in Kew Bull. **17**: 119 (1963) continue to apply in the Flora Zambesiaca area, i.e. although the distribution of the two varieties overlaps considerably, the geography is not the same. Excluding cultivated specimens in central Zimbabwe, only at Gurué in Mozambique do the two varieties occur in the same general locality, where var. *macrantha* is represented by *Torre* 1510, a specimen with very few prickles.

S. macrantha var. *levis* has often been confused with *Sesbania coerulescens* in fruit, but the sutures of the pod are of equal width and non-angular in the former whereas in the latter the lower suture is wider than the upper one and is distinctly angular. The K and SRGH sheets of *Bingham* 184 are mixed collections. The flowering portions belong to *S. macrantha* var. *levis*, the fruit to *S. tetraptera* subsp. *rogersii*.

7. **Sesbania cinerascens** Baker [in F.T.A. **2**: 134 (1871) for most part] emend. E.P. Sousa in C.F.A. **3**: 183 (1962). —White, F.F.N.R.: 433 (1962). —Gillett in Kew Bull. **17**: 121, fig.5/1 (1963). —Schreiber in Merxmüller, Prodr. Fl. SW Afrika, fam. 60: 109 (1970). —Lewis in Kirkia **13**: 31 (1988). —K. Coates Palgrave, Trees Sthn. Africa, ed.2: 313 (1988). —M. Coates Palgrave, Trees Sthn. Africa: 380 (2002). Type: Angola, Pungo Andongo, near Lombe R., iii.1857, *Welwitsch* 1999 (LISU holotype, BM). FIGURES 3.3.**32**/6, 3.3.**33**/6.

Sesbania kapangensis Cronquist in Bull. Jard. Bot. État **22**: 229 (1952); in F.C.B. **5**: 83 (1954). —Lock, Leg. Afr. Check-list: 461 (1989). Type from Congo.

Short-lived suffrutescent perennial herb 2.5–4 m tall. Stems pithy, glabrous, glabrescent or less commonly densely shortly woolly, particularly on the young branches. Leaves 7–36 cm long; petioles 0.3–1.2(4.5) cm long; leaflets in 20–65 pairs, 0.6–3.5 cm × 1.5–6.5 mm, oblong, obtuse or rarely emarginate at the apex, apiculate, glabrous or sparsely spreading pubescent especially on the midvein and margins beneath; stipules up to 7 mm long, marginally or totally woolly, caducous. Raceme 7–33 cm long, 13–23(26)-flowered; peduncle (1)2–7.5 cm long, densely woolly to sparingly pubescent or glabrescent; pedicels 1–2.2 cm long, densely pubescent, glabrescent or glabrous; bracts caducous; bracteoles 2.5–3 mm long, linear-lanceolate, marginally pubescent to glabrous, caducous. Calyx 6.5–9 × 6–9 mm, oblique, the tube glabrous or sparsely pubescent, the teeth 0.75–1 mm long, acuminate, slightly pubescent marginally. Standard yellow, usually streaked or with dark purple speckles, 2.2–2.7 × 1.7–2.2 cm, broad, rounded, apically emarginate, the appendages with blunt strongly incurved free tips up to 1.5 mm long; wings yellow, 2.2–2.6 cm × 5–8 mm, with an acute basal tooth variable in length or without an obvious tooth, and with lamellate sculpturing in the upper basal one-third to one-half on the outer face; keel yellowish-green, 2.2 cm × 6–11 mm, with a broad subquadrate limb and a basal tooth projecting upwards. Ovary glabrous or rarely pubescent on the lower suture; style glabrous. Pod dark reddish-brown or purplish when mature, area over septa darker than area over seeds when immature, 11.5–33(36.5) cm long (including a stipe of 3–10 mm and beak of 2–3 mm), 2.75–3.5 mm wide, pendulous, linear, very slightly falcate, very occasionally constricted between the seeds along the sutures but usually parallel-sided, thicker above the seed cavities than above the septa, glabrous, (17)30–44(50)-seeded; septa 6–7 mm apart; exocarp of mature pods often fracturing at right-angles to the sutures. Seeds olive-green, brown or black, sometimes mottled, 3.5–5 × 1.5–2 mm, 1.5–2 mm thick, subcylindrical; hilum in a subcentral circular pit surrounded by an obscure rim aril. Seedlings with first foliage leaf simple, ± equalling the cotyledons in size, second foliage leaf paripinnate.

Botswana. N: Ngamiland Dist., Kwando R., main stream bank, 18°12'S, 23°21'E, fl. 1.ii.1978, *P.A. Smith* 2330 (K, PRE, SRGH); Chobe Dist., Linyanti R. at Shaile, fr. 28.x.1972, *Gibbs Russell* 2423 (K, PRE, SRGH). **Zambia**. B: Sesheke Dist., Kwando R., between Imusho and Sinjembela, fl.& fr. ix.1959, *Guy* in GHS 98856 (SRGH). N: Nchelenge Dist., L. Mweru, near Nchelenge, fl.& fr. 26.iv.1951, *Bullock* 3825 (K, PRE). W: Kitwe, fl., fr. & seedl., 4.vii.1960, *Fanshawe* 3049A (K, LISC, PRE, SRGH).

Also in Chad, Congo, Angola and Namibia. In colonies in swampy areas and seasonally flooded lake shores, river-sides and dambos, in peaty soil or in mud; 900–1400 m.

Also recorded from Mozambique T: by Gonçalves in Garcia de Orta, sér.Bot. **5**: 110 (1982).

The degree of pubescence is very variable, although all specimens tend to be glabrescent on the older parts. Specimens from N Botswana (together with those from Namibia) show a pronounced increase in pubescence so that, particularly on the young shoots, there is a dense grey tomentum. These specimens also possess calyces that have sparsely appressed-pubescent tubes, ovaries that are sparsely pubescent along the lower suture and leaflets that are pubescent beneath and sparsely pubescent or glabrous above. By contrast, material from Zambia (together with that from Congo and Chad) is generally glabrous or glabrescent, the calyx tubes are usually totally glabrous, the ovaries always so and the leaflets if not glabrous on both surfaces have, at the most, a few scattered hairs on the margins and the midvein beneath. The Angolan type collection, *Welwitsch* 1999 (fruiting specimen at BM), has a pubescence somewhat intermediate between these two groups, although it is closest to the less pubescent material from Zambia, etc. Given further collections it may prove desirable to recognize the material from Botswana and Namibia as a distinct subspecies.

Verboom 214 from Zambia, a specimen with fruit only, whilst possessing a pod characteristic of *S. cinerascens* has obscure prickles on the undersurface of the leaf rachis and may represent a link with *S. bispinosa*.

8. **Sesbania rostrata** Bremek. & Oberm. in Ann. Transvaal Mus. **16**: 419 (1935). — Gillett in Bol. Soc. Brot., sér.2, **35**: 8 (1961); in Kew Bull. **17**: 122, fig.3/8 (1963). —Schreiber in Merxmüller, Prodr. Fl. SW Afrika, fam. 60: 111 (1970). —Gillett

in F.T.E.A., Legum., Pap.: 343 (1971). —Corby in Kirkia **9**: 326 (1974). —Lewis in Kirkia **13**: 32 & 21 (1988). Type: Botswana, Ngami Flats, *van Son* in *Transvl. Mus.* 28895 (PRE holotype, BM, K photo). FIGURES 3.3.**32**/7, 3.3.**33**/7 & 21, 3.3.**34**/3.

  Sesbania pachycarpa DC., Prodr. **2**: 265 (1825) in part. —Baker in F.T.A. **2**: 134 (1871) for most part. —Phillips & Hutchinson in Bothalia **1**: 50 (1921) in part. —Berhaut in Bull. Soc. Bot. France **99**: 297–301(1953). —Hepper in F.W.T.A., ed.2, **1**: 532 (1958).

  Sesbania hirticalyx Cronquist in Bull. Jard. Bot. État **22**: 228 (1952); in F.C.B. **5**: 82 (1954). Type: Congo, Katanga, Mabwe near L. Upemba, *Van Meel* in *de Witte* 4719 (BR holotype, K).

An erect, robust, softly woody, non-aculeate annual or biennial herb, 1–3 m tall. Stems pithy, sparsely pilose at first, later glabrescent with vertical rows of pustules usually evident above the leaf axils and producing warty outgrowths on older stems, submerged portions clothed with matted fibrous roots. Leaves (4.6)7–25.5 cm long, the rachides sparsely pilose; petioles 3–8 mm long, sparsely to densely pilose; leaflets opposite or sometimes alternate in (6)12–24(27) pairs, 0.9–3.5 cm × 2–10 mm, the basal pair usually smaller than the rest, sometimes 5 × 3 mm, oblong, rounded to obtuse to slightly emarginate at the apex, apiculate, glabrous above or rarely with an occasional hair on the midrib, usually sparsely pilose on margins and midrib beneath; stipules 5–10 mm long, linear-lanceolate to lanceolate, reflexed, pilose, rather persistent. Racemes axillary, shorter than the subtending leaf, (0.5)1–5.9 cm long, the rachis pilose, (1)3–12(15)-flowered; peduncle (0.2)0.4–1.5(2.2) cm long, pilose especially at the base; pedicels 0.4–1.5(1.9) cm long, sparsely pilose; bracts and bracteoles linear to linear-lanceolate, sparsely pilose, caducous, the bracts 5–8 mm long, the bracteoles c.5 mm long. Calyx 5–7.5 × 4–5 mm, sparsely pilose, the teeth 1–2 mm long, subulate, sparsely pilose. Standard yellow or orange, speckled dark purple or reddish, 1.2–1.6(1.8) × 1.1–1.4(1.5) cm, the blade suborbicular, apically emarginate, the appendages with short triangular upward-pointing or slightly incurved free tips less than 1 mm long; wings yellow, 1.3–1.6(1.7) cm × 3.5–5 mm, the small, triangular tooth and the upper margin of the basal half of the blade together characteristically inrolled, lamellate sculpturing in upper basal section on outer face; keel yellow to pale greenish, 1.2–1.5(1.7) cm × 6.5–9 mm, basal tooth short, triangular, slightly upward-pointing with a small, 3–4 mm long, pocket below it on inner face of the blade. Ovary sparsely pilose on upper margin or glabrous; style glabrous. Pod 15–22 cm long (including a slender beak 1.5–3.5 cm long), 3.5–5 mm wide, erect, falcate, distinctly thicker at the centre than at the sutures, the central areas on mature fruits with prominent venation, this sometimes interspersed with red blotching or dark red warts, glabrous, (13)24–51-seeded; septa 3.5–4.5(5) mm apart. Seeds brown, greenish or dark reddish-brown, 3–3.5 × 2.5–3 mm, 2–2.5 mm thick, ± cube-shaped to subcylindrical, the hilum in a small central circular pit.

Botswana. N: Ngamiland Dist., Matlapaneng, Thamalakane R. bridge, 8 km NE of Maun, fl.& fr. 19.iii.1965, *Wild & Drummond* 7169 (K, LISC). **Zambia**. S: Monze Dist., Lochinvar Nat. Park (Ranch), Kafue Flats, fr. 25.iv.1962, *Mitchell* 13/96 (SRGH). **Zimbabwe**. E: Chipinge Dist., Save (Sabi) escarpment behind Dotts Drift, fl.& fr. 9.iv.1959, *Savory* 405 (K, PRE, SRGH). S: Chiredzi Dist., Gonarezhou, between Chitsa's store and Save/Runde junction (Sabi/Lundi), Tambahata (Tamboharta) Pan, fl.& fr. 31.v.1971, *Grosvenor* 586 (K, LISC, PRE, SRGH). **Malawi**. N: Rumphi Dist., Nyika Plateau, Mwenembwe (Mwanemba), fl.& imm.fr. iii.1903, *McClounie* 83 (K). S: Nsanje Dist., James Lagoon, 10 km N of Chiromo, fl. 22.iii.1960, *Phipps* 2613 (K, PRE, SRGH). **Mozambique**. Z: Nicoadala Dist., S.A. Madal Estates, Mussuluga floodplains, fl.& fr. 8.iv.1958, *Pedro* 65 (PRE). MS: Gorongosa Dist., Gorongosa Nat. Park, Urema floodplains, fl.& fr. vi.1972, *Tinley* 2626 (K, PRE, SRGH).

Also from Senegal to Ethiopia, Congo (Katanga), Tanzania, Namibia (Okavango area) and Madagascar. In wetter parts of floodplains, on river banks and in seasonally inundated rainwater pans, on alluvium and black clays, in wet times of the year can be found in standing water up to 1 m deep; 500–1200 m.

The inrolled upper margin of the wing petals and the persistent pocket on the inner face of the keel petals, together with the almost cube-shaped, closely packed seeds and the long beak to the pod, separate this from all other species.

De Winter & Marais 4869 from the Okavango Native Territory (E Caprivi) may be the basis for Gillett's inclusion of the Caprivi Strip in the distributions cited in Kew Bull. **17**: 122 (1963) and F.T.E.A., Legum., Pap.: 344 (1971). His reference to the pod as "25–30-seeded" in F.T.E.A. is a mistake as specimens seen by him for that flora have up to 40 seeds per pod.

9. **Sesbania bispinosa** (Jacq.) W.F. Wight in U.S. Dept. Agric. Bur. Pl. Industr. Bull.
   137: 15 (1909). —Sprague & Milne-Redhead in Bull. Misc. Inform., Kew **1939**: 159 (1939). —Gillett in Kew Bull. **17**: 129, figs.2/11 & 12, fig. 3/12 (1963) for most part, but not *Obermeyer* 2445. —Schreiber in Merxmüller, Prodr. Fl. SW Afrika, fam. 60: 109 (1970). —Gillett in F.T.E.A., Legum., Pap.: 349 (1971). — Lewis in Kirkia **13**: 34 & map 2 (1988). Type: plant of unknown origin, probably Asiatic, cultivated in Vienna before 1788 and figured in Jacq., Icon. Pl. Rar. **3**: t.564 (1792). FIGURES 3.3.**32**/8, 3.3.**33**/8.

   Aeschynomene bispinosa Jacq., Ic. Pl. Rar. **3**: 13, t. 564 (1792).
   Coronilla aculeata Willd., Sp. Pl. **3**: 1147 (1802) nom. illegit. Type as for species.
   Sesbania aculeata (Willd.) Pers., Syn. Pl. **2**: 316 (1807). —Harvey in F.C. **2**: 212 (1862). — Baker in F.T.A. **2**: 134 (1871) in part. —Phillips & Hutchinson in Bothalia **1**: 50 (1921) in part. —Baker f., Legum. Trop. Africa: 262 (1929) in part, non sensu Baker in J. Linn. Soc., Bot. **37**: 143 (1905) nom. illegit.
   Sesbania aculeata var. *micrantha* Chiov., Fl. Somala **2**: 164 (1932). Type from Somalia.
   Sesbania bispinosa var. *micrantha* (Chiov.) J.B. Gillett in Kew Bull. **17**: 130 (1963).

Erect annual, biennial or short-lived perennial slightly woody herb, (0.6)1–3 m tall. Stems glabrous, or very sparsely pilose when young, usually aculeate although often obscurely so. Leaves (5.3)9.5–29.5(35) cm long, the rachis aculeate beneath especially towards the base, sparsely pilose when young, later glabrous; petioles 2–15 mm long; leaflets in (11)20–50(55) pairs, 0.75–2(2.6) cm × 1.5–3(5) mm, oblong to oblong-linear, apically emarginate, usually apiculate, sparsely pilose when young on margins and midvein beneath, later glabrous; stipules 5–11 mm long, lanceolate to linear-lanceolate, ± erect, pilose on margins and inner face, at length falling. Raceme (0.8)2.3–15(16.5) cm long, 1–12(14)-flowered; peduncle (0.5)1.5–4(6) cm long, glabrous, nearly always aculeate especially at the base; pedicels (2)4–8(11) mm long, glabrous; bracts 3–5 mm long, linear-lanceolate, caducous; bracteoles 2–3 mm long, filiform, caducous. Calyx 3–4 × 3–4 mm; tube glabrous except for the woolly margins, the teeth 0.5–1 mm long, triangular, ± acuminate. Standard pale yellowish or cream-coloured, flecked or spotted brownish or purplish, 0.9–1.5 × 0.8–1.4 cm, rounded, apically emarginate, the appendages wedge-shaped and truncate or slightly emarginate apically; wings yellow, 0.9–1.25 cm × 2.5–3 mm, with a short, blunt, unhooked upper basal tooth and with lamellate sculpturing in the upper basal one-half to two-thirds; keel 0.9–1.3 cm × 3.5–5 mm, the blade distinctly longer than wide, the upper edge gently sloping from the apex to the short blunt or slightly acute basal tooth parallel to the claw or slightly downward pointing. Ovary and style glabrous. Pod brownish mottled with reddish-brown patches, 12.5–25(28) cm long (including a beak 1–1.5 cm long), 2–3 mm wide, curved, the central area sometimes ridged parallel to the sutures, glabrous, 28–45-seeded; septa 4–6 mm apart. Seeds pale brown, olive-green or greenish-black, 3 × 1.5 mm, 1.2 mm thick, elliptical in cross-section, the hilum in a small subcentral circular pit surrounded by a white rim aril.

Botswana. N: Ngamiland Dist., 83 km W of Nokaneng, fl.& imm.fr. 12.iii.1965, *Wild & Drummond* 6909 (K, LISC, PRE, SRGH); Toromoja, Boteti (Botletle) R., fr. 22.iv.1971, *Campbell* 6 (K, SRGH). SE: Kweneng Dist., Gaborone, Broadhurst Farm, fl. 19.iii.1977, *Hansen* 3088 (K, PRE). **Zimbabwe**. N: Kariba, fr. iii.1960, *Goldsmith* 39/60 (K, PRE, SRGH). W: Hwange Dist., Matetsi Safari Area, fl.& fr. 19.iii.1979, *Gonde* 120

(K, PRE, SRGH). C: Hwedza Dist., Ziyambe Purchase Area, Dorowa road, fl.& imm.fr. 28.ii.1967, *Corby* 1782 (K, LISC, PRE, SRGH). E: Chimanimani Dist., road between Mutare and Junction Gate, fl.& imm.fr. 28.ii.1968, *Corby* 1999 (K, LISC, PRE, SRGH). S: Gwanda Dist., Mkezi R., fr. v.1955, *Davies* 1312 (K). **Malawi**. S: Machinga Dist., Liwonde Nat. Park, fl.& fr. 17.iv.1980, *Blackmore, Banda & Brummitt* 1279 (K, MAL). **Mozambique**. N: Montepuez Dist., base of Mt Matuta, c.5 km S of Rio Messalo (M'salo) near Nairoto (Nantulo), c.350 m, fr. 9.iv.1964, *Torre & Paiva* 11794 (LISC). Z: Lugela Dist., Mocuba, Namagoa, fl.& imm.fr. 18.iii.1949, *Faulkner* Kew 411a (K). T: Cahora Bassa Dist., c.6 km from Marueira, fl. 5.ii.1972, *Macêdo* 4801 (LISC). MS: Beira Dist., Manga, bud & fr. ii.1923, *Honey* 697 (K, PRE). GI: Limpopo, between Caniçado and Mejinge, fl.& fr. 18.v.1948, *Torre* 7859 (LISC). M: Machanecke, Incomáti R. mouth, fl.& fr. 21.ii.1961, *Mogg* 29685 (K, LISC).

Also in Somalia, Kenya, Tanzania, Namibia, South Africa (Limpopo, Mpumalanga and KwaZulu-Natal Provinces), Swaziland and Lesotho; and in China, India, SE Asia, Fiji, Oman, Madagascar, Mauritius, Jamaica and Guyana, probably as an introduced weed in many of these areas. Floodplains and river banks, in seasonally inundated rainwater pans and depressions, and in swampy ground, on sandy soils and alluvial clays. In wet times of the year can be found in standing water up to 1 m deep; 0–1100 m.

Gillett in Kew Bull. **17**: 130 (1963) discusses the distribution of *S. bispinosa* var. *micrantha*, and comments that "it may represent merely a stunted state of whatever form of the species occurs in a given area". This is supported by material from the Flora area, e.g. *P.A. Smith* 1815 & 1977 and *Biggs* M293 from Botswana, *Grosvenor* 610, *Drummond* 5612 and *Rushworth* 147 from Zimbabwe and *Torre* 2593 & 7859 from Mozambique, all with 1–2-flowered inflorescences. The filament sheaths of the flowers of these collections range in length from 8 mm in *Torre* 7859 to 11 mm in *Rushworth* 147 and *P.A. Smith* 1815. The specimens either represent young or, more commonly, scrappy depauperate plants and it would appear that such plants have a tendency to produce few-flowered racemes. *Grosvenor* 610 is an extreme of this condition with reduced leaves and the flowers single in the leaf axils. It was collected in a dried pan where stressful growing conditions could well have caused stunting. Occasionally, normal specimens displaying 3–12-flowered inflorescences also produce 1–2-flowered inflorescences in the upper axils. In light of these observations var. *micrantha* is considered as a synonym of *S. bispinosa*. The second sheet of *Faulkner* 411a is part of a mixed collection, the other specimen 411b is *S. greenwayi*. The altitudinal range of the species is far greater in the Flora Zambesiaca area than in East Africa; *Gonde* 120 from the Hwange District of Zimbabwe is recorded from c.1100 m.

S. bispinosa has been confused with many other species, notably *S. leptocarpa* and *S. microphylla*, from which it is separable in flower by the truncate shape of the unopened standard, the keel which is longer than wide with a short fairly blunt tooth, and by the apically truncate or slightly emarginate appendages. The pod is straight-sided and never truly torulose.

10. **Sesbania leptocarpa** DC., Prodr. **2**: 265 (1825). —Baker in F.T.A. **2**: 135 (1871). —Phillips & Hutchinson in Bothalia **1**: 50 (1921). —Hepper in F.W.T.A., ed.2, **1**: 532 (1958) for most part. —Gillett in Kew Bull. **17**: 142, fig.4/7 (1963) in part excl. specimen *Dubber* 10. —Lewis in Kirkia **13**: 37 (1988). Type: Senegal, no locality, *Bacle* s.n. (G holotype).

    Sesbania arabica E. Phillips & Hutch. in Bothalia **1**: 51 (1921) for most part.

    Sesbania mossambicensis Klotzsch in Peters, Naturw. Reise Mossamb. **6**(1): 45 (1861). —Phillips & Hutchinson in Bothalia **1**: 52 (1921). —Baker f., Legum. Trop. Africa: 263 (1929) for type only. —Gillett in Kew Bull. **17**: 141, fig.5/4 (1963). —Corby in Kirkia **9**: 326

(1974). Types: Mozambique, Nampula Prov., Cabaceira Peninsula near Ilha de Moçambique, *Peters* s.n. (B† syntype, K, P); Zambezia Province, near Quelimane, *Peters* s.n. (B† syntype).

Annual herb 1–3 m tall. Stems glabrous, often slightly aculeate (occasionally the prickles very evident). Leaves 2.8–24.5 cm long; petioles 0.3–1.1(1.8) cm long; leaflets in 8–44 pairs, 0.45–2.5(2.8) cm × 1.5–4.5(5.5) mm, oblong, apiculate, distinctly asymmetrical at the base, entire, glabrous except when very young; stipels slightly hooked, usually persistent; stipules 2–5(8) mm long, narrowly triangular to linear-lanceolate, caducous, sparsely hairy on the margins. Raceme 13–14.3 cm long, 1–17-flowered; peduncle 2.1–4.3 cm long, sparsely pubescent near the base only, rarely totally glabrous, sometimes very slightly aculeate; pedicels 3–5(8) mm long, glabrous; bracts and bracteoles linear to linear-lanceolate, sparsely pubescent on the margins or glabrous, caducous, the bracts c.2.5 mm long, the bracteoles 1.5 mm long. Calyx 2.5–5 × 3–4.5 mm, the teeth 0.5–1 mm long, acute, somewhat woolly at the margins. Standard yellow speckled with dark purple on the outer face, 7–12 × 6–9 mm, rounded, the blade usually wider than long, apically emarginate, sometimes distinctly cordate at the base, the appendages narrowly wedge-shaped without free tips and ± truncate at the apex; wings 6.5–11 × 2.5–3.5 mm, the upper basal tooth rather variable, short, blunt, and horizontal or acute and down-turned, occasionally almost lacking, the blades with lamellate sculpturing in the upper basal one-third to one-half on the outer face; keel 6.5–10 × 3–5 mm, the claw usually longer than the blade, the tooth narrow, acuminate and horizontal or wider, acute and ± downward pointing. Ovary 17–29(31)-ovulate, glabrous; style glabrous (or very rarely with a few hairs). Pod pale green to straw-coloured, usually with a well-marked dark blotch above the septal area, (5.4)8–17.2 cm × 2–3.5 mm, held ± erect, linear to slightly falcate, torulose, shortly beaked, glabrous, (7)12–29(31)-seeded; septa 5–7 mm apart. Seeds pale to dark brown, unmottled, 3.5–4 × 1.5–2 mm, 1–1.25 mm thick, ellipsoid; subcentral hilum in small circular pit.

Var. **leptocarpa** —Lewis in Kirkia **13**: 38 & map 2 (1988). FIGURES 3.3.**32**/9, 3.3.**33**/9.

Filament sheath 6 mm long or longer.

Zimbabwe. E: Chipinge Dist., Save (Sabi) Valley, fl.& imm.fr. 23.i.1964, *Corby* 1070 (K, SRGH), specimen verging towards *S. bispinosa*. S: Beitbridge Dist., Shashe–Limpopo confluence, fl.& imm.fr. 22.iii.1959, *Drummond* 5954 (K, LISC, PRE), specimen verging towards *S. bispinosa*. **Malawi**. S: Chikwawa Dist., Lengwe Nat. Park (Game Reserve), fl. 1.ii.1970, *Hall-Martin* 529 (K, SRGH). **Mozambique**. N: Nampula Prov., Mossuril Dist., Cabaceira Peninsula near Ilha de Moçambique, fl.& imm.fr. no date, *Peters* s.n. (K). Z: Chinde Dist., between Chinde and Luabo, fl.& fr. 13.x.1941, *Torre* 3642 (LISC). T: Cahora Bassa Dist., between Chicoa and Chetima, 19.2 km from Chicoa, fl.& fr. 30.vi.1949, *Barbosa & Carvalho* 3398 (K, LMA). MS: Chemba Dist., Chiou, Expt. Station of C.I.C.A., fl.& fr. 12.iv. 1960, *Lemos & Macuácua* 72 (K, LISC). GI: Inhambane, Massinga, between Funhalouro and Saúte, fl.& fr. 19.v.1941, *Torre* 2689 (LISC). M: Matutuíne, between Salamanga and Quinta da Pedra, fl.& imm.fr. 4.ii.1948, *Gomes e Sousa* 3662 (COI, K, LISC, SRGH).

Also in the Yemen, Senegal, N Nigeria, Sudan, Eritrea, Ethiopia and Somalia. Coastal dunes, floodplains, riverine forest margins and seasonally inundated grassland, often as a weed of irrigated farmland at low altitudes; on sandy soil; 0–700 m.

Sesbania leptocarpa DC. (1825), being the older name, is broadened to include *S. mossambicensis* Klotzsch (1861). The *S. leptocarpa* "complex" is undoubtedly a confusing one, showing tendencies towards both *S. microphylla* and *S. bispinosa* within its morphological range. To recognize *S. mossambicensis* as a SE African subspecies of *S. leptocarpa* was rejected on two counts. Firstly, whilst plants found at either end of the geographical range of the "complex" do show differences in flower number and inflorescence length, there exists a morphological gradation from the Mozambique coast to W Africa and the Yemen. Many specimens in Mozambique occur on coastal sand dunes, and these are usually 5–17-flowered and have longer inflorescences than plants growing further inland. Collections from the Yemen and W Africa are fewer (1–4, rarely 5) flowered and often occur as weeds of cultivation on sandy soils. However, *Westphal & Westphal-Stevels* 1218, from Ethiopia, shows greater affinity to material from coastal Mozambique than to other Ethiopian or Yemeni material. Similarly, *Brummitt* 8893 and *Hall-Martin* 529 from southern Malawi, and *Barbosa & Carvalho* 3222 & 3398 from the Tete Province of Mozambique, show greater affinity to material from the Yemen, NE Africa and W Africa. Secondly, some individual plants show variable flower and fruit numbers per inflorescence and correspondingly variable inflorescence lengths on the same shoot.

The length of the claw of the keel petal relative to the length of the blade, the angle, width, length and apex of the basal tooth of both keel and wing petals, and the distance between the pedicels are all rather variable characters. Specimens are often aculeate and, in the absence of mature fruits, are difficult to separate from *S. bispinosa*. The blade of the keel petal in *S. bispinosa* is always longer than wide and the basal tooth is short and blunt. Although most collections of *S. leptocarpa* have keel petals with the blade as wide as, or wider than, long and the basal tooth is usually long, narrow and acuminate, some specimens show a tendency towards *S. bispinosa*. *Corby* 1070, *Obermeyer* 2445, *Kelly* 527, *Drummond* 5954 from SE Zimbabwe and *Torre* 7161 & 7316 from S Mozambique are such examples. Whether these are the result of hybridization between the two species is not known. It is more likely they are the result of continuing speciation that has also produced the closely related species *S. microphylla*, *S. leptocarpa* and *S. bispinosa*.

Var. **minimiflora** (J.B. Gillett) G.P. Lewis in Kirkia **13**: 39, map 2 (1988). Type: Zambia S: Zambezi Valley, Gwembe area, *Trapnell* 1478 (K holotype).

    Sesbania mossambicensis subsp. *minimiflora* J.B. Gillett in Kew Bull. **17**: 141 (1963). — Drummond in Kirkia **8**: 226 (1972). —Gonçalves in Garcia de Orta, sér.Bot. **5**: 111 (1982).

Filament sheath 5 mm long or less.

Zambia. S: Gwembe Dist., Zambezi Valley near Chirundu, fl. 8.iii.1958, *Angus* 1860 (K, PRE). **Zimbabwe**. N: Mount Darwin Dist., Muzarabani C.L. (Mzarabani Tribal Trust Land), Musingwa R., fl.& fr. 1.v.1972, *Mavi* 1367 (K, PRE, SRGH). **Mozambique**. T: Luenha (Changara), Tete–Harare road, fl. 1.iii.1961, *Richards* 14511 (K, SRGH). MS: Gorongosa Dist., Gorongosa Nat. Park, Urema floodplain, fl.& fr. iii.1969, *Tinley* 1713 (K, SRGH).

Not known outside the Flora Zambesiaca area. Floodplain grasslands, sometimes as a weed of cultivation; on sandy soils; 430–830 m.

Although the overall appearance of var. *minimiflora* is different from var. *leptocarpa*, only the length of the filament sheath consistently separates one from the other. It is with some hesitation that the variety is recognized here. Both varieties occur in coastal Mozambique near Beira, and at Luenha in Tete Province, Mozambique. On the other hand, var. *minimiflora* occurs also in S Zambia and N Zimbabwe where var. *leptocarpa* has not been recorded. In S Zambia in the area around Mazabuka, var. *minimiflora* overlaps with the geographical range of *S. microphylla*, but the latter is otherwise never found within the range of *S. leptocarpa*, preferring moister environments.

As in var. *leptocarpa*, specimens of var. *minimiflora* sometimes display few-flowered inflorescences. *Torre & Correia* 18542 from Luenha is an example with the lower infructescences with only a single fruit (although the upper inflorescences have up to 7 flowers). The blade of the standard is sometimes very cordate at the base.

11. **Sesbania microphylla** E. Phillips & Hutch. in Bothalia **1**: 52 (1921). —Baker f., Legum. Trop. Africa: 263 (1929). —White, F.F.N.R.: 165 (1962). —Sousa in C.F.A. **3**: 186 (1962). —Gillett in Kew Bull. **17**: 135, fig. 5/3 (1963); in F.T.E.A., Legum., Pap.: 348 (1971). —Schreiber in Merxmüller, Prodr. Fl. SW Afrika, fam. 60: 110 (1970). —Corby in Kirkia **9**: 326 (1974). —Lewis in Kirkia **13**: 39 & map 2 (1988). Type: Angola, junction of Longo and Cuito rivers, *Baum* 569 (K holotype, E). FIGURES 3.3.**32**/10, 3.3.**33**/10.

   Sesbania microphylla Harms in Warburg, Kunene-Samb.-Exped. Baum: 260 (1903) nom. nud.

   Sesbania mossambicensis sensu Phillips & Hutchinson in Bothalia **1**: 52 (1921) in part, non Klotzsch.

   Sesbania leptocarpa sensu Cronquist in F.C.B. **5**: 80 (1954) non DC.

Erect, slightly woody annual herb 0.5–3(4) m tall, glabrous except for the woolly calyx margin and a few caducous hairs on the very young stems and the margins, midvein beneath and petiolules of the young leaflets, very rarely with a few obscure scattered prickles. Stems longitudinally ribbed. Leaves (0.8)2.5–18.5(26.5) cm long; petioles 3–7(17) mm long; leaflets in (5)10–40(48) pairs, 0.2–1.1(1.8) cm × 0.75–2.5 mm, oblong, often apically emarginate, apiculate, sometimes broadening towards the apex; stipules 2–5(9) mm long, narrowly triangular, caducous. Raceme (0.7)1.6–10.5(14) cm long, 1–9-flowered; peduncle (0.4)1–4.2 cm long; pedicels 3–11 mm long; bracts 1.5–3 mm long, narrowly triangular, caducous; bracteoles 1.5–3 mm long, ± filiform, caducous or rarely persistent. Calyx 3–4(6) × 3–4(5) mm, the tube with 5 reddish-brown bands, each running into a tooth, the teeth 0.5–1.2 mm long, triangular-acuminate, subulate at the tip. Standard sometimes pure yellow or more commonly lightly to heavily dotted or flecked purplish, (1.1)1.5–2(2.8) × 1–1.6(1.95) cm, with appendages narrow, rounded near the top and without free tips; wings 1.1–1.7(2.6) cm × 3–6(7) mm, the basal tooth ± straight or slightly hooked, blades with lamellate sculpturing in the upper basal one-half to two-thirds; keel 1.1–1.6(2.2) cm × 5–7.5(11) mm, the basal tooth straight, at an angle of 40°–60° to the claw or ± parallel to it, less often bent over and pointing downwards. Ovary 25–45-ovulate, glabrous, slightly thickened just below the glabrous style. Pod (2.9)9.5–25.5 cm long (including a beak of 4–10 mm), 1.5–3.25 mm wide, slightly curved, usually somewhat torulose, sometimes regularly constricted between the seeds so as to be moniliform, less often straight-edged, ± stipitate, (4)12–42-seeded; septa 4.5–6.5 mm apart, the septal cavities elliptic or oblong, the area above the sutural ridge darker than the area above the seed. Seeds dark olive-brown or so heavily mottled purplish as to appear black, 2.5–4 × 1.5–1.8 mm, 1–1.2 mm thick, oblong-ellipsoid, with a small subcentral hilum in a circular pit.

Caprivi Strip. Shitangadimba Camp at Andara Mission Station, fl. 22.ii.1956, *de Winter & Marais* 4793 (PRE). **Botswana**. N: Central Dist., Boteti (Botletle) R., 9 km E of Makalamabedi, fl.& fr. 22.iii.1965, *Wild & Drummond* 7214 (K, LISC, PRE). **Zambia**. B: Mongu Dist., Mongu-Lealui, fl. 15.i.1960, *Gilges* 873 (K, SRGH). N: Mbala Dist., Mpulungu, fl. 6.iii.1952, *Richards* 1032 (K). W: Mufulira, bank of Kafue R., Government Expt. Farm, fl. 18.iii.1961, *Linley* 121 (K, SRGH). C: Lusaka Dist., Chikupi Plantation (Estate), fl.& fr. 12.iv.1963, *van Rensburg* KBS 1897 (K, SRGH). S: 12 km W of Mazabuka, fl.& fr. 24.iii.1952, *White* 2334 (K). **Zimbabwe**. N: Zvimba Dist., near Rodcamp Mine, fl.& fr. 20.x.1960, *Rutherford-Smith* 322 (K, LISC, PRE, SRGH). W: Hwange Dist., Kazungula, fl.& imm.fr. iv.1955, *Davies* 1195 (SRGH). C: Harare, Salisbury Expt. Station, fl.& imm.fr. 17.iii.1943, *Arnold* in GHS 9781 (K, LISC).

Also in Central African Republic, Congo, Sudan, Uganda, Tanzania, Angola and Namibia. Floodplains, river banks, lake shores and dambos usually in swampy ground, in grassland and *Acacia* woodland on black alluvial clays and on sandy shores, in wet times of the year can be found in standing water up to 1 m deep; 900–1400 m.

Gillett in Kew Bull. **17**: 135 (1963) commented that although *S. microphylla* showed "a great deal of variation especially in the basin of the Upper Zambezi", it seemed better "not to name these intraspecific variants until more material is available". Contrary to Gillett's expectations, material collected in the last 40 years does not greatly clarify the taxonomy of this species. What is needed is concentrated collecting and in-field population studies within the Barotse region of Zambia and along the Okavango River in N Botswana. Nevertheless, some general observations seem worthy of note. The standard ranges from pure yellow to, more commonly, mottled or flecked purplish. On herbarium specimens some confusion may arise as the standards sometimes lose their purple markings on drying, thus *de Winter & Marais* 4793 from the Caprivi Strip, appearing to have unmottled standards, originally possessed standards "flecked and stippled with purplish-brown" according to the field-notes. Although all yellow-flowered specimens occur to the west of 25° E longitude they do not constitute a geographically distinct population. *P.A. Smith* 1136 and *Richards* 14821, for example, both from N Botswana, are collections made up of some plants with mottled and some with unmottled standards.

Flower size is another variable character. *Milne-Redhead* 4572 has the standard 2.8 cm long and 1.95 cm wide, far larger than any other specimen. The keel petals may dry with a purplish blotch near the apex; the basal tooth is usually straight and at an angle of 40°–60° to the claw, but it is occasionally down-turned as in *S. transvaalensis*. Most specimens are totally without prickles but, rarely, plants are obscurely aculeate. The SRGH sheet of *P.A. Smith* 1136 is more evidently prickly and is included here with some hesitation; it is also similar to the second SRGH sheet of *Wild & Drummond* 6925, a pure yellow-flowered form of *S. macowaniana*. When growing in water, plants may develop a matted fibrous root system above the more usual taproot and secondary branching system (e.g. *P.A. Smith* 1433). Leaves vary in length and in number and size of leaflets; in *P.A. Smith* 1433 they are greatly reduced.

Pods are sometimes straight-edged (e.g. *Kruger* 'A' from the Caprivi Strip), although these may represent fruits in the early stages of development. Mature pods are always somewhat torulose and are usually regularly constricted between the seeds. The SRGH sheet of *Fagan* BMF 20 displays one straight-edged (?juvenile) pod and all the rest torulose.

Eyles 695 with persistent, paired bracteoles below the calyx hypanthium is similar to *S. transvaalensis* but the standard is dorsally mottled and not veined purple. *S. wildmannii* from Congo and *S. dalzielii* from W Africa seem very close to *S. microphylla* sensu lato, but both have pubescent leaflets. They would perhaps be better treated as subspecies of *S. microphylla*.

Dubber 10 from the Kafue Flats in S Zambia was placed in *S. leptocarpa* "with some hesitation" by *Gillett* in 1963. After further study, I have replaced it in *S. microphylla*, as originally identified, but the specimen is slightly aculeate and undoubtedly shows some affinity with *S. leptocarpa*. Only in S Zambia in the area around Mazabuka, does the geographical range of *S. microphylla* overlap with that of *S. leptocarpa* (var. *minimiflora*); otherwise the two species are geographically separate. It is exactly in this area of overlap that *Dubber* 10 was collected and a plant of hybrid origin is suggested. This is supported by an analysis of the pollen where it was found that pollen from *Dubber* 10 possessed 30% sterile grains. Samples of *S. leptocarpa* and *S. microphylla*, the putative parents of this collection, possessed less than 1% sterile pollen. The flower shape, i.e. the blade of the standard longer than wide and widest above the middle, is constant for the species. This, together with the rounded appendages without free tips, the ovary slightly thickened below the style and the reddish-brown bands on the calyx, separates *S. microphylla* from its closest allies in the Flora Zambesiaca area, *S. leptocarpa* and *S. bispinosa*.

12. **Sesbania sericea** (Willd.) Link, Enum. Hort. Berol. Alt., pt. 2: 244 (1822). — Hutchinson in Bull. Misc. Inform., Kew **1920**: 252, fig.1 (1920). —Sousa in C.F.A. **3**: 182 (1962). —Gillett in Kew Bull. **17**: 133, fig. 3/13 (1963); in F.T.E.A., Legum., Pap.: 350 (1971). —Lewis in Kirkia **13**: 42 (1988). Type: plant cultivated at Berlin from seeds perhaps supplied by Thonning from Ghana; neotype from Sri Lanka, Colombo, *Ferguson* in *Thwaites* c.Pl. 3850 (K), chosen by Gillett (1963). FIGURES 3.3.**32**/11 & 3.3.**33**/11.

> *Coronilla sericea* Willd., Enum. Pl.: 773 (1809).
>
> *Sesbania pubescens* DC., Prodr. **2**: 265 (1825). —Baker in F.T.A. **2**: 135 (1871) for most part. —Hiern, Cat. Afr. Pl. Welw. **1**: 231 (1896). —Eyles in Trans. Roy. Soc. S. Africa **5**: 376 (1916). —Phillips & Hutchinson in Bothalia **1**: 48 (1921). —Baker f., Legum. Trop. Africa: 261 (1929). —Brenan, Check-list For. Trees Shrubs Tang. Terr.: 441 (1949). —Cronquist in F.C.B. **5**: 83 (1954). —Hepper in F.W.T.A., ed.2, **1**: 532 (1958). Type: Ghana, *Thonning* (G holotype).

Erect annual or biennial herb to subshrub, 1–3 m tall, pubescent throughout except for the flower and fruit, silky when young. Stems often (although not in the Flora area) with minute prickles hidden amongst the hairs but not obviously aculeate, exuding bluish, slightly milky juice when cut. Leaves 10.5–16.5 cm long; petioles 5–10 mm long; leaflets in 18–26 pairs, up to 2 cm × 4 mm, oblong to linear-oblong, rounded at apex, apiculate; stipules 5–6 mm long, linear-lanceolate, caducous. Racemes axillary, 1–5.6 cm long, 2–7-flowered; peduncle 0.8–2 cm long, softly silky or pilose; pedicels 3–6 mm long, sparsely silky pilose; bracts and bracteoles linear-lanceolate, acuminate, sparsely silky pilose, caducous, the bracts up to 5 mm long, the bracteoles 3 mm long. Calyx 3–4.2 × 3 mm; tube glabrous; teeth 0.5–0.75 mm long, triangular, subulate, pubescent on the margins and inner face. Standard pale cream slightly flecked with violet or purple, 7.5–9 × 8–10 mm, broader than long, apically emarginate, the appendages very narrowly wedge-shaped and truncate at the apex; wings 7.5–9 × 3–4.5 mm, with a small upper basal triangular tooth and lamellate sculpturing in upper basal one-third on outer face; keel 7–8 × 4–6 mm, the limb subquadrate, ± incurved, with a small acute basal tooth. Ovary and style glabrous. Pod up to 16 cm × 2.5–3.5 mm, linear to slightly falcate, acuminate, not torulose, glabrous, 15–30-seeded; septa 5 mm apart, thicker at the centre than at the sutures. Seeds brown to reddish-brown, minutely spotted blackish, 3–3.5 × 1.75–2 mm, 1.5 mm thick, the hilum in a small, central circular pit.

Malawi. S: Mangochi, at E end of bridge over Shire R., fl.& imm.fr. 24.ii.1979, *Brummitt & Patel* 15473 (K, SRGH).

Also from Senegal to Somalia, and southwards through Congo, Uganda, Kenya and Tanzania to Angola; and from S Arabia and Sri Lanka (introduced), the West Indies and northern South America (possibly introduced). River banks; elsewhere known from marshy ground and old rice-fields; c.450 m.

13. **Sesbania greenwayi** J.B. Gillett in Kew Bull. **17**: 138, fig.3/11 (1963); in F.T.E.A., Legum., Pap.: 347, fig.49/11 (1971). —Lewis in Kirkia **13**: 43 (1988). Type: Tanzania, Rufiji Dist., Mafia Is., *Greenway* 5188 (K holotype, EA). FIGURES 3.3.**32**/12, 3.3.**33**/12, 3.3.**34**/2.

Erect slightly woody annual herb, 1–5.5 m tall. Stems glabrous or sparsely pilose when young, rarely with a few obscure prickles. Leaves 1.2–10 cm long (37.5 cm long on lower parts of plant), the rachis sometimes with a few prickles; petioles 0.15–0.9 cm long (6.5 cm on larger leaves); leaflets in (1)3–32 subopposite pairs, 0.5–1.4(4) cm × 2–5(6.5) mm, oblong to narrowly obovate, rounded to emarginate at the apex, apiculate, entire, glabrous although slightly pubescent when young; stipules 2–7 mm long, lanceolate, long acuminate, erect or spreading, sparsely pilose, usually caducous. Raceme (1)2–12 cm long, (1)2–13-flowered; peduncle 0.8–2.8(4) cm long, glabrous or sparsely pilose especially near the base, sometimes with small prickles intermixed with the hairs; pedicels (0.2)0.6–1 cm long, glabrous or with an occasional hair, sometimes with very small prickles; bracts and bracteoles linear-lanceolate, sparsely pilose on the margins, caducous, the bracts 3–4 mm long, the bracteoles 2–3 mm long. Calyx 4–7 × 4–7 mm, the tube glabrous; teeth 0.7–1.3 mm long, slightly acuminate, shortly woolly on the margins. Standard yellow, speckled with purplish-black on the outer face, 1.4–1.65 × 1.1–1.4 cm, broad, rounded, apically emarginate, the appendages 1–1.5 mm long abruptly rounded and sometimes emarginate at the top; wings 1.4–1.75 cm × 3–5.5 mm, usually with a small, slightly hooked upper basal tooth and with lamellate sculpturing in the upper basal half on outer face; keel 1.3–1.6 cm × 6–7.5 mm, with basal tooth acute, the apex 2 mm long, forming an angle of 40°–70° with the claw. Ovary glabrous, often slightly thickened towards the base of the style, the style pubescent except at the base. Pod straw-coloured, mottled dull reddish-brown when mature, (4)7.5–29.5 cm × 2.5–4.5 mm, slightly falcate, torulose, stipitate, shortly beaked, glabrous, (2)5–40-seeded; septa 6–8 mm apart. Seeds pale brown to olive-greenish-brown to almost black, not mottled, 4–4.5 × 2–2.5 mm, 1–2 mm thick, elliptical; hilum in a small, subcentral circular pit.

Zambia. N: Mpika Dist., Mfuwe Lagoon, South Luangwa Nat. Park (Game Reserve), fl. 27.iv.1965, *Mitchell* 2702 (K, SRGH). E: Chipata Dist., Jumbe and Munkanya (Mkhania) area, fl.& fr. 24.iii.1963, *Verboom* 691 (K, SRGH). **Zimbabwe**. N: Hurungwe Dist., E of Chewore–Zambezi confluence, fl.& fr. 21.iv.1967, *Cleghorn* 1632 (SRGH). **Mozambique**. N: Angoche Dist., between Angoche (António Enes) and Missão Católica de Malatane, fl.& imm.fr. 8.xi.1936, *Torre* 987 (COI, LISC). Z: Lugela Dist., Mocuba, Namagoa Estate, fl. 18.iii.1949 & fr. 16.iv.1949, *Faulkner* Kew 411/B (K, PRE). T: Magoe Dist., Msusa, Zambezi R., fl.& fr. 22.vii.1950, *Chase* 2717 (K, LISC, SRGH). MS: Chemba Dist., c.5 km from Chemba, on road to Nhacolo (Tambara), fl.& imm.fr. 23.iv.1960, *Lemos & Macuácua* 137 (COI, K, LISC, PRE, SRGH).

Also from Somalia and Tanzania. Seasonally flooded areas, swamps, dambos, river banks, roadside ditches and savanna margins; on heavy clay and alluvial soils; 0–1000 m.

As noted by Gillett, *Faulkner* Kew 411B was part of a mixed gathering, 411A being *S. bispinosa*. Phiri, with respect to his specimen 174, states that the plant is edible but does not specify which parts. The LISC sheet of *Chase* 2717 shows a degree of fasciation on the flowering portion. *E.A. Robinson* 819 is a confusing sheet and led Gillett to suggest that it may represent a distinct variety or subspecies. The specimen is aculeate, having small, scattered prickles on the stems and occasionally on the peduncles. However, prickles are found on other specimens of the species, being most pronounced on the peduncles of *Macêdo* 4802 from Tete Province in Mozambique. It is the flowers of *Robinson* 819 which throw some light on the problem. The first of two flowers dissected had 6 petals (3 of which are wing-like) and 10 stamens plus 2 short staminodes. The second had the normal number of petals and stamens but two bracteoles unusually fused half-way up the calyx tube. The specimen is undoubtedly abnormal, perhaps the result of insect damage, and is not here considered as a distinct taxon.

14. **Sesbania brevipeduncula** J.B. Gillett in Kew Bull. **17**: 143, fig. 7/1 (1963). — Corby in Kirkia **9**: 326 (1974). —Lewis in Kirkia **13**: 44 (1988). —Lock, Leg. Afr. Check-list: 460 (1989) as "*brevipedunculata*". Type: Botswana, NE margin of Makgadikgadi (Makarikari) Pan, mouth of Nata R., *Drummond & Seagrief* 5184 (K holotype, LISC, PRE, SRGH). FIGURES 3.3.**32**/13, 3.3.**33**/13, 3.3.**34**/1.
     Sesbania punctata var. sensu N.E. Brown in Bull. Misc. Inform., Kew **1909**: 104 (1909) in part, non DC.
     Sesbania aculeata sensu Phillips & Hutchinson in Bothalia **1**: 50 (1921) in lesser part, non Pers.

A much-branched annual herb, 1–2.5 m tall. Stems glabrous, usually sparsely aculeate. Leaves 3.5–20 cm long; rachis glabrous, occasionally aculeate; petioles 0.3–1.7 cm long; leaflets in 8–41 pairs, 4–16 × 1.75–4.5 mm, oblong, apiculate, slightly broader and asymmetrical at the base, glabrous; stipules 3–7 mm long, narrowly triangular to lanceolate, caducous to ± persistent. Racemes 0.3–3.5(9) cm long, (1)2–8(17)-flowered; peduncle 0.1–0.9(1.8) cm long, glabrous; pedicels 0.3–1.5 cm long, glabrous; bracts 2–3 mm long, linear-lanceolate, caducous; bracteoles 1–2 mm long, linear, located slightly below the hypanthium, caducous. Calyx 3–4.5 × 3–4 mm, the tube glabrous with 5 vertical dark brown bands each running into a tooth; teeth 0.5–0.75 mm long and well-spaced often giving the calyx a ± truncate appearance, acuminate, woolly at the margins. Standard yellow finely speckled brown or purplish, 1.1–1.4 × 1–1.3 cm, the blade wider than long, suborbicular, apically emarginate, basally rounded, truncate or slightly cordate, the appendages small, triangular, divergent and without free tips; wings 1.1–1.4 cm × 3–5.5 mm, with a 1 mm long upper basal, finely acuminate ± upward-pointing tooth and with open lamellate sculpturing in the upper basal section; keel 1–1.3 cm × 5–7 mm, the limb with an erect basal tooth 0.3–0.5 mm long and a secondary widely triangular tooth above it. Ovary 15–26(32)-ovulate, glabrous; style glabrous. Pod (2.4)7.2–11.5 (rarely 13.5) cm × (2)2.5–3(5.5) mm, erect, moniliform, shortly stipitate and beaked, (2)8–18(rarely 26)-seeded; septa 5–6 mm apart; sutures 1.5 mm wide, purplish-green, distinctly darker than the valves which, particularly when immature, have a secondary sutural ridge. Seeds brown or blackish, 2–3.3 × 2 mm, 1.5–2 mm thick, ellipsoid or ± spherical with a subcentral hilum in a very small circular pit.

Botswana. N: Central Dist., NE margin of Makgadikgadi (Makarikari) Pan, mouth of Nata R., fl.& fr. 22.iv.1957, *Drummond & Seagrief* 5184 (K, LISC, PRE, SRGH). **Zambia.** S: between Kalomo and Livingstone, fl.& imm.fr. 20.ii.1963, *van Rensburg* 1374 (K, SRGH). **Zimbabwe.** N: Gokwe South Dist., c.1 km S of Tari (Tare) R. on Gokwe–Binga road, fl.& fr. 31.iii.1962, *Bingham* 218 (SRGH). W: Bulawayo, fl.& fr. 8.i.1909, *MacDonald in Tvl. Mus.* 4942 (K; PRE). C: Gweru Dist., Gweru (Gwelo) Kopje, fl.& fr. 5.iii.1967, *Biegel* 1968 (K, LISC, SRGH).

Also in Angola and South Africa (North-West, Limpopo and Gauteng Provinces). Swampy ground, damp grassland and sandy river banks; on black clays and granitic soils; 900–1450 m.

Denny in NRSH 187 from Zimbabwe has longer, narrower pods than usual with up to 26 seeds per pod. For further discussion on the species, reference should be made to Gillett's comments in Kew Bull. **17**: 146 (1963).

15. **Sesbania macowaniana** Schinz in Verh. Bot. Vereins Prov. Brandenburg **30**: 165 (1888). —Dinter in Repert. Spec. Nov. Regni Veg. **23**: 234 (1926). —Gillett in Kew Bull. **17**: 146, fig. 7/2 (1963). —Schreiber in Merxmüller, Prodr. Fl. SW Afrika, fam. 60: 110 (1970). —Lewis in Kirkia **13**: 45 (1988). Type: Namibia, Ovamboland, Olukonda, c.18°S, 16°10'E, *Schinz* (Z holotype). FIGURES 3.3.**33**/14, 3.3.**36**.

  Sesbania mossambicensis sensu Phillips & Hutchinson in Bothalia **1**: 52 (1921). —sensu Baker f., Legum. Trop. Africa: 263 (1929) in lesser part, non Klotzsch.

Erect annual herb (20)35–50(150) cm tall. Stems sometimes single, glabrous, obscurely to clearly aculeate. Leaves 1.5–6.5(11) cm long; rachis sometimes aculeate; petioles 0.2–0.8(1) cm long; leaflets in (4)6–19(26) pairs, 0.3–1.2(13) cm × 1.5–3.5 mm, oblong, apically obtuse, apiculate, asymmetrical at the base, with a prominent purple to reddish-brown midvein on the undersurface, young leaflets with a few hairs on the margins and midvein below, becoming glabrous; petiolules invested with a few hairs; stipules linear-lanceolate to narrowly triangular, 2–5 mm long with a few obscure marginal hairs. Inflorescence a reduced axillary raceme, 2–6(7) mm long, 1–8(9)-flowered; peduncle 1–2 mm long, glabrous to slightly spreading hairy; pedicels 2–5 mm long, glabrous; bracts 2 mm long, linear to linear-lanceolate, caducous; bracteoles 0.5–0.75 mm long, filiform. Calyx 3–3.5(4) × 3–3.5 mm; tube glabrous with 5 dark brown bands running into the teeth; teeth acuminate, 1–1.75 mm long, obscurely hairy on the margins. Standard yellow with strong purple veining on the outer face, rarely pure yellow, 1–1.3 × 1–1.1 cm, rounded, apically emarginate, without appendages; wings yellow,

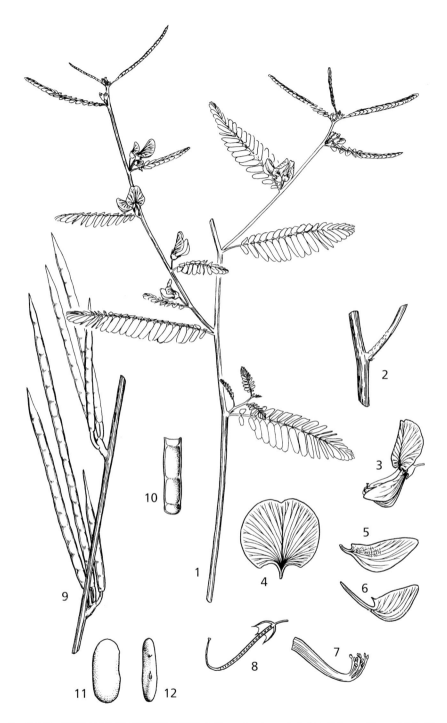

Fig. 3.3.**36**. SESBANIA MACOWANIANA. 1, flowering branch (× ²/₃); 2, base of branchlet with obscure prickles (× 1¹/₂); 3, flower (× 1); 4, standard (× 2); 5, wing (× 2); 6, keel petal (× 2); 7, stamens (× 2); 8, gynoecium (× 2), 1–8 from *Wild & Drummond* 6925; 9, infructescences (× ²/₃); 10, inside pod (× 1); 11 & 12, seed, side and front views (× 4), 9–12 from *Dinter* 7394. Drawn by Eleanor Catherine. From Kirkia (1988).

(0.95)1.1–1.2(1.25) cm × 3.5–4 mm, with an upper basal tooth and with lamellate sculpturing in the upper basal section on outer face; keel yellow, slightly striped purple or brownish on the beak, (1)1.2(1.35) cm × (5)7 mm, the limb subtriangular with a basal tooth pointing ± upwards. Ovary and style glabrous. Pod (4.5)8–9.5(10.5) cm × 2–3(4) mm, linear to slightly falcate, acuminate, held erect, glabrous, (4)12–17(18)-seeded; septa 5–6 mm apart; valves with prominent raised venation and occasional small scattered prickles (visible with a ×10 lens). Seeds olive-green with fine black speckling or brownish, 3.5–4 × 1–1.5 mm, 1.5–2 mm thick; small hilum in central circular pit.

Botswana. N: Ngamiland Dist., Qangwa R. (Kangwa, Xanwe), 27 km NE of Aha Hills, fl.& fr. 12.iii.1965, *Wild & Drummond* 6925 (K, LISC, PRE, SRGH). SE: Ghanzi Dist., Kobe Pan, Farm No.107, fl. 21.ii.1970, *R.C. Brown & D.C. Brown* 8680 (PRE, SRGH).

Also in Namibia. Brackish flats, calcareous pans and cracks in limestone pavements near springs; on grey calcareous or brownish clayey-loam soils; 900–1200 m.

This handsome species with its strikingly attractive flowers just enters the Flora area, being more widely found in Namibia. Habitat data suggests that it is a calcicole preferring limestone pavements and pans. Rarely, the standard is pure yellow, as seen on the second SRGH sheet of *Wild & Drummond* 6925. The shorter inflorescences and straighter-edged pods separate it from its closest relatives, *S. notialis* and *S. transvaalensis*.

16. **Sesbania notialis** J.B. Gillett in Kew Bull. **17**: 146 (1963). —Lewis in Kirkia **13**: 47, figs. 2/14, 3/15, 4/5 (1988). Type: South Africa, Northern Cape Prov., Modder R. near Kimberley, clay vlei soils, 17.ii.1948, *Rodin* 3680 (K holotype). FIGURES 3.3.**32**/14, 3.3.**33**/15, 3.3.**34**/5.

Erect annual herb 0.3–1 m tall. Stems with or occasionally without prickles, glabrous or sometimes the prickles with a terminal hair. Leaves (2)6–8(12) cm long, the rachis glabrous or almost so, with or occasionally without prickles; petioles 4–7 mm long; leaflets in (4)12–21(25) pairs, 4–7(9) × 1.75–3 mm, oblong to oblong-elliptic, rounded to slightly emarginate at the apex, apiculate, asymmetrical at the base, glabrous; stipules 2.5–3 mm long, narrowly triangular, acuminate, margins sometimes with a few sparse hairs, usually persistent. Raceme axillary, 1.1–5 cm long, 1–5-flowered, the rachis usually aculeate; peduncle 1.1–2.5 cm long; pedicels 3–8 mm long, glabrous; bracts 1.5–2 mm long, linear-lanceolate, somewhat persistent; bracteoles 1–1.5 mm long, linear, located at base of the hypanthium, at length falling. Calyx 3–4.2 × 3 mm; tube glabrous; teeth 0.6–1 mm long, acuminate, sparsely woolly on the margins. Standard yellow with blackish-purple veins on the outer face, 1.1 cm × 8–9 mm, orbicular, apically emarginate, basally attenuate, the appendages small, triangular, divergent and without free tips; wings yellow, 1.1 cm × 3.5–4 5 mm, with an upper basal tooth and with lamellate sculpturing in upper basal half on outer face; keel yellow with some purple striping on the beak, 1.1–1.2 cm × 5–6 mm, the limb with a ± horizontal to slightly upward-pointing, acuminate tooth. Ovary and style glabrous, 16–19(20)-ovulate. Pod 8–10 cm × 2–2.5 mm, linear to very slightly falcate, torulose when young, hardly so when mature, rostrate, slightly stipitate, erect, glabrous, 10–19(20)-seeded; septa 5 mm apart; valves with reticulate venation in central area above seeds. Seeds brownish-green, with or without blackish-purple spotting, 3 × 1.5 mm, 1.3 mm thick; small hilum in central circular pit.

Botswana. SE: Kgatleng Dist., Mochudi, fl.& imm.fr. iii.1974, *N. Mitchison* 118 (K).

Also in South Africa (Northern Cape, North-West and Free State Provinces) where it is recorded from limestone hollows, vleis and grasslands; on clay soils; 1100 m.

Mitchison 118, the only specimen of *S. notialis* in the Flora area, extends the northerly range of the species by about 200 km. The specimen is unusual in being non-aculeate, although a few obscure warty outgrowths on the stems may represent rudimentary prickles. The lack of armature makes it difficult to separate this from *S.*

transvaalensis (a species also represented by only one collection in Botswana, and this being geographically close to *Mitchison* 118). It is, however, separable by its 1–3-flowered inflorescences and 16–19-seeded pods which are not moniliform.

17. **Sesbania transvaalensis** J.B. Gillett in Kew Bull. **17**: 147, fig. 7/3 (1963). —Lewis in Kirkia **13**: 47 (1988). Type: South Africa, Pretoria Dist., Apies R., 1500 m, marshes, 15.i.1894, *Schlechter* 4179 (K holotype, BM). FIGURES 3.3.**32**/15, 3.3.**33**/16, 3.3.**34**/4.

> *Sesbania mossambicensis* sensu Phillips & Hutchinson in Bothalia **1**: 52 (1921). —sensu Burtt Davy, Fl. Pl. Ferns Transvaal, pt. 2: 381 (1932) for most part, but not type.

Erect annual herb or spindly subshrub (0.6)1.5(2) m tall. Stems glabrous, without prickles. Leaves (2)6.4–9.5(12.3) cm long, the rachis glabrous or occasionally with scattered hairs; petioles 3–7 mm long; leaflets in 5–20(29) pairs, 3.5–10 × 1.5–3 mm, oblong, rounded to slightly emarginate at the apex, apiculate, asymmetrical at the base, entire, glabrous or with minute hairs scattered along the margins, on the midvein beneath and on the petiolules, punctate above with minute dark dots; stipules 3–5 mm long, linear-triangular to linear-lanceolate, caducous to persistent. Racemes 1.6–9 cm long, (2)5–7(10)-flowered; peduncle 1.5–3.9 cm long, glabrous; pedicels 4–10 mm long, glabrous; bracts 1.5–3(4) mm long, linear, persistent; bracteoles 1.5–3 mm long, linear, persistent, located below the receptacle. Calyx 3–4 × 2.5–4 mm; tube glabrous; teeth 0.5–1 mm long, subulate, sparsely woolly on the inner surface and at the tips. Standard yellow with strong purple to reddish veining on the outer face, 1–1.2 × 0.85–1.1 cm, suborbicular, apically emarginate, the appendages small, triangular, divergent and without free tips; wings yellow, 1–1.1 cm × 3.25–4 mm, with an upper basal tooth and with lamellate sculpturing in upper basal half on outer face; keel yellow (slightly paler than wings) with some purple striping on the beak, 1.1–1.2 cm × 4–6 mm, the limb with a ± horizontal acuminate tooth. Ovary and style glabrous, 10–13-ovulate, ovules widely spaced. Pod (4.6)6–8.5 cm × 1.5–2.75 mm, linear to slightly falcate, moniliform, rostrate, stipitate, held ± erect, glabrous, 7–12-seeded; septa 5–7 mm apart; valves with reticulate venation in central area above seeds, the margins developing a slightly raised ridge on drying which tapers out towards the septa. Seeds brown or olive-green mottled blackish, 3 × 1.75 mm, 1.75–2 mm thick; small hilum in central circular pit.

Botswana. SE: Bangwaketse (Banquaketse) territory, Masupa R., fl.& immr.fr. *Holub* s.n. (K).

A species of South Africa (North-West, Limpopo, Gauteng and Free State Provinces) and just entering into SE Botswana, where it is only known from one collection. In South Africa it is recorded from brackish flats, river banks and wet grassland; on black loam and coarse sandy soils; 1230–1450 m.

Easily confused with *S. notialis*, some specimens of which seem to be non-aculeate, it is nevertheless separable by its fewer-seeded moniliform pods and by usually having more flowers per inflorescence.

18. **Sesbania tetraptera** Baker in F.T.A. **2**: 136 (1871). —Phillips & Hutchinson in Bothalia **1**: 53 (1921). —Baker f., Legum. Trop. Africa: 263 (1929). —Gillett in Kew Bull. **17**: 149, fig. 3/14 (1963); in F.T.E.A., Legum., Pap.: 351 (1971). — Lewis in Kirkia **13**: 48 (1988). Type: Sudan, Kordofan, Mt Arashkol, Kotschy 131 (K holotype, BM).

> *Sesbania kirkii* E. Phillips & Hutch. in Bothalia **1**: 54 (1921). —Baker f., Legum. Trop. Africa: 264 (1929). Types: Malawi, Lower Shire, Elephant Marsh, ii.1887, *Scott* s.n. (K syntype); Mozambique, Tete, ii.1859, *Kirk* s.n.; Sena, vi.1859, *Kirk* s.n. (K syntypes).
> *Sesbania hamata* E. Phillips & Hutch. in Bothalia **1**: 54 (1921). —Burtt Davy, Fl. Pl. Ferns Transvaal, pt. 2: 381 (1932). Type: South Africa, Limpopo Prov., Soutspansberg Dist., *Schlechter* 4620 (BOL holotype, K).

Erect annual or biennial herb 1–4 m tall, sometimes with wide regular branching. Stems greenish or tinged purplish-blue, sparsely pilose when young, glabrous later, sometimes with a few to many scattered prickles. Leaves (5.5)7–38 cm long, the rachis sparsely pilose on the upper surface, sometimes sparsely aculeate below; petioles (0.2)0.3–1.8(3.5) cm long; leaflets in (9)16–42 pairs, 0.8–4.3 cm × 2–10.5 mm, often purplish beneath, oblong, apiculate, occasionally obscurely pilose on margins and midvein beneath, otherwise glabrous; stipules 5–10 mm long, narrowly lanceolate, pilose, usually spreading and persistent, often with a curled apex. Raceme 3.4–16.2 cm long, 2–13-flowered; peduncle 1.8–7 cm long, beset, often densely so, with flattened, sometimes fused prickles especially at the base, the prickles occasionally tipped with hairs; pedicels 0.4–2.1 cm long, glabrous; bracts 3–5.5 mm long, linear to narrowly lanceolate, spreading, caducous or persisting until after the corolla has fallen off; bracteoles 2–4 mm long, linear-lanceolate, caducous. Calyx 4–7 × 4.5–8 mm; tube glabrous; teeth 0.5–1.5 mm long, acuminate, marginally hairy. Standard pale yellow with dark purple speckling on the outer face, 0.9–2 × 0.8–1.8 cm, broad, rounded, apically emarginate, the blade often wider than long, sometimes prominently cordate basally, the appendages gently rounded and without free tips; wings yellow, 1–2 cm × 3–6 mm, with an upper basal very slightly downward-pointing tooth and with lamellate sculpturing in the upper basal section on outer face; keel pale yellow to greenish-yellow, 0.7–1.6 cm × 5–9 mm; basal tooth usually at an angle of 10–30° with the claw but occasionally at 45° (e.g. *Richards* 14469). Ovary winged, 15–32-ovulate, glabrous; style pubescent except at the base. Pod (8.3)12–23.3 × (0.9)1–1.4 cm, linear or slightly curved, shortly beaked, stipitate, 4-winged, (0.9)19–28-seeded, glabrous; wings darker than the central window-like areas above the seeds except in very old fruit; septa 6–8 mm apart. Seeds ochre, olive-green sometimes over-coloured bluish-black, brown or reddish-brown, 4–5.5 × 2–4 mm, 1.5–2.5 mm thick, ellipsoid to brick-shaped; hilum in small subcentral circular pit surrounded by a rim aril.

Subsp. **tetraptera**

Stems greenish. Peduncle to 7 cm long; pedicels to 2.1 cm long; bracts persistent until the corolla has fallen. Standard to 1.2 cm long. Septa of pod 6–7(7.5) mm apart.

Zimbabwe. S: Mwenezi Dist., between Palfrey's store and Mwenezi (Nuanetsi) Drift, fl.& fr. 29.iv.1961, *Drummond & Rutherford-Smith* 7597 (K, LISC, SRGH). **Malawi**. S: Chikwawa Dist., Lengwe Nat. Park (Game Reserve), NE corner, fl.& fr. 5.iii.1970, *Brummitt* 8886 (K, PRE, SRGH). **Mozambique**. T: Tete (Tette), fl. ii.1859, *Kirk* s.n. (K); Cahora Bassa Dist., Chitima (Estima), fl.& fr. 22.iv.1972, *Macêdo* 5235 (K, LISC, SRGH). MS: Chemba Dist., Chiou, Estação Experimental do C.I.C.A., fr. 12.iv.1960, *Lemos & Macuácua* 86 (COI, K, LISC, PRE). GI: between Mapai and Mabalane, 88 km from Mapai, fr. 5.iv.1952, *Barbosa & Balsinhas* 5124 (K). M: Magude Dist., in area of Mapulanguene, fl.& fr. 23.i.1948, *Torre* 7191 (LISC).

Also in Sudan, Ethiopia, Tanzania and South Africa (Limpopo Province). Riverside pans and grassland; on heavy black clay and sandy soils; occasionally cultivated; 60–200 m.

Although further collections of *S. tetraptera* subsp. *tetraptera* from the Flora area, together with a collection from Ethiopia (*Parker* E467) extend the geographical range of this taxon, its distribution remains essentially the same as that given for *S. tetraptera* by Gillett in Kew Bull. **17**: 149 (1963). Gillett notes that "the extraordinary distribution of *S. tetraptera* (i.e. subsp. *tetraptera*) in a series of well-separated localities on a North-South axis" is very likely due to distribution by migrating birds.

Subsp. **rogersii** (E. Phillips & Hutch.) G.P. Lewis in Kirkia **13**: 50 (1988). Types: Zimbabwe, Victoria Falls, *Flanagan* 3109 (BOL syntype) and, probably in same region, *F.A. Rogers* 8747 (BOL syntype). FIGURES 3.3.**32**/16, 3.3.**33**/17 & 20, 3.3.**34**/6 & 7.

Sesbania rogersii E. Phillips & Hutch. in Bothalia **1**: 55 (1921). —Baker f., Legum. Trop. Africa: 264 (1929). —White, F.F.N.R.: 165 (1962). —Gillett in Kew Bull. **17**: 150 (1963). — Corby in Kirkia 9: 326 (1974).

    Sesbania tetraptera sensu Eyles in Trans. Roy. Soc. S. Africa **5**: 376 (1916) non Baker.

Stems nearly always tinged purplish-blue. Peduncle 1.8–3(3.7) cm long; pedicels 0.5–1.2(1.4) cm long; bracts caducous. Standard 1.6–2 cm long. Septa of pod 7–8 mm apart.

Botswana. N: Central Dist., junction of Sibanini (Sebanini) and Nata rivers, fr. 21.iv.1931, *Pole Evans* 3320 (K, PRE). **Zambia**. S: Kalomo Dist., c.12 km W of Kazungula, fl.& imm.fr. 14.iii.1952, *White* 2261 (K, PRE). **Zimbabwe**. N: Gokwe Dist., fl.& imm.fr. 20.iv.1973, *Chiparawasha* 726 (SRGH). W: Hwange Dist., Gwayi (Gwaai) R., Ngamo Forest Reserve, fl.& imm.fr. iii.1960, *Armitage* 117/60 (K, SRGH). C: Harare, Salisbury Agric. Expt. Station (?cultivated), fl.& imm.fr. 4.iii.1943, *Arnold* in *GHS* 9745 (K, SRGH). E: Chipinge Dist., Save (Sabi) Expt. Station, fl.& fr. 16.iii.1965, *Corby* 1284 (K, SRGH). **Malawi**. S: Mangochi Dist., 4 km N of Mpale R. Bridge, 32 km from Mangochi (Fort Johnston) towards Liwonde, fl.& imm.fr. 2.iii.1970, *Brummitt* 8849 (K, SRGH).

Not known from outside the Flora Zambesiaca area, mostly occurring west of the range of subsp. *tetraptera*, but both subspecies occur in the area south of Lake Malawi. Seasonal swamps and river margins, in mopane and mixed deciduous woodlands and roadside ditches; on heavy black clays and sandy soils; 375–1100 m.

S. tetraptera subsp. *rogersii*, unlike subsp. *tetraptera*, is confined to the Flora area, being found at higher altitudes and generally further inland. Malawi S: is the only division from where both subspecies are recorded, although not from the same locality. Both are occasionally cultivated and records from experimental stations tend to obscure the geography.

Referring to *S. rogersii*, Gillett (Kew Bull. **17**: 150, 1963) states that "this species is closely related to *S. tetraptera*, but seems sufficiently distinct; specimens without flowers can, however, often not be determined". His unpublished notes on a Kew species folder of *S. rogersii* outline the differences he found between the two taxa. With further collections now available certain characters lose their usefulness as species separators and measurements are found to overlap. Although the two taxa recognized here do grade into one another to some extent they seem best treated as subspecies based particularly on differing geographical distributions and flower-size. The possibility of treating all the material as one taxon is ruled out by comparing, for example, *Brummitt* 8886 (subsp. *tetraptera*) and *Brummitt & Patel* 15478 (subsp. *rogersii*), both from Malawi and evidently different. However, many collections fall somewhere between the two. On the whole, subsp. *rogersii* is a more robust plant, nearly always with a purplish-blue tinge to the stems. In flower it is quite distinct from subsp. *tetraptera* with all flower parts being larger and with caducous bracts. Naming fruiting material remains difficult and the only character that seems reliable is the septal distance, being consistently greater in subsp. *rogersii*.

Some similarities between and variation within the subspecies may be due to hybridization between subsp. *tetraptera* and perhaps *S. macrantha* sensu lato, although the geographical distribution of the two does not make this a convincing hypothesis. Certainly subsp. *rogersii* possesses many of the characteristics of *S. macrantha* var. *macrantha* and introgressive hybridization between the latter and subsp. *tetraptera* possibly helps to explain this. The above speculation needs experimental testing, and field observations and further collecting would undoubtedly throw light on the problem. The K and SRGH sheets of *Bingham* 184 are mixed collections. The fruits belong to subsp. *rogersii*, the flowering portions to *S. macrantha* var. *levis*.

P.A. Smith 2426, from Botswana, is an abnormal specimen. Of two flowers dissected the first had no true standard, only one true keel petal and four wing-like petals and the ten stamens fused into a sheath. The second possessed a strange, almost two-lipped, calyx. The leaves are reduced with few leaflets. Similar abnormalities were observed in *S. greenwayi* and in both cases insect damage may be the cause or some sort of root nematode may be involved. *Richards* 14469 has a keel petal with the tooth at 45° to the claw, a greater angle than is usual in the subspecies.

INDEX TO BOTANICAL NAMES